TECHNICAL INFORMATION IN CHEMICAL ENGINEERING

TECHNICAL INFORMATION IN CHEMICAL ENGINEERING

Manoj Verma

RANDOM

RANDOM PUBLICATIONS

NEW DELHI - 110 002 (INDIA)

Technical Information in Chemical Engineering

ISBN 978-93-51116-64-6

Published in 2015 in India by

RANDOM PUBLICATIONS

4376-A/4B, Gali Murari Lal, Ansari Road
New Delhi-110 002
Phone: +9111-43580356, 23289044
E-mail: randomexports@gmail.com; sales@randompublications.com; info@randompublications.com
Reprinted 2026

Type Setting by: Friends Media, Delhi-110089

Preface

Chemical Engineering is the branch of engineering that deals with the technology of large-scale chemical production and the manufacture of products through chemical processes. Chemical Engineering is a dynamic discipline and it has been evolving continuously since the days of craft innovations of prehistoric times. Chemical engineering involves managing plant processes and conditions to ensure optimal plant operation. Chemical reaction engineers construct models for reactor analysis and design using laboratory data and physical parameters, such as chemical thermodynamics, to solve problems and predict reactor performance. A chemical plant is an industrial process plant that manufactures (or otherwise processes) chemicals, usually on a large scale. The general objective of a chemical plant is to create new material wealth via the chemical or biological transformation and or separation of materials. Chemical plants use specialized equipment, units, and technology in the manufacturing process. Other kinds of plants, such as polymer, pharmaceutical, food, and some beverage production facilities, power plants, oil refineries or other refineries, natural gas processing and biochemical plants, water and wastewater treatment, and pollution control equipment use many technologies that have similarities to chemical plant technology such as fluid systems and chemical reactor systems.

Chemical engineering design concerns the creation of plans, specification, and economic analyses for pilot plants, new plants or plant modifications. Design engineers often work in a consulting role, designing plants to meet clients' needs. Design is limited by a number of factors, including funding, government regulations and safety standards. These constraints dictate a plant's choice of process, materials and equipment. A unit operation is a physical step in an individual chemical engineering process. Unit operations (such as crystallization, drying and evaporation) are used to prepare reactants, purifying and separating its products, recycling unspent reactants, and controlling energy transfer in reactors. On the other hand, a unit process is the chemical equivalent of a unit operation. Along with unit operations, unit processes constitute a process operation. Unit processes (such as nitration and oxidation) involve the conversion of material by biochemical, thermochemical and other means. Chemical engineers responsible for these are called process engineers. Transport phenomena occur frequently in industrial problems. These include fluid dynamics, heat transfer and mass transfer, which mainly concern momentum transfer, energy transfer and transport of chemical species respectively. Basic equations for describing the three phenomena in the macroscopic, microscopic and molecular levels are very similar. Thus, understanding transport phenomena requires thorough understanding of mathematics. Advancements in computer science found applications designing and managing plants, simplifying calculations and drawings that previously had to be done manually. The completion of the Human Genome Project is also seen as a major development, not only advancing chemical engineering but genetic engineering and genomics as well. Chemical engineering principles were used to produce DNA sequences in large quantities.

The present resource book is the most complete book for chemical and process engineers incorporates the latest information for engineers and practitioners who depend on it as a working tool. Intended primarily for graduate chemical-engineering students, this book also includes material which bridges the gap between graduate and postgraduate requirements.

I thank all members of my team who have helped in the preparation of the book. My special thanks go to "Random Publications" who have published the book.

— Manoj Verma

Contents

Chapter 1

Introduction

Chemical engineering is a branch of engineering that applies the natural (or experimental) sciences (e.g. chemistry and physics) and life sciences (e.g. biology, microbiology and biochemistry) together with mathematics and economics to production, transformation, transportation and proper usage of chemicals, materials and energy. It essentially deals with the engineering of chemicals, energy and the processes that create and/or convert them. Modern chemical engineers are concerned with processes that convert raw-materials or (cheap) chemicals into more useful or valuable forms. In addition, they are also concerned with pioneering valuable materials and related techniques – which are often essential to related fields such as nanotechnology, fuel cells and bioengineering. Within chemical engineering, two broad subgroups include design, manufacture, and operation of plants and machinery in industrial chemical and related processes ("chemical process engineers") and development of new or adapted substances for products ranging from foods and beverages to cosmetics to cleaners to pharmaceutical ingredients, among many other products ("chemical product engineers").

History of Chemical Engineering

Chemical engineering as a discipline that was developed out of those practicing "industrial chemistry" in the late 19th century. Before the Industrial Revolution (18th century), industrial chemicals and other consumer products such as soap were mainly produced through batch processing. Batch processing is labour-intensive and individuals mix predetermined amounts of ingredients in a vessel, heat, cool or pressurize the mixture for a predetermined length of time. The product may then be isolated, purified and tested to achieve a saleable product. Batch processes are still performed today on higher value products, such as pharmaceutical intermediates, speciality and formulated products such as perfumes and paints, or in food manufacture such as pure maple syrups, where a profit can still be made despite batch methods being slower and inefficient in terms of labour and equipment usage. Due to the application of Chemical Engineering techniques during manufacturing process development, larger volume chemicals are now produced through a continuous "assembly line" chemical processes. The Industrial Revolution was when a shift from batch to more continuous processing began to occur. Today commodity chemicals and petrochemicals are predominantly made using continuous manufacturing processes whereas speciality chemicals, fine chemicals and pharmaceuticals are made using batch processes.

Origin

The Industrial Revolution led to an unprecedented escalation in demand, both with regard to quantity and quality, for bulk chemicals such as sulfuric acid and soda ash. This meant two things: one, the size of the activity and the efficiency of operation had to be enlarged, and two, serious alternatives to batch processing, such as continuous operation, had to be examined.

The First Chemical Engineer

Industrial Chemistry was being practiced in the mid 1800s, but it was not until the 1880s that the engineering elements required to control chemical processes were being recognised as a distinct professional activity. Chemical engineering was first established as a profession in the United Kingdom when the first chemical engineering course was given at the University of Manchester in 1887 by George E. Davis in the form of twelve lectures covering various aspects of industrial chemical practice. As a consequence George E. Davis is regarded as the world's first chemical engineer. Today, chemical engineering is a highly regarded profession. Chemical engineers with experience can become licensed Professional Engineers in the United States, aided by the National Society of Professional Engineers, or gain "Chartered" chemical-engineer status through the UK-based Institution of Chemical Engineers.

Professional Associations

Early programmes simply married industrial chemistry with mechanical engineering. For the profession to be born, proponents needed to clearly define their activity as something more than a mishmash of chemistry and engineering. To emphasize their identity and thus help the growth of their profession, chemical engineers formed the American Institute of Chemical Engineers in 1908. The Institution of Chemical Engineers was founded in 1922 and awarded a Royal Charter in 1957. In 1959, the Instituto Mexicano de Ingenieros Quimicos (IMIQ) was founded in Mexico.

Definitions

For the other established branches of engineering, there were ready associations in the mind of the common man: Mechanical Engineering meant machines, Electrical Engineering meant circuitry, and Civil Engineering meant structures. So chemical engineering can be symbolised as chemicals production.

Unit Operation

The answer, provided by Arthur D. Little to the President of MIT, was to emphasize the approach chemical engineers took to the design and analysis of processes rather than a process or a product. The concept of Unit operations was developed to emphasize the underlying unity among seemingly different operations. For example, the principles are the same whether one is concerned about separating alcohol from water in a fermenter, or separating gasoline from diesel in a refinery, as long as the basis of separation is generation of a vapor of a different composition from the liquid. Therefore such separation processes can be studied together as a unit operation (in this case called distillation). The concept has stood the profession in good stead in its phase of growth, and has even been used to understand the way the human body functions.

Unit Processes

In the early part of the last century, a parallel concept called Unit Processes was used to classify reactive processes. Thus oxidations, reductions, alkylations, etc. formed separate unit processes and were studied as such. This was natural considering the close affinity of chemical engineering to industrial chemistry at its inception. Gradually however, the subject of chemical reaction engineering has largely replaced the unit process concept. This subject looks at the entire body of chemical reactions as having a personality of its own, independent of the particular chemical species or chemical bonds involved. The latter does contribute to this personality in no small measure, but to design and operate chemical reactors, a knowledge of characteristics such as rate behaviour, thermodynamics, single or multiphase nature, etc. are more important. The emergence of chemical reaction engineering as a discipline truly signaled the severance of the umbilical cord connecting chemical engineering to industrial chemistry, and served to cement the truly unique character of this discipline.

Chemical Engineering Today

Sustainable Process Development

Chemical Engineers are today focusing their disciplines on the use of alternative feedstocks such as biomass and societal waste to produce chemicals and energy. Their aim is to use less carbon in the manufacturing processes by making them more carbon efficient and these lower carbon processes is sometimes termed "Green Chemistry" or sustainable processing. An example of this is the work of the Northeast of England Bioresources Group (NEBR) created by the members of the Northeast of England Process Industry Cluster (NEPIC). Their work is aimed at creating the future biorefinery and other opportunities for energy and products to be manufactured from biomass and waste materials. An example where such work has brought results is with Coca Cola who have already required the commodity chemical polyethylene terephthalate (PET) to be produced in a more environmentally friendly way for use in their soft drinks bottles and their product "PlantBottle" was introduced in 2009.

Chemical Engineering Locations

In the 21st century most professional Chemical Engineers work in contract engineering companies or in the process industry companies that such engineering contractors serve. These are Commodity-speciality-fine chemical, petrochemical, pharmaceutical, biotechnology, polymer and renewable energy and sustainable materials companies. Many such companies can be found clustered around the large scale infrastructure that these industries need, such as ports, power stations, storage facilities, pipelines etc. This infrastructure is particularly important to commodity and petrochemical companies. An example of such a location is on Teesside in the Northeast of the United Kingdom where the process industry companies and the engineering companies collaborate through membership of the Northeast of England Process Industry Cluster (NEPIC). In this region, as well as the process industry companies themselves, there is a particularly high number of engineering suppliers supporting the sector. This includes offices of global suppliers like ABB, Jacobs, AMEC and Foster Wheeler as well as many local firms that also serve the industry internationally, such as KHome International, Cordell, Hertel etc. Chemical Engineers also contribute to industry via specialist technology companies again examples in the NEPIC Cluster are companies such as DRD Power, and CompactGTL. One report has suggested that the highest concentration of such knowledge intensive businesses (KIBS) in Europe can be found in Northeast England and a significant number of their 14000 employees are chemical engineers. In locations such as this Chemical Engineers share best practice through local sub-groups of their Professional Institution. The Teesside sub-group of the Institution of Chemical Engineers regularly meet to present updates on technical and operational issues and to share best practice.

New Concepts and Innovations

By the 1940s, it became clear that unit operations alone was insufficient in developing chemical reactors. While the predominance of unit operations in chemical engineering courses in Britain and the United States continued until the 1960s, transport phenomena started to experience greater focus. Along with other novel concepts, such process systems engineering (PSE), a "second paradigm" was defined. Transport phenomena gave an analytical approach to chemical engineering while PSE focused on its synthetic elements, such as control system and process design. Developments in chemical engineering before and after World War II were mainly incited by the petrochemical industry, however, advances in other fields were made as well. Advancements in biochemical engineering in the 1940s, for example, found application in the pharmaceutical industry, and allowed for the mass production of various antibiotics, including penicillin and streptomycin. Meanwhile, progress in polymer science in the 1950s paved way for the "age of plastics".

Safety and Hazard Developments

Concerns regarding the safety and environmental impact of large-scale chemical manufacturing facilities were also raised during this period. *Silent Spring*, published in 1962, alerted its readers to the harmful effects of DDT, a potent insecticide. The 1974 Flixborough disaster in the United Kingdom resulted in 28 deaths, as well as damage to a chemical plant and three nearby villages. The 1984 Bhopal disaster in India resulted in almost 4,000 deaths. These incidents, along with other incidents, affected the reputation of the trade as industrial safety and environmental protection were given more focus. In response, the IChemE required safety to be part of every degree course that it accredited after 1982. By the 1970s, legislation and monitoring agencies were instituted in various countries, such as France, Germany, and the United States.

Recent Progress

Advancements in computer science found applications designing and managing plants, simplifying calculations and drawings that previously had to be done manually. The completion of the Human Genome Project is also seen as a major development, not only advancing chemical engineering but genetic engineering and genomics as well. Chemical engineering principles were used to produce DNA sequences in large quantities.

While the application of chemical engineering principles to these fields only began in the 1990s, Rice University researchers see this as a trend towards biotechnology. The latest book 'Chemical Process Technology and Simulation' gives a very good comprehensive account of the various products and methods of its production and also deals with computer based process simulation of complex processes as practiced in industry.

Chemical engineering involves the application of several principles. Key concepts are presented below.

Chemical Reaction Engineering

Chemical engineering involves managing plant processes and conditions to ensure optimal plant operation. Chemical reaction engineers construct models for reactor analysis and design using laboratory data and physical parameters, such as chemical thermodynamics, to solve problems and predict reactor performance.

Plant Design

Chemical engineering design concerns the creation of plans, specification, and economic analyses for new plants or plant modifications. Design engineers often work in a consulting role, designing plants to meet clients' needs. Design is limited by a number of factors, including funding, government regulations and safety standards. These constraints dictate a plant's choice of process, materials and equipment.

Process Design

A unit operation is a physical step in an individual chemical engineering process. Unit operations (such as crystallization, drying and evaporation) are used to prepare reactants, purifying and separating its products, recycling unspent reactants, and controlling energy transfer in reactors. On the other hand, a unit process is the chemical equivalent of a unit operation. Along with unit operations, unit processes constitute a process operation. Unit processes (such as nitration and oxidation) involve the conversion of material by biochemical, thermochemical and other means. Chemical engineers responsible for these are called process engineers.

Process Design is the most challenging field of chemical engineering. Overall process simulation is to be done using various software. The recent book "Chemical Process Technology and Simulation" by Srikumar Koyikkal gives many classical examples. It is also a text book of Chemical Process Technology.

Transport Phenomena

Transport phenomena occur frequently in industrial problems. These include fluid dynamics, heat transfer and mass transfer, which mainly concern momentum transfer, energy transfer and transport of chemical species respectively. Basic equations for describing the three transport phenomena in the macroscopic, microscopic and molecular levels are very similar. Thus, understanding transport phenomena requires thorough understanding of mathematics.

Applications and Practice

Chemical engineers "develop economic ways of using materials and energy". Chemical engineers use chemistry and engineering to turn raw materials into usable products, such as medicine, petrochemicals and plastics on a large-scale, industrial setting. They are also involved in waste management and research. Both applied and research facets could make extensive use of computers.

A chemical engineer may be involved in industry or university research where they are tasked in designing and performing experiments to create new and better ways of production, controlling pollution, conserving resources and making these processes safer. They may be involved in designing and constructing plants as a project engineer. In this field, the chemical engineer uses their knowledge in selecting plant equipment and the optimum method of production to minimize costs and increase profitability.

After its construction, they may help in upgrading its equipment. They may also be involved in its daily operations. Chemical engineers may be permanently employed at chemical plants to manage operations. Alternatively, they may serve in a consultant role to troubleshoot problems, manage process changes and otherwise assist plant operators.

Today, the field of chemical engineering is a diverse one, covering areas from biotechnology and nanotechnology to mineral processing.

Biochemical Engineering

Biochemistry, sometimes called biological chemistry, is the study of chemical processes within and relating to living organisms. By controlling information flow through biochemical signaling and the flow of chemical energy through metabolism, biochemical processes give rise to the complexity of life. Over the last 40 years, biochemistry has become so successful at explaining living processes that now almost all areas of the life sciences from botany to medicine are engaged in biochemical research. Today, the main focus of pure biochemistry is in understanding how biological molecules give rise to the processes that occur within living cells, which in turn relates greatly to the study and understanding of whole organisms. Biochemistry is closely related to molecular biology, the study of the molecular mechanisms by which genetic information encoded in DNA is able to result in the processes of life. Depending on the exact definition of the terms used, molecular biology can be thought of as a branch of biochemistry, or biochemistry as a tool with which to investigate and study molecular biology.

Much of biochemistry deals with the structures, functions and interactions of biological macromolecules, such as proteins, nucleic acids, carbohydrates and lipids, which provide the structure of cells and perform many of the functions associated with life. The chemistry of the cell also depends on the reactions of smaller molecules and ions. These can be inorganic, for example water and metal ions, or organic, for example the amino acids which are used to synthesize proteins. The mechanisms by which cells harness energy from their environment via chemical reactions are known as metabolism. The findings of biochemistry are applied primarily in medicine, nutrition, and agriculture. In medicine, biochemists investigate the causes and cures of disease. In nutrition, they study how to maintain health and study the effects of nutritional deficiencies. In agriculture, biochemists investigate soil and fertilizers, and try to discover ways to improve crop cultivation, crop storage and pest control.

Starting Materials: the Chemical Elements of Life

Around two dozen of the 92 naturally occurring chemical elements are essential to various kinds of biological life. Most rare elements on Earth are not needed by life (exceptions being selenium and iodine), while a few common ones (aluminum and titanium) are not used. Most organisms share element needs, but there are a few differences between plants and animals. For example ocean algae use bromine but land plants and animals seem to need none. All animals require sodium, but some plants do not. Plants need boron and silicon, but animals may not (or may need ultra-small amounts).

Just six elements—carbon, hydrogen, nitrogen, oxygen, calcium, and phosphorus—make up almost 99% of the mass of a human body. In addition to the six major elements that compose most of the human body, humans require smaller amounts of possibly 18 more.

Biomolecules

The four main classes of molecules in biochemistry (often called biomolecules) are carbohydrates, lipids, proteins, and nucleic acids. Many biological molecules are polymers: in this terminology, *monomers* are relatively small micromolecules that are linked together to create large macromolecules known as *polymers*. When monomers are linked together to synthesize a biological polymer, they undergo a process called dehydration synthesis. Different macromolecules can assemble in larger complexes, often needed for biological activity.

Carbohydrates

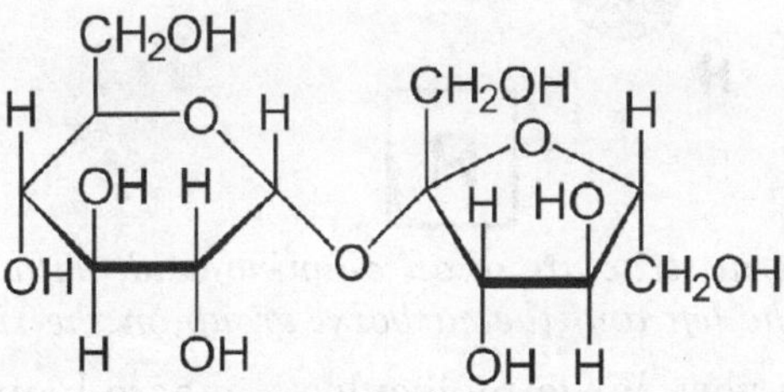

Figure: *A molecule of sucrose (glucose + fructose), a disaccharide.*

Carbohydrates are made from monomers called *monosaccharides*. Some of these monosaccharides include glucose ($C_6H_{12}O_6$), fructose ($C_6H_{12}O_6$), and deoxyribose ($C_5H_{10}O_4$). When two monosaccharides undergo dehydration synthesis, water is produced, as two hydrogen atoms and one oxygen atom are lost from the two monosaccharides' hydroxyl group.

Lipids

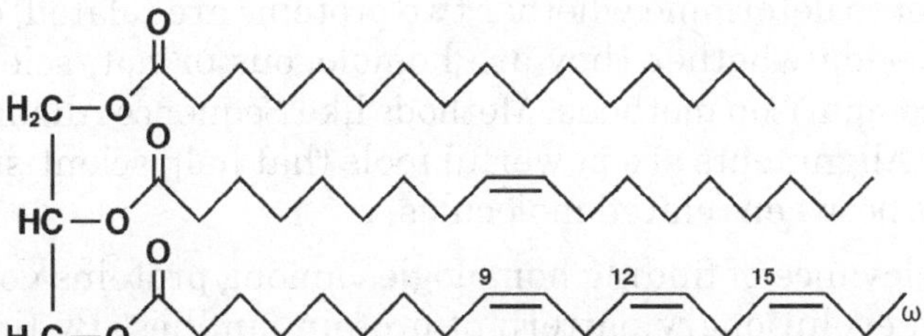

Figure: *A triglyceride with a glycerol molecule on the left and three fatty acids coming off it.*

Lipids are usually made from one molecule of glycerol combined with other molecules. In triglycerides, the main group of bulk lipids, there is one molecule of glycerol and three fatty acids. Fatty acids are considered the monomer in that case, and may be saturated (no double bonds in the carbon chain) or unsaturated (one or more double bonds in the carbon chain).

Lipids, especially phospholipids, are also used in various pharmaceutical products, either as co-solubilisers (e.g., in parenteral infusions) or else as drug carrier components (e.g., in a liposome or transfersome).

Proteins

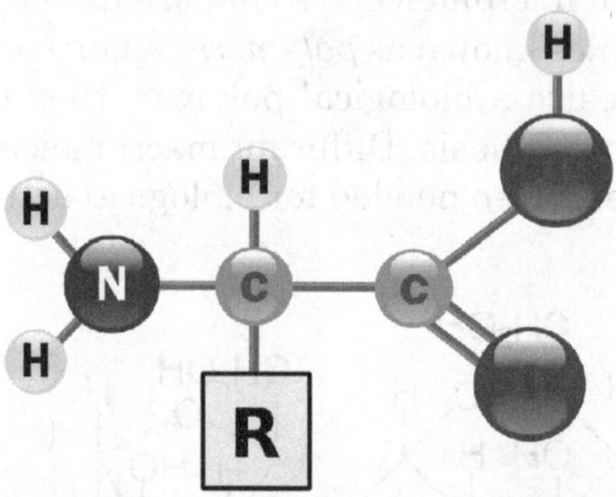

Figure: *The general structure of an α-amino acid, with the amino group on the left and the carboxyl group on the right.*

Proteins are very large molecules – macro-biopolymers – made from monomers called *amino acids*. There are 20 standard amino acids, each containing a carboxyl group, an amino group, and a side-chain (known as an "R" group). The "R" group is what makes each amino acid different, and the properties of the side-chains greatly influence the overall three-dimensional conformation of a protein. When amino acids combine, they form a special bond called a peptide bond through dehydration synthesis, and become a *polypeptide*, or protein.

In order to determine whether two proteins are related, or in other words to decide whether they are homologous or not, scientists use sequence-comparison methods. Methods like Sequence Alignments and Structural Alignments are powerful tools that help scientists identify homologies between related molecules.

The relevance of finding homologies among proteins goes beyond forming an evolutionary pattern of protein families. By finding how similar two protein sequences are, we acquire knowledge about their structure and therefore their function.

Nucleic Acids

Nucleic acids are the molecules that make up DNA, an extremely important substance that all cellular organisms use to store their genetic information. The most common nucleic acids are deoxyribonucleic acid (DNA) and ribonucleic acid (RNA). Their monomers are called nucleotides.

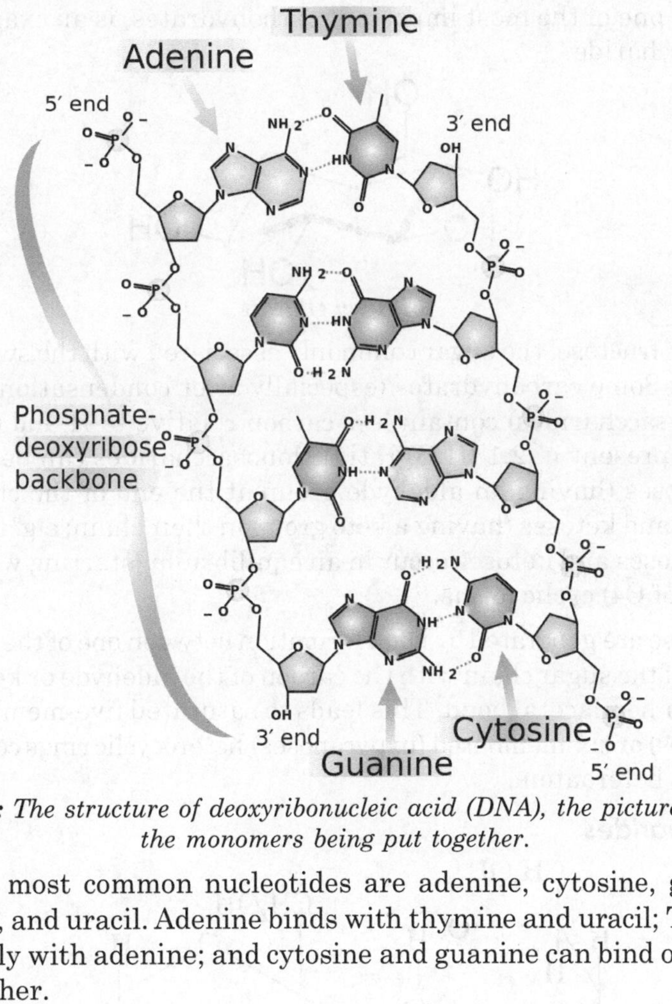

***Figure:** The structure of deoxyribonucleic acid (DNA), the picture shows the monomers being put together.*

The most common nucleotides are adenine, cytosine, guanine, thymine, and uracil. Adenine binds with thymine and uracil; Thymine binds only with adenine; and cytosine and guanine can bind only with one another.

Carbohydrates

The function of carbohydrates includes energy storage and providing structure. Sugars are carbohydrates, but not all carbohydrates are sugars. There are more carbohydrates on Earth than any other known type of biomolecule; they are used to store energy and genetic information, as well as play important roles in cell to cell interactions and communications.

Monosaccharides

The simplest type of carbohydrate is a monosaccharide, which between other properties contains carbon, hydrogen, and oxygen, mostly in a ratio of 1:2:1 (generalized formula $C_nH_{2n}O_n$, where n is at least 3). Glucose, one of the most important carbohydrates, is an example of a monosaccharide.

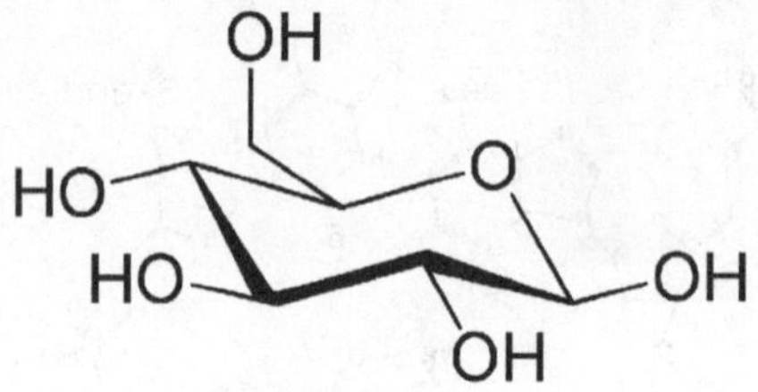

Figure: *Glucose*

So is fructose, the sugar commonly associated with the sweet taste of fruits. Some carbohydrates (especially after condensation to oligo- and polysaccharides) contain less carbon relative to H and O, which still are present in 2:1 (H:O) ratio. Monosaccharides can be grouped into aldoses (having an aldehyde group at the end of the chain, e.g. glucose) and ketoses (having a keto group in their chain; e.g. fructose). Both aldoses and ketoses occur in an equilibrium (starting with chain lengths of C4) cyclic forms.

These are generated by bond formation between one of the hydroxyl groups of the sugar chain with the carbon of the aldehyde or keto group to form a hemiacetal bond. This leads to saturated five-membered (in furanoses) or six-membered (in pyranoses) heterocyclic rings containing one O as heteroatom.

Disaccharides

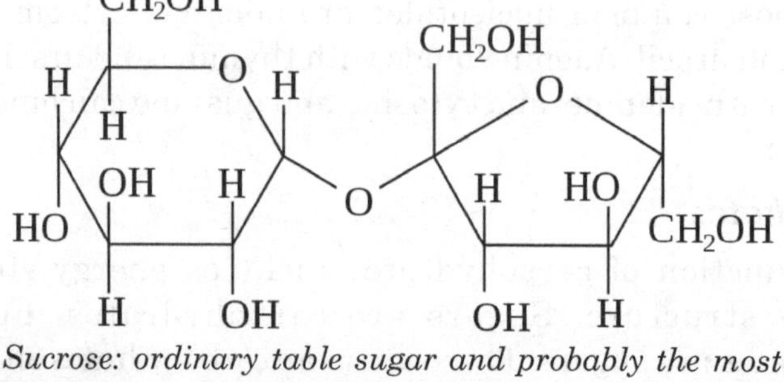

Figure: *Sucrose: ordinary table sugar and probably the most familiar carbohydrate.*

Two monosaccharides can be joined using dehydration synthesis, in which a hydrogen atom is removed from the end of one molecule and a hydroxyl group (—OH) is removed from the other; the remaining residues are then attached at the sites from which the atoms were removed. The H—OH or H_2O is then released as a molecule of water, hence the term *dehydration*. The new molecule, consisting of two monosaccharides, is called a *disaccharide* and is conjoined together by a glycosidic or ether bond. The reverse reaction can also occur, using a molecule of water to split up a disaccharide and break the glycosidic bond; this is termed *hydrolysis*. The most well-known disaccharide is sucrose, ordinary sugar (in scientific contexts, called *table sugar* or *cane sugar* to differentiate it from other sugars). Sucrose consists of a glucose molecule and a fructose molecule joined together. Another important disaccharide is lactose, consisting of a glucose molecule and a galactose molecule. As most humans age, the production of lactase, the enzyme that hydrolyzes lactose back into glucose and galactose, typically decreases. This results in lactase deficiency, also called *lactose intolerance*.

Sugar polymers are characterized by having reducing or non-reducing ends. A reducing end of a carbohydrate is a carbon atom that can be in equilibrium with the open-chain aldehyde or keto form. If the joining of monomers takes place at such a carbon atom, the free hydroxy group of the pyranose or furanose form is exchanged with an OH-side-chain of another sugar, yielding a full acetal. This prevents opening of the chain to the aldehyde or keto form and renders the modified residue non-reducing. Lactose contains a reducing end at its glucose moiety, whereas the galactose moiety form a full acetal with the C4-OH group of glucose. Saccharose does not have a reducing end because of full acetal formation between the aldehyde carbon of glucose (C1) and the keto carbon of fructose (C2).

Oligosaccharides and Polysaccharides

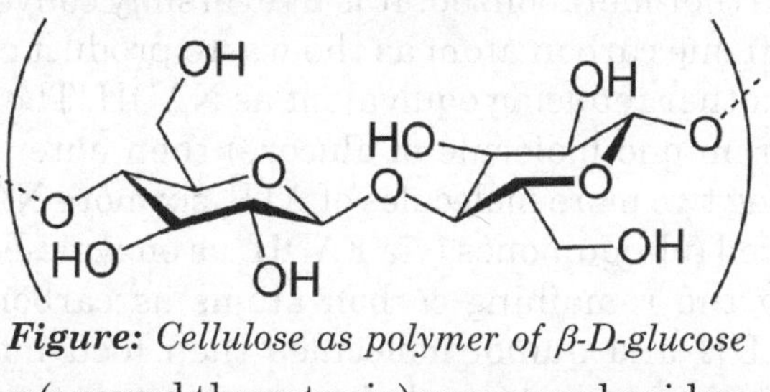

Figure: *Cellulose as polymer of β-D-glucose*

When a few (around three to six) monosaccharides are joined, it is called an *oligosaccharide* (*oligo*- meaning "few"). These molecules tend to be used as markers and signals, as well as having some other uses.

Many monosaccharides joined together make a polysaccharide. They can be joined together in one long linear chain, or they may be branched. Two of the most common polysaccharides are cellulose and glycogen, both consisting of repeating glucose monomers.

- *Cellulose* is made by plants and is an important structural component of their cell walls. Humans can neither manufacture nor digest it.
- *Glycogen*, on the other hand, is an animal carbohydrate; humans and other animals use it as a form of energy storage.

Use of Carbohydrates as an Energy Source

Glucose is the major energy source in most life forms. For instance, polysaccharides are broken down into their monomers (glycogen phosphorylase removes glucose residues from glycogen). Disaccharides like lactose or sucrose are cleaved into their two component monosaccharides.

Glycolysis (Anaerobic)

Glucose is mainly metabolized by a very important ten-step pathway called glycolysis, the net result of which is to break down one molecule of glucose into two molecules of pyruvate; this also produces a net two molecules of ATP, the energy currency of cells, along with two reducing equivalents as converting NAD^+ to NADH. This does not require oxygen; if no oxygen is available (or the cell cannot use oxygen), the NAD is restored by converting the pyruvate to lactate (lactic acid) (e.g., in humans) or to ethanol plus carbon dioxide (e.g., in yeast). Other monosaccharides like galactose and fructose can be converted into intermediates of the glycolytic pathway.

Aerobic

In aerobic cells with sufficient oxygen, as in most human cells, the pyruvate is further metabolized. It is irreversibly converted to acetyl-CoA, giving off one carbon atom as the waste product carbon dioxide, generating another reducing equivalent as NADH. The two molecules acetyl-CoA (from one molecule of glucose) then enter the citric acid cycle, producing two more molecules of ATP, six more NADH molecules and two reduced (ubi)quinones (via $FADH_2$ as enzyme-bound cofactor), and releasing the remaining carbon atoms as carbon dioxide. The produced NADH and quinol molecules then feed into the enzyme complexes of the respiratory chain, an electron transport system transferring the electrons ultimately to oxygen and conserving the released energy in the form of a proton gradient over a membrane (inner mitochondrial membrane in eukaryotes). Thus, oxygen is reduced to water and the original electron acceptors NAD^+ and quinone are regenerated. This is why humans breathe in oxygen and breathe out carbon dioxide. The energy released from transferring the electrons from high-energy states in NADH and quinol is conserved first as proton gradient and converted to ATP via ATP synthase. This generates an additional *28* molecules of ATP (24 from the 8 NADH + 4 from the 2 quinols), totaling to 32 molecules of ATP conserved per degraded glucose (two from glycolysis + two from the citrate cycle). It is clear that using oxygen to completely oxidize glucose provides an organism with far more energy than any oxygen-independent metabolic feature, and this is thought to be the reason why complex life appeared only after Earth's atmosphere accumulated large amounts of oxygen.

Gluconeogenesis

In vertebrates, vigorously contracting skeletal muscles (during weightlifting or sprinting, for example) do not receive enough oxygen to meet the energy demand, and so they shift to anaerobic metabolism, converting glucose to lactate. The liver regenerates the glucose, using a process called gluconeogenesis. This process is not quite the opposite of glycolysis, and actually requires three times the amount of energy gained from glycolysis (six molecules of ATP are used, compared to the two gained in glycolysis). Analogous to the above reactions, the glucose produced can then undergo glycolysis in tissues that need energy, be stored as glycogen (or starch in plants), or be converted to other monosaccharides or joined into di- or oligosaccharides. The combined pathways of glycolysis during exercise, lactate's crossing via the bloodstream to the liver, subsequent gluconeogenesis and release of glucose into the bloodstream is called the Cori cycle.

Proteins

Like carbohydrates, some proteins perform largely structural roles. For instance, movements of the proteins actin and myosin ultimately are responsible for the contraction of skeletal muscle. One property many proteins have is that they specifically bind to a certain molecule or class of molecules—they may be *extremely* selective in what they bind. Antibodies are an example of proteins that attach to one specific type of molecule. In fact, the enzyme-linked immunosorbent assay (ELISA), which uses antibodies, is one of the most sensitive tests modern medicine uses to detect various biomolecules. Probably the most important proteins, however, are the enzymes. These molecules recognise specific reactant molecules called *substrates*; they then catalyze the reaction between them. By lowering the activation energy, the enzyme speeds up that reaction by a rate of 10^{11} or more: a reaction that would normally take over 3,000 years to complete spontaneously might take less than a second with an enzyme. The enzyme itself is not used up in the process, and is free to catalyze the same reaction with a new set of substrates. Using various modifiers, the activity of the enzyme can be regulated, enabling control of the biochemistry of the cell as a whole.

In essence, proteins are chains of amino acids. An amino acid consists of a carbon atom bound to four groups. One is an amino group, —NH_2, and one is a carboxylic acid group, —COOH (although these exist as —NH_3^+ and —COO^- under physiologic conditions). The third is a simple hydrogen atom. The fourth is commonly denoted "—R" and is different for each amino acid. There are 20 standard amino acids. Some of these have functions by themselves or in a modified form; for instance, glutamate functions as an important neurotransmitter.

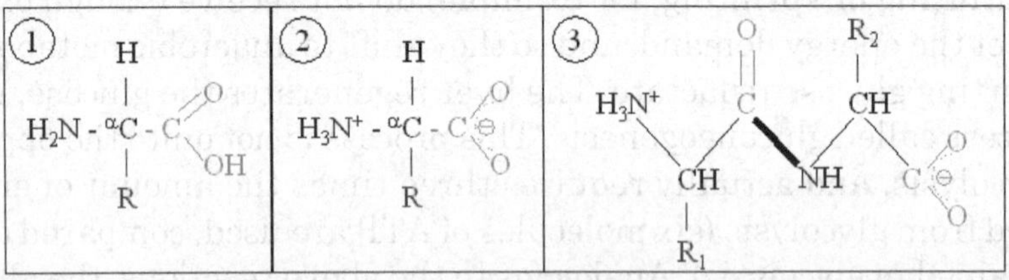

***Figure:** Generic amino acids (1) in neutral form, (2) as they exist physiologically, and (3) joined together as a dipeptide.*

Amino acids can be joined via a peptide bond. In this dehydration synthesis, a water molecule is removed and the peptide bond connects the nitrogen of one amino acid's amino group to the carbon of the other's carboxylic acid group. The resulting molecule is called a *dipeptide*, and short stretches of amino acids (usually, fewer than thirty) are called *peptides* or polypeptides. Longer stretches merit the title *proteins*. As an example, the important blood serum protein albumin contains 585 amino acid residues.

The structure of proteins is traditionally described in a hierarchy of four levels. The primary structure of a protein simply consists of its linear sequence of amino acids; for instance, "alanine-glycine-tryptophan-serine-glutamate-asparagine-glycine-lysine-...". Secondary structure is concerned with local morphology (morphology being the study of structure). Some combinations of amino acids will tend to curl up in a coil called an α-helix or into a sheet called a β-sheet; some α-helixes can be seen in the hemoglobin schematic above. Tertiary structure is the entire three-dimensional shape of the protein. This shape is determined by the sequence of amino acids. In fact, a single change can change the entire structure. The alpha chain of hemoglobin contains 146 amino acid residues; substitution of the glutamate residue at position 6 with a valine residue changes the behaviour of hemoglobin so much that it results in sickle-cell disease. Finally, quaternary structure is concerned with the structure of a protein with multiple peptide subunits, like hemoglobin with its four subunits. Not all proteins have more than one subunit.

Ingested proteins are usually broken up into single amino acids or dipeptides in the small intestine, and then absorbed. They can then be joined to make new proteins. Intermediate products of glycolysis, the citric acid cycle, and the pentose phosphate pathway can be used to make all twenty amino acids, and most bacteria and plants possess all the necessary enzymes to synthesize them. Humans and other mammals, however, can synthesize only half of them. They cannot synthesize isoleucine, leucine, lysine, methionine, phenylalanine, threonine, tryptophan, and valine. These are the essential amino acids, since it is essential to ingest them. Mammals do possess the enzymes to synthesize alanine, asparagine, aspartate, cysteine, glutamate, glutamine, glycine, proline, serine, and tyrosine, the nonessential amino acids. While they can synthesize arginine and histidine, they cannot produce it in sufficient amounts for young, growing animals, and so these are often considered essential amino acids.

If the amino group is removed from an amino acid, it leaves behind a carbon skeleton called an α-keto acid. Enzymes called transaminases can easily transfer the amino group from one amino acid (making it an α-keto acid) to another α-keto acid (making it an amino acid). This is important in the biosynthesis of amino acids, as for many of the pathways, intermediates from other biochemical pathways are converted to the α-keto acid skeleton, and then an amino group is added, often via transamination. The amino acids may then be linked together to make a protein.

A similar process is used to break down proteins. It is first hydrolyzed into its component amino acids. Free ammonia (NH_3), existing as the ammonium ion (NH_4^+) in blood, is toxic to life forms. A suitable method for excreting it must therefore exist. Different tactics have evolved in different animals, depending on the animals' needs. Unicellular organisms, of course, simply release the ammonia into the environment. Likewise, bony fish can release the ammonia into the water where it is quickly diluted. In general, mammals convert the ammonia into urea, via the urea cycle.

Lipids

The term lipid composes a diverse range of molecules and to some extent is a catchall for relatively water-insoluble or nonpolar compounds of biological origin, including waxes, fatty acids, fatty-acid derived phospholipids, sphingolipids, glycolipids, and terpenoids (e.g., retinoids and steroids). Some lipids are linear aliphatic molecules, while others have ring structures. Some are aromatic, while others are not. Some are flexible, while others are rigid.

Most lipids have some polar character in addition to being largely nonpolar. In general, the bulk of their structure is nonpolar or hydrophobic ("water-fearing"), meaning that it does not interact well with polar solvents like water. Another part of their structure is polar or hydrophilic ("water-loving") and will tend to associate with polar solvents like water. This makes them amphiphilic molecules (having both hydrophobic and hydrophilic portions). In the case of cholesterol, the polar group is a mere -OH (hydroxyl or alcohol). In the case of phospholipids, the polar groups are considerably larger and more polar, as described below.

Lipids are an integral part of our daily diet. Most oils and milk products that we use for cooking and eating like butter, cheese, ghee etc., are composed of fats. Vegetable oils are rich in various polyunsaturated fatty acids (PUFA). Lipid-containing foods undergo digestion within the body and are broken into fatty acids and glycerol, which are the final degradation products of fats and lipids.

Nucleic Acids

A nucleic acid is a complex, high-molecular-weight biochemical macromolecule composed of nucleotide chains that convey genetic information. The most common nucleic acids are deoxyribonucleic acid (DNA) and ribonucleic acid (RNA). Nucleic acids are found in all living cells and viruses. Aside from the genetic material of the cell, nucleic acids often play a role as second messengers, as well as forming the base molecule for adenosine triphosphate (ATP), the primary energy-carrier molecule found in all living organisms.

Nucleic acid, so called because of its prevalence in cellular nuclei, is the generic name of the family of biopolymers. The monomers are called nucleotides, and each consists of three components: a nitrogenous heterocyclic base (either a purine or a pyrimidine), a pentose sugar, and a phosphate group. Different nucleic acid types differ in the specific sugar found in their chain (e.g., DNA or deoxyribonucleic acid contains 2-deoxyriboses). Also, the nitrogenous bases possible in the two nucleic acids are different: adenine, cytosine, and guanine occur in both RNA and DNA, while thymine occurs only in DNA and uracil occurs in RNA.

Relationship to Other "Molecular-Scale" Biological Sciences

Researchers in biochemistry use specific techniques native to biochemistry, but increasingly combine these with techniques and ideas developed in the fields of genetics, molecular biology and biophysics. There has never been a hard-line among these disciplines in terms of content and technique. Today, the terms *molecular biology* and *biochemistry* are nearly interchangeable. The following figure is a schematic that depicts one possible view of the relationship between the fields:

- *Biochemistry* is the study of the chemical substances and vital processes occurring in living organisms. Biochemists focus heavily on the role, function, and structure of biomolecules. The study of the chemistry behind biological processes and the synthesis of biologically active molecules are examples of biochemistry.

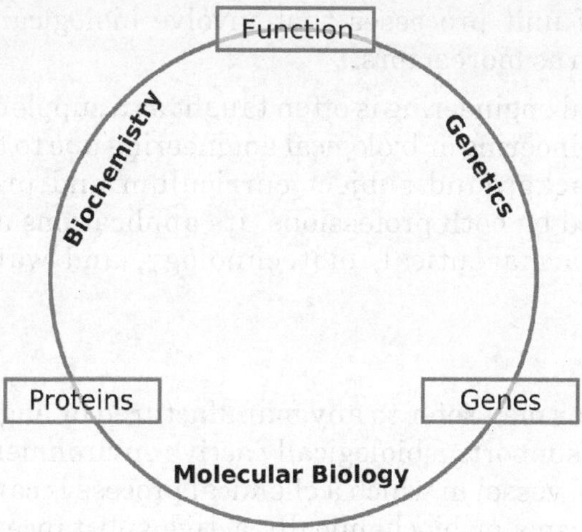

Figure: *Schematic relationship between biochemistry, genetics, and molecular biology*

- *Genetics* is the study of the effect of genetic differences on organisms. Often this can be inferred by the absence of a normal component (e.g., one gene). The study of "mutants" – organisms with a changed gene that leads to the organism being different with respect to the so-called "wild type" or normal phenotype. Genetic interactions (epistasis) can often confound simple interpretations of such "knock-out" or "knock-in" studies.

- *Molecular biology* is the study of molecular underpinnings of the process of replication, transcription and translation of the genetic material.

  The central dogma of molecular biology where genetic material is transcribed into RNA and then translated into protein, despite being an oversimplified picture of molecular biology, still provides a good starting point for understanding the field. This picture, however, is undergoing revision in light of emerging novel roles for RNA.
- *Chemical biology* seeks to develop new tools based on small molecules that allow minimal perturbation of biological systems while providing detailed information about their function. Further, chemical biology employs biological systems to create non-natural hybrids between biomolecules and synthetic devices (for example emptied viral capsids that can deliver gene therapy or drug molecules).

Biochemical engineering is a branch of chemical engineering or biological engineering that mainly deals with the design and construction of unit processes that involve biological organisms or molecules, such as bioreactors.

Biochemical engineering is often taught as a supplementary option to chemical engineering or biological engineering due to the similarities in both the background subject curriculum and problem-solving techniques used by both professions. Its applications are used in the food, feed, pharmaceutical, biotechnology, and water treatment industries.

Bioreactor

A bioreactor may refer to any manufactured or engineered device or system that supports a biologically active environment. In one case, a bioreactor is a vessel in which a chemical process is carried out which involves organisms or biochemically active substances derived from such organisms. This process can either be aerobic or anaerobic. These bioreactors are commonly cylindrical, ranging in size from litres to cubic metres, and are often made of stainless steel.

A bioreactor may also refer to a device or system meant to grow cells or tissues in the context of cell culture. These devices are being developed for use in tissue engineering or biochemical engineering. On the basis of mode of operation, a bioreactor may be classified as batch, fed batch or continuous (e.g. a continuous stirred-tank reactor model). An example of a continuous bioreactor is the chemostat.

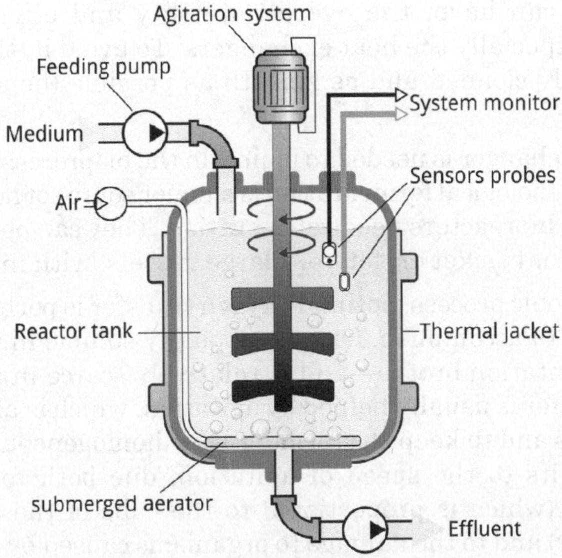

Figure: *General structure of batch type bioreactor*

Organisms growing in bioreactors may be suspended or immobilized. Immobilization is a general term describing a wide variety of the cell or the particle attachment or entrapment. It can be applied to basically all types of biocatalysts including enzymes, cellular organelles, animal and plant cells.

Large scale immobilized cell bioreactors are:

- moving media, also known as Moving Bed Biofilm Reactor (MBBR)
- packed bed
- fibrous bed
- membrane

Bioreactor Design

Bioreactor design is a relatively complex engineering task, which is studied in the discipline of biochemical engineering. Under optimum conditions, the microorganisms or cells are able to perform their desired function with a 100 percent rate of success. The bioreactor's environmental conditions like gas (i.e., air, oxygen, nitrogen, carbon dioxide) flow rates, temperature, pH and dissolved oxygen levels, and agitation speed/circulation rate need to be closely monitored and controlled.

Most industrial bioreactor manufacturers use vessels, sensors and a control system networked together.

Fouling can harm the overall sterility and efficiency of the bioreactor, especially the heat exchangers. To avoid it, the bioreactor must be easily cleaned and as smooth as possible (hence the round shape)..

A heat exchanger is needed to maintain the bioprocess at a constant temperature. Biological fermentation is a major source of heat, therefore in most cases bioreactors need refrigeration. They can be refrigerated with an external jacket or, for very large vessels, with internal coils.

In an aerobic process, optimal oxygen transfer is perhaps the most difficult task to accomplish. Oxygen is poorly soluble in water—even less in fermentation broths—and is relatively scarce in air (20.95%). Oxygen transfer is usually helped by agitation, which is also needed to mix nutrients and to keep the fermentation homogeneous. There are, however, limits to the speed of agitation, due both to high power consumption (which is proportional to the cube of the speed of the electric motor) and to the damage to organisms caused by excessive tip speed. In practice, bioreactors are often pressurized; this increases the solubility of oxygen in water.

Photobioreactor

Figure: *Moss photobioreactor with* Physcomitrella patens

A photobioreactor (PBR) is a bioreactor which incorporates some type of light source. Virtually any translucent container could be called a PBR, however the term is more commonly used to define a closed system, as opposed to an open tank or pond. Photobioreactors are used to grow small phototrophic organisms such as cyanobacteria, algae, or moss plants. These organisms use light through photosynthesis as their energy source and do not require sugars or lipids as energy source.

Consequently, risk of contamination with other organisms like bacteria or fungi is lower in photobioreactors when compared to bioreactors for heterotroph organisms.

Sewage Treatment

Bioreactors are also designed to treat sewage and wastewater. In the most efficient of these systems, there is a supply of a free-flowing, chemically inert medium which acts as a receptacle for the bacteria that break down the raw sewage. Examples of these bioreactors often have separate, sequential tanks and a mechanical separator or cyclone to speed the separation of water and biosolids. Aerators supply oxygen to the sewage and medium, further accelerating breakdown. Submersible mixers provide agitation in anoxic bioreactors to keep the solids in suspension and thereby ensure that the bacteria and the organic materials "meet".

In the process, the liquid's Biochemical Oxygen Demand (BOD) is reduced sufficiently to render the contaminated water fit for reuse. The biosolids can be collected for further processing, or dried and used as fertilizer. An extremely simple version of a sewage bioreactor is a septic tank whereby the sewage is left in situ, with or without additional media to house bacteria. In this instance, the biosludge itself is the primary host (activated sludge) for the bacteria. Septic systems are best suited where there is sufficient landmass, and the system is not subject to flooding or overly saturated ground, and where time and efficiency are not prioritized.

Because they are the engine that drives biological wastewater treatment, it is critical to closely monitor the quantity and quality of microorganisms in bioreactors. One method for this is via 2nd Generation ATP tests.

NASA Tissue Cloning Bioreactor

In bioreactors in which the goal is to grow cells or tissues for experimental or therapeutic purposes, the design is significantly different from industrial bioreactors. Many cells and tissues, especially mammalian ones, must have a surface or other structural support in order to grow, and agitated environments are often destructive to these cell types and tissues. Higher organisms, being auxotrophic, also require highly specialised growth media. NASA has developed a new type of bioreactor that artificially grows tissue in cell cultures. NASA's tissue bioreactor can grow heart tissue, skeletal tissue, ligaments, cancer tissue for study, and other types of tissue.

ATP Test

The ATP test is a process of rapidly measuring actively growing microorganisms through detection of adenosine triphosphate, or ATP.

ATP Testing Method

ATP is a molecule found in and around living cells, and as such it gives a direct measure of biological concentration and health. ATP is quantified by measuring the light produced through its reaction with the naturally occurring firefly enzyme luciferase using a luminometre. The amount of light produced is directly proportional to the amount of living organisms present in the sample.

ATP tests can be used to:

- Control biological treatment reactors
- Guide biocide dosing programs
- Determine drinking water cleanliness
- Manage fermentation processes
- Assess soil activity
- Determine corrosion / deposit process type
- Measure equipment or product sanitation

1st Generation Testing vs. 2nd Generation Testing

1st generation ATP tests are derived from hygiene monitoring uses where samples are relatively free of interferences. 2nd Generation tests are specifically designed for water, wastewater and industrial applications where, for the most part, samples contain a variety of components that can interfere with the ATP assay.

How ATP is Measured

ATP is a molecule found only in and around living cells, and as such it gives a direct measure of biological concentration and health. ATP is quantified by measuring the light produced through its reaction with the naturally-occurring firefly enzyme Luciferase using a Luminometre. The amount of light produced is directly proportional to the amount of biological energy present in the sample.

Within a water sample containing microorganisms, there are two types of ATP:

- Intracellular ATP – ATP contained within living biological cells.
- Extracellular ATP – ATP located outside of biological cells that has been released from dead or stressed organisms.

Accurate measurement of these two types of ATP is critical to utilising ATP-based measurements. Being able to accurately measure these different types of ATP offers the ability to assess biological health and activity, and subsequently control water and wastewater processes.

Biological Hydrogen Production (Algae)

The biological hydrogen production with algae is a method of photobiological water splitting which is done in a closed photobioreactor based on the production of hydrogen by algae. Algae produce hydrogen under certain conditions. In 2000 it was discovered that if *C. reinhardtii* algae are deprived of sulfur they will switch from the production of oxygen, as in normal photosynthesis, to the production of hydrogen.

History

In 1939 a German researcher named Hans Gaffron, while working at the University of Chicago, observed that the algae he was studying, *Chlamydomonas reinhardtii* (a green-algae), would sometimes switch from the production of oxygen to the production of hydrogen. He never discovered the cause for this change and for many years other scientists failed in their attempts at its discovery.

In the late 1990s, professor Anastasios Melis a researcher at the University of California at Berkeley discovered that if the algae culture medium is deprived of sulfur it will switch from the production of oxygen (normal photosynthesis), to the production of hydrogen. He found that the enzyme responsible for this reaction is hydrogenase, but that the hydrogenase lost this function in the presence of oxygen.

Melis found that depleting the amount of sulfur available to the algae interrupted its internal oxygen flow, allowing the hydrogenase an environment in which it can react, causing the algae to produce hydrogen. *Chlamydomonas moewusii* is also a good strain for the production of hydrogen. Scientists at the U.S. Department of Energy's Argonne National Laboratory are currently trying to find a way to take the part of the hydrogenase enzyme that creates the hydrogen gas and introduce it into the photosynthesis process. The result would be a large amount of hydrogen gas, possibly on par with the amount of oxygen created.

Milestones

1997- Professor Anastasios Melis discovered, after following Hans Gaffron's work, that the deprivation of sulfur will cause the algae to switch from producing oxygen to producing hydrogen. The enzyme, hydrogenase, he found was responsible for the reaction.

2006 - Researchers from the University of Bielefeld and the University of Queensland have genetically changed the single-cell green alga *Chlamydomonas reinhardtii* in such a way that it produces an especially large amount of hydrogen. The Stm6 can, in the long run, produce five times the volume made by the wild form of alga and up to 1.6-2.0 percent energy efficiency.

2007 - It was discovered that if copper is added to block oxygen generation algae will switch from the production of oxygen to hydrogen.

2007 - Anastasios Melis studying solar-to-chemical energy conversion efficiency in *tlaX* mutants of *Chlamydomonas reinhardtii*, achieved 15% efficiency, demonstrating that truncated Chl antenna size would minimize wasteful dissipation of sunlight by individual cells This solar-to-chemical energy conversion process could be coupled to the production of a variety of bio-fuels including hydrogen.

2008 - Anastasios Melis studying solar-to-chemical energy conversion efficiency in *tlaR* mutants of *Chlamydomonas reinhardtii*, achieved 25% efficiency out of a theoretical maximum of 30%.

2009 - A team from the University of Tennessee, Knoxville and Oak Ridge National Laboratory stated that the process was more than 10 times more efficient as the temperature increased.

2011 - Adding a bioengineered enzyme increases the rate of algal hydrogen production by about 400 percent.

2011 - A team at Argonne's Photosynthesis Group demonstrated how platinum nanoparticles can be linked to key proteins in algae to produce hydrogen fuel five times more efficiently.

Research

As of 2009, HydroMicPro is testing plate reactors. As of 2013, Grow Energy has developed novel system for the large-scale production of hydrogen from structural bioreactors.

Economics

It would take about 25,000 square kilometres to be sufficient to displace gasoline use in the US. To put this in perspective, this area represents approximately 10% of the area devoted to growing soya in the US. In 2004, the US Department of Energy issued a selling price of $2.60 per kilogram ($1.18/lb) as a goal for making renewable hydrogen economically viable. 1 kg is approximately the energy equivalent to a gallon of gasoline. To achieve this, the efficiency of light-to-hydrogen conversion must reach 10% while 2004 achieved-efficiency is only 1% and the 2004 actual selling price is estimated at $13.53 per kilogram

($6.14/lb). According to a 2004 DOE cost estimate, for a refueling station to supply 100 cars per day, it would need 300 kg. With current technology, a 300 kg per day stand-alone system will require 110,000 m^2 of pond area, 0.2 g/l cell concentration, a truncated antennae mutant and 10 cm pond depth. Areas of research to increase efficiency include developing oxygen-tolerant FeFe-hydrogenases and increased hydrogen production rates through improved electron transfer.

Truncated Antenna

The chlorophyll (Chl) antenna size in green algae is minimized, or truncated, to maximize photobiological solar conversion efficiency and H_2 production. The truncated Chl antenna size minimizes absorption and wasteful dissipation of sunlight by individual cells, resulting in better light utilisation efficiency and greater photosynthetic productivity by the green alga mass culture.

Moving Bed Biofilm Reactor (MBBR)

Moving Bed Biofilm Reactor (MBBR) processes improve reliability, simplify operation, and require less space than traditional wastewater treatment systems. MBBR technology employs thousands of polyethylene biofilm carriers operating in mixed motion within an aerated wastewater treatment basin. Each individual biocarrier increases productivity through providing protected surface area to support the growth of heterotrophic and autotrophic bacteria within its cells. It is this high-density population of bacteria that achieves high-rate biodegradation within the system, while also offering process reliability and ease of operation. This technology provides cost-effective treatment with minimal maintenance since MBBR processes self-maintain an optimum level of productive biofilm. Additionally, the biofilm attached to the mobile biocarriers within the system automatically responds to load fluctuations.

Process Benefits

Compact Design: A fraction of the size of conventional systems

Expandable: Capacity can be easily upgraded by simply increasing the fill fraction of biofilm carriers

Single Pass Process: No return activated sludge stream required

Load Responsive: Actively sloughed biofilm automatically responds to load fluctuations

Minimal Maintenance: No F/M ratios or MLSS levels to maintain

MBBR processes are an excellent solution for common wastewater applications including:

BOD Reduction: Hydrogen peroxide (H2O2) has been used to reduce the BOD and COD of industrial wastewaters for many years. While the cost of removing BOD and COD through chemical oxidation with hydrogen peroxide is typically greater than that through physical or biological means, there are nonetheless specific situations which justify the use of hydrogen peroxide.

These include: Predigestion of wastewaters which contain moderate to high levels of compounds that are toxic, inhibitory, or recalcitrant to biological treatment (e.g., pesticides, plasticizers, resins, coolants, and dyestuffs);

- Pretreatment of high strength / low flow wastewaters – where biotreatment may not be practical – prior to discharge to a Publicly Owned Treatment Works (POTW);
- Enhanced separation of entrained organics by flotation and settling processes; and
- Supply of supplemental Dissolved Oxygen (DO) when biological treatment systems experience temporary overloads or equipment failure.

As indicated by these examples, H2O2 can be used as a stand-alone treatment or as an enhancement to existing physical or biological treatment processes, depending on the situation.

Discussion of H_2O_2 Applications and Mechanisms

Direct Chemical Oxidation Using Hydrogen Peroxide

Hydrogen peroxide can be used alone or with catalysts – such as iron (Fe2+ or Fe3+), UV light, ozone (O_3) and alkali – to oxidize BOD/COD contributing compounds in wastewaters. The type of oxidation needed depends on the type of BOD/COD present. This relationship is present in the figure below.

Oxidant System	*Chemical Oxygen Demand*		
	Type A (Sulfide, Thiosulfate, Sulfite)	*Type B (Phenols, Cyanides, Amines)*	*Type C (BTEX, TOCl, Paraffins)*
Type A			
H2O2	X		
Type B			
H2O2 / OH-	X	X	
H2O2 / M+	X	X	
H2O2 / H+	X	X	
Type C			
H2O2 / Fe	X	X	X
H2O2 / O3	X	X	X
H2O2 / UV	X	X	X

Note: Whether an oxidant system will degrade a specific pollutant (i.e., affect its COD) will depend on the oxidant system and the pollutant. Type A oxidants react only with Type A pollutants; whereas, Type C oxidants, being more reactive, react with most any pollutant. However, Type C oxidants generally react preferentially with Type A pollutants.

If a large fraction of the BOD and COD is contributed by inorganic reduced sulfur compounds such as sulfides, sulfides, or thiosulfate, then hydrogen peroxide alone is typically effective.

Depending on the wastewater pH, the oxidation of these compounds by H2O2 yields sulfate or colloidal sulfur, neither which contribute to BOD and COD. If the primary contributors to BOD and COD are dissolved organics, then a more reactive oxidation system is needed. Moderate activation of hydrogen peroxide can be achieved by:

1) alkali (generating the perhydroxyl ion , OOH- – the active agent in peroxide bleaching systems);
2) certain transition metals (e.g., tungstate, vanadate, molybdate) which form reactive peroxometal complexes in-situ; and
3) certain mineral acids (e.g., sulfuric) which form reactive peroxyacid derivatives such as peroxymonosulfuric acid (Caro's Acid) ex-situ.

For the more recalcitrant organics such as chlorinated solvents, extremely reactive free radical systems (termed Advanced Oxidation Processes) are needed. A generalized reaction using Fenton's Reagent for reducing BOD and COD can be expressed as follows:

With Fe+2

Step-1: BOD/COD + H2O2 —> partially oxidized species With Fe+2

Step-2: partially oxidized species + H_2O_2 —> CO_2 + H_2O + inorganic salts

The extent of oxidation (and therefore the degree of direct BOD/COD reduction) typically depends on the amount of hydrogen peroxide used. The theoretical hydrogen peroxide requirement is about 2.1 lbs (as 100%) per lb-BOD and COD oxidized.

In many cases, however, complete digestion of the organic compounds to carbon dioxide and water is not needed. Partial oxidation to intermediate compounds minimizes chemical consumption and often results in substantial reductions in BOD and COD and toxicity.

Degrees of Oxidation

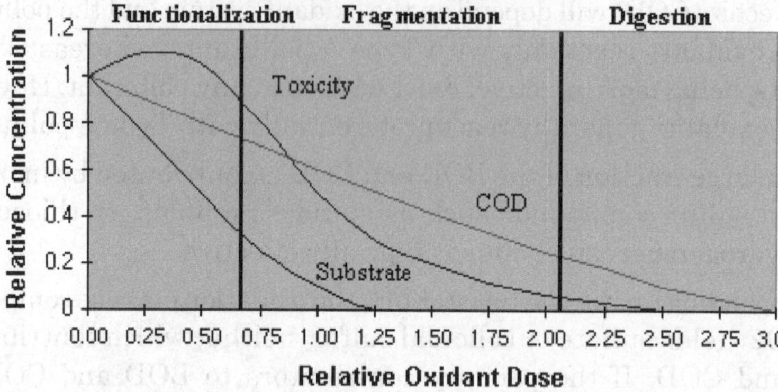

Enhanced Physical Separation of BOD and COD with Hydrogen Peroxide

Enhanced physical separation of BOD and COD with hydrogen peroxide may occur is two ways. In the first case, partial oxidation of organic contaminants results in more polar (charged) substances which are more amenable to adsorption onto coagulants and flocculants. As illustrated in the example below, this allows BOD and COD removal efficiencies with less than stoichiometric hydrogen peroxide doses.

In the second case, enhanced physical separation (flotation) of fats, oils and greases (FOG) is provided by H2O2. This occurs by the natural decomposition of hydrogen peroxide to oxygen and water, i.e., hydrogen peroxide will supersaturate the wastewater with oxygen, which results in the formation of evenly dispersed microbubbles which scavenge FOG constituents as they rise to the surface of the water. In some cases, this can increase BOD removal through dissolved air flotation cells from e.g., 50% to 90-95%. Typical doses are 25-100 mg/L H_2O_2, the cost for which can often be offset against savings in coagulant use – a polyelectrolyte polymer is generally still needed.

Enhanced Flocculation by Pre-Oxidation

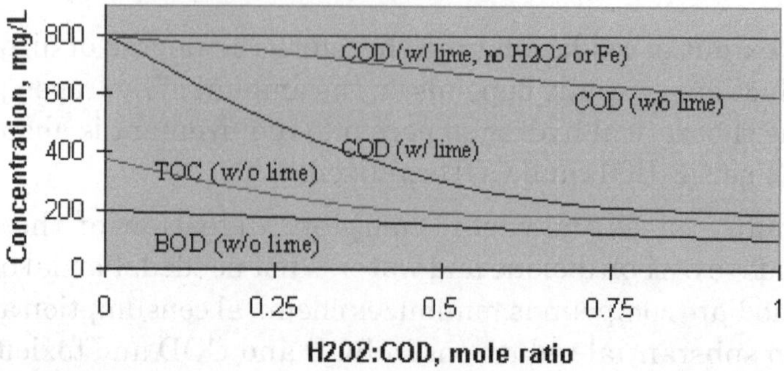

Hydrogen Peroxide (H_2O_2) as a Supplemental Oxygen Source

The BOD removal efficiency of aerobic biological treatment processes depends on a number of factors including (but not limited to): influent BOD loading, F:M ratio, temperature, nutrient levels, and dissolved oxygen (DO) concentrations.

Many biological treatment facilities use hydrogen peroxide to supplement DO levels when oxygen limited conditions in aeration basins or lagoons result in poor BOD removal. These conditions can be brought about by unexpected peaks in influent BOD loading; seasonal variations in BOD loading (e.g., fruit and vegetable processing); and hot weather – which reduces the efficiency of oxygen transfer by mechanical aeration equipment (i.e., O2 solubility decreases as temperature increases). These conditions may or may not be accompanied by filamentous bulking.

When hydrogen peroxide is used to supplement DO, it is metred directly into the aeration basin of a biological treatment system to provide an immediate source of DO. The conversion of hydrogen peroxide to DO in an activated sludge mixed liquor proceeds according to the following reaction:

(Catalase enzyme)

$$2\ H2O2 \longrightarrow O2 + 2\ H2O$$

Theoretical hydrogen peroxide requirement: 0.48 lbs H2O2 (100%) per mg/L DO

Catalase enzyme is a natural decomposition catalyst for hydrogen peroxide, and is found in all activated sludge mixed liquors, being produced by most aerobic organisms. Because this enzymatic decomposition of hydrogen peroxide is very rapid, the oxygen supplied by hydrogen peroxide is immediately available for uptake by the aerobic organisms. The above reaction shows that two parts of hydrogen peroxide will yield one part of DO.

Therefore, the amount of hydrogen peroxide required to oxygenate the wastewater is surprisingly small. For example, the theoretical amount of hydrogen peroxide required to increase the DO by 1 mg/L in a treatment plant that averages 5 MGD flow is about 17 gpd-50%. In actual practice, the requirement may be higher due to inefficiencies in oxygen uptake and side reactions with oxidizable compounds. Moving Bed Biofilm Reactor systems deliver a flexible, cost-effective, and easy-to-operate means to address current wastewater requirements and the expandability to meet future loads or more stringent discharge requirements within a compact design.

Septic Tank

A septic tank is a key component of the septic system, a small-scale sewage treatment system common in areas with no connection to main sewage pipes provided by local governments or private corporations. Other components, typically mandated and/or restricted by local governments, optionally include pumps, alarms, sand filters, and clarified liquid effluent disposal means such as a septic drain field, ponds, natural stone fibre filter plants or peat moss beds.

Septic systems are a type of onsite sewage facility (OSSF). In North America, approximately 25% of the population relies on septic tanks; this can include suburbs and small towns as well as rural areas. Indianapolis is an example of a large city where many of the city's neighbourhoods are still on separate septic systems. In Europe, septic systems are generally limited to rural areas. Since a septic system requires a drainfield that uses a lot of land area, they are not suitable for densely built cities. The term "septic" refers to the anaerobic bacterial environment that develops in the tank which decomposes or mineralizes the waste discharged into the tank. Septic tanks can be coupled with other onsite wastewater treatment units such as biofilters or aerobic systems involving artificial forced aeration.

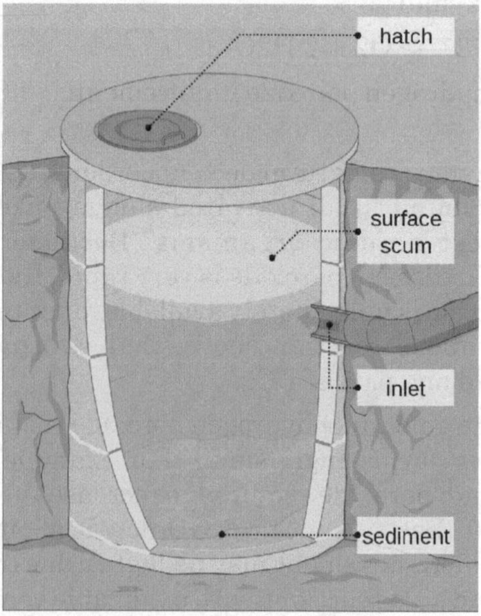

Figure: *Cesspit scheme*

Periodic preventive maintenance is required to remove the irreducible solids that settle and gradually fill the tank, reducing its efficiency. In most jurisdictions this maintenance is required by law, yet often not enforced. According to the U.S Environmental Protection Agency, in the United States it is the home owners' responsibility to maintain their septic system. Those who disregard the requirement will eventually be faced with extremely costly repairs when solids escape the tank and clog the clarified liquid effluent disposal system. A properly maintained system, on the other hand, can last for decades or possibly even a lifetime.

Description

A septic tank generally consists of a tank (or sometimes more than one tank) of between 4000 and 7500 litres (1,000 and 2,000 gallons) connected to an inlet wastewater pipe at one end and a septic drain field at the other. In general, these pipe connections are made via a T pipe, which allows liquid to enter and exit without disturbing any crust on the surface. Today, the design of the tank usually incorporates two chambers (each equipped with a manhole cover), which are separated by means of a dividing wall that has openings located about midway between the floor and roof of the tank. Waste water enters the first chamber of the tank, allowing solids to settle and scum to float. The settled solids are anaerobically digested, reducing the volume of solids. The liquid component flows through the dividing wall into the second chamber, where further settlement takes place, with the excess liquid then draining in a relatively clear condition from the outlet into the leach field, also referred to as a drain field or seepage field, depending upon locality. A percolation test is required to establish the porosity of the local soil conditions for the drain field design.

The remaining impurities are trapped and eliminated in the soil, with the excess water eliminated through percolation into the soil (eventually returning to the groundwater), through evaporation, and by uptake through the root system of plants and eventual transpiration. A piping network, often laid in a stone-filled trench, distributes the wastewater throughout the field with multiple drainage holes in the network. The size of the leach field is proportional to the volume of wastewater and inversely proportional to the porosity of the drainage field. The entire septic system can operate by gravity alone or, where topographic considerations require, with inclusion of a lift pump. Certain septic tank designs include siphons or other methods of increasing the volume and velocity of outflow to the drainage field. This helps to load all portions of the drainage pipe more evenly and extends the drainage field life by preventing premature clogging.

An Imhoff tank is a two-stage septic system where the sludge is digested in a separate tank. This avoids mixing digested sludge with incoming sewage. Also, some septic tank designs have a second stage where the effluent from the anaerobic first stage is aerated before it drains into the seepage field.

Waste that is not decomposed by the anaerobic digestion eventually has to be removed from the septic tank, or else the septic tank fills up and wastewater containing undecomposed material discharges directly to the drainage field. Not only is this detrimental for the environment but, if the sludge overflows the septic tank into the leach field, it may clog the leach field piping or decrease the soil porosity itself, requiring expensive repairs.

How often the septic tank has to be emptied depends on the volume of the tank relative to the input of solids, the amount of indigestible solids, and the ambient temperature (because anaerobic digestion occurs more efficiently at higher temperatures), as well as usage, system characteristics and the requirements of the relevant authority. Some health authorities require tanks to be emptied at prescribed intervals, while others leave it up to the determination of an inspector.

Some systems require pumping every few years or sooner, while others may be able to go 10–20 years between pumpings. An older system with an undersize tank that is being used by a large family will require much more frequent pumping than a new system used by only a few people. Anaerobic decomposition is rapidly restarted when the tank re-fills. A properly designed and normally operating septic system is odor-free and, besides periodic inspection and pumping of the septic tank, should last for decades with no maintenance. A well designed and maintained concrete, fibreglass, or plastic tank should last about 50 years.

Potential Problems

Like any system, a septic system requires maintenance. Although septic systems generally require less maintenance than others, in contrast to communally connected sewage systems, the maintenance of a septic system is often the responsibility of the resident or property owner. For this reason they are sometimes perceived to require higher maintenance than other systems. Neglected or abused systems can pose the following problems:

- Excessive dumping of cooking oils and grease can cause the inlet drains to block. Oils and grease are often difficult to degrade and can cause odor problems and difficulties with the periodic emptying.

- Flushing non-biodegradable items such as cigarette butts and hygiene products such as sanitary napkins, tampons, and cotton buds/swabs will rapidly fill or clog a septic tank, so as in other systems, those materials should not be disposed of in that way.
- As with all drainage systems, the use of garbage disposals for disposal of waste food can cause a rapid overload of the system with solids and early failure.
- Certain chemicals may damage the components of a septic tank, especially pesticides, herbicides, materials with high concentrations of bleach or caustic soda (lye) or any other inorganic materials such as paints or solvents.
- Certain chemicals can kill the septic bacteria needed for the system to operate. Most notably, even very small quantities of silver nitrate will kill an entire culture.
- As with communal sewage systems, roots from trees and shrubbery growing above the tank or the drainfield may clog and/or rupture them.
- Playgrounds and storage buildings may cause damage to a tank and the drainage field. In addition, covering the drainage field with an impermeable surface, such as a driveway or parking area, will seriously affect its efficiency and possibly damage the tank and absorption system.
- The flushing of salted water into the septic system can lead to sodium binding in the drainfield. The clay and fine silt particles bind together and effectively waterproof the leach field, rendering it ineffective.
- Like communal sewage systems, excessive water entering the system will overload it and cause it to fail. Checking for plumbing leaks and practising water conservation will help optimize the system's operation.
- Very high rainfall, rapid snowmelt, and flooding from rivers or the sea can all prevent a drain field from operating, and can cause flow to back up, interfering with the normal operation of the tank. High winter water tables can also result in groundwater flowing back into the septic tank.
- Over time, biofilms develop on the pipes of the drainage field, which can lead to blockage. Such a failure can be referred to as "biomat failure".
- Septic tanks by themselves are ineffective at removing nitrogen compounds that have potential to cause algal blooms in

waterways into which affected water from a septic system finds its way. This can be remedied by using a nitrogen-reducing technology, or by simply ensuring that the leach field is properly sited to prevent direct entry of effluent into bodies of water.

Environmental Issues

Whilst a properly maintained septic tank poses no more environmental problems than centralised municipal sewage treatment, certain problems could potentially arise from a neglected installation. Some pollutants, especially sulfates, under the anaerobic conditions of septic tanks, are reduced to hydrogen sulfide, a pungent and toxic gas. Likewise, methane, a potent greenhouse gas, is another by-product. Nitrates and organic nitrogen compounds are reduced to ammonia.

Because of the anaerobic conditions, fermentation processes take place, which ultimately generate carbon dioxide and methane. The fermentation processes cause the contents of a septic tank to be anaerobic with a low redox potential, which keeps phosphates in a soluble and, thus, mobilized form. Phosphates discharged from a septic tank into the environment can trigger prolific plant growth including algal blooms, which can also include blooms of potentially toxic cyanobacteria.

The soil's capacity to retain phosphorus is usually large enough to cope with the load through a normal residential septic tank. An exception occurs when septic drain fields are located in sandy or coarser soils on property adjoining a water body. Because of limited particle surface area, these soils can become saturated with phosphates. Phosphates will progress beyond the treatment area, posing a threat of eutrophication to surface waters.

In areas with high population density, groundwater pollution levels often exceed acceptable limits. Some small towns are facing the costs of building very expensive centralized wastewater treatment systems because of this problem, owing to the high cost of extended collection systems. To slow residential development which might increase the need to construct an expensive centralised sewerage system, building moratoriums and limits on the subdivision of property are often imposed. Ensuring existing septic tanks are functioning properly can also be helpful for a limited time, but becomes less effective as a primary remediation strategy as population density increases.

In areas adjacent to water bodies with fish or shellfish intended for human consumption, improperly maintained and failing septic systems contribute to pollution levels that can force harvest restrictions and/or commercial or recreational harvest closures. In Washington State, USA, a “shellfish protection district” or “clean water district” is a geographic service area designated by a county to protect water quality and tideland resources. The district provides a mechanism to generate local funds for water quality services to control non-point sources of pollution, such as septic system maintenance. The district also serves as an educational resource, calling attention to the pollution sources that threaten shellfish growing waters.

Trees in the vicinity of a concrete septic tank have the potential to penetrate the tank as the system ages and the concrete begins to develop cracks and small leaks. Tree roots can cause serious flow problems due to plugging and blockage of drain pipes, added to which the trees themselves tend to grow extremely vigorously due to the ready supply of nutrients from the septic system.

Chapter 2

Petrochemical Engineering

Petrochemicals are chemical products derived from petroleum. Some chemical compounds made from petroleum are also obtained from other fossil fuels, such as coal or natural gas, or renewable sources such as corn or sugar cane.

***Figure:** Petrochemical plant in The Kingdom of Saudi Arabia*

The two most common petrochemical classes are olefins (including ethylene and propylene) and aromatics (including benzene, toluene and xylene isomers). Oil refineries produce olefins and aromatics by fluid catalytic cracking of petroleum fractions.

Chemical plants produce olefins by steam cracking of natural gas liquids like ethane and propane. Aromatics are produced by catalytic reforming of naphtha. Olefins and aromatics are the building-blocks for a wide range of materials such as solvents, detergents, and adhesives. Olefins are the basis for polymers and oligomers used in plastics, resins, fibres, elastomers, lubricants, and gels.

Global ethylene and propylene production are about 115 million tonnes and 70 million tonnes per annum, respectively. Aromatics production is approximately 70 million tonnes. The largest petrochemical industries are located in the USA and Western Europe; however, major growth in new production capacity is in the Middle East and Asia. There is substantial inter-regional petrochemical trade.

Primary petrochemicals are divided into three groups depending on their chemical structure:

- Olefins includes ethylene, propylene, and butadiene. Ethylene and propylene are important sources of industrial chemicals and plastics products. Butadiene is used in making synthetic rubber.
- Aromatics includes benzene, toluene, and xylenes. Benzene is a raw material for dyes and synthetic detergents, and benzene and toluene for isocyanates MDI and TDI used in making polyurethanes. Manufacturers use xylenes to produce plastics and synthetic fibres.
- Synthesis gas is a mixture of carbon monoxide and hydrogen used to make ammonia and methanol. Ammonia is used to make the fertilizer urea and methanol is used as a solvent and chemical intermediate.

The prefix "petro-" is an arbitrary abbreviation of the word "petroleum"; since "petro-" is Ancient Greek for "rock" and "oleum" means "oil". Therefore, the etymologically correct term would be "oleochemicals". However, the term oleochemical is used to describe chemicals derived from plant and animal fats.

Sources

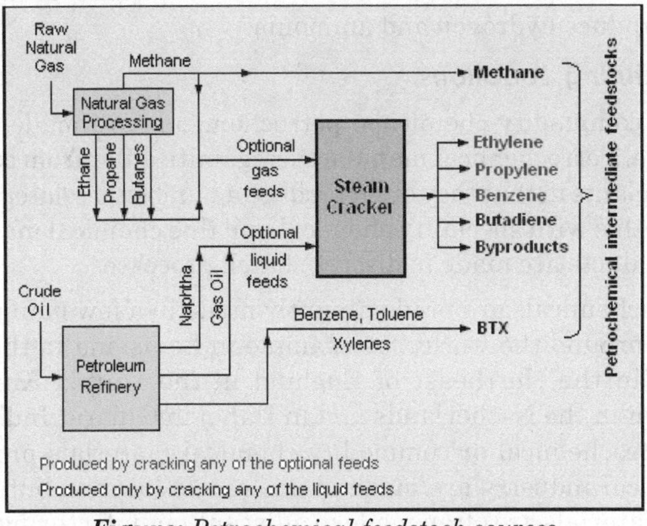

***Figure:** Petrochemical feedstock sources*

The adjacent diagram schematically depicts the major hydrocarbon sources used in producing petrochemicals are:

- Methane, ethane, propane and butanes: Obtained primarily from natural gas processing plants.
- Naphtha obtained from petroleum refineries.
- Benzene, toluene and xylenes, as a whole referred to as BTX and primarily obtained from petroleum refineries by extraction from the reformate produced in catalytic reformers.
- Gas oil obtained from petroleum refineries.

Methane and BTX are used directly as feedstocks for producing petrochemicals. However, the ethane, propane, butanes, naphtha and gas oil serve as optional feedstocks for steam-assisted thermal cracking plants referred to as *steam crackers* that produce these intermediate petrochemical feedstocks:

- Ethylene
- Propylene
- Butenes and butadiene
- Benzene

In 2007, the amounts of ethylene and propylene produced in steam crackers were about 115 Mt (megatonnes) and 70 Mt, respectively. The output ethylene capacity of large steam crackers ranged up to as much as 1.0 – 1.5 Mt per year.

Steam crackers are not to be confused with steam reforming plants used to produce hydrogen and ammonia.

Manufacturing Locations

Like commodity chemicals, petrochemicals are made on a very large scale. Petrochemical manufacturing units differ from commodity chemical plants in that they often produce a number of related products. Compare this with speciality chemical and fine chemical manufacture where products are made in discrete batch processes.

Petrochemicals are predominantly made in a few manufacturing locations around the world, for example in Louisiana in the USA, in Teesside in the Northeast of England in the United Kingdom, in Rotterdam in the Netherlands and in Dahej in Gujarat India. Not all of the petrochemical or commodity chemical materials produced by the chemical industry are made in one single location but groups of related materials are often made in adjacent manufacturing plants to induce industrial symbiosis as well as material and utility efficiency and other economies of scale. This is known in chemical engineering terminology as integrated manufacturing.

Speciality and fine chemical companies are sometimes found in similar manufacturing locations as petrochemicals but, in most cases, they do not need the same level of large scale infrastructure (e.g., pipelines, storage, ports and power, etc.) and therefore can be found in multi-sector business parks.

The large scale petrochemical manufacturing locations have clusters of manufacturing units that share utilities and large scale infrastructure such as power stations, storage tanks, port facilities, road and rail terminals.

In the United Kingdom for example, there are 4 main locations for such manufacturing: near the River Mersey in Northwest England, on the Humber on the East coast of Yorkshire, in Grangemouth near the Firth of Forth in Scotland and in Teesside as part of the Northeast of England Process Industry Cluster (NEPIC). To demonstrate the clustering and integration, some 50% of the United Kingdom's petrochemical and commodity chemicals are produced by the NEPIC industry cluster companies in Teesside.

List of Significant Petrochemicals and Their Derivatives

The following is a partial list of the major commercial petrochemicals and their derivatives:

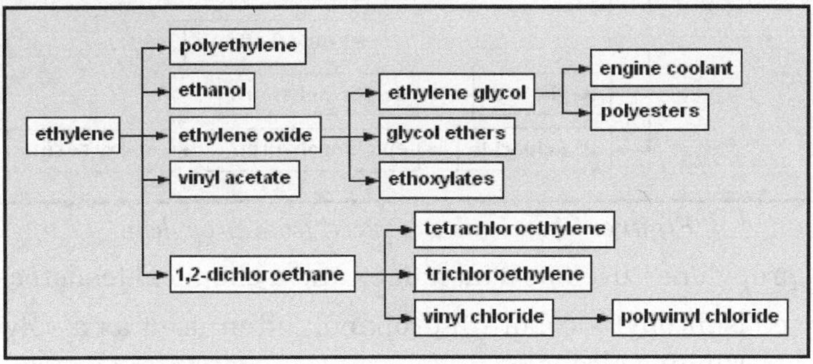

Figure: *Chemicals produced from ethylene*

- ethylene - the simplest olefin; used as a chemical feedstock and ripening stimulant
  - o polyethylene - polymerized ethylene
  - o ethanol - via ethylene hydration (chemical reaction adding water) of ethylene

- o ethylene oxide - via ethylene oxidation
  - – ethylene glycol - via ethylene oxide hydration
  - – engine coolant - ethylene glycol, water and inhibitor mixture
  - – polyesters - any of several polymers with ester linkages in the backbone chain
  - – glycol ethers - via glycol condensation
  - – ethoxylates
- o vinyl acetate
- o 1,2-dichloroethane
  - – trichloroethylene
  - – tetrachloroethylene - also called perchloroethylene; used as a dry cleaning solvent and degreaser
  - – vinyl chloride - monomer for polyvinyl chloride
  - – polyvinyl chloride (PVC) - type of plastic used for piping, tubing, other things

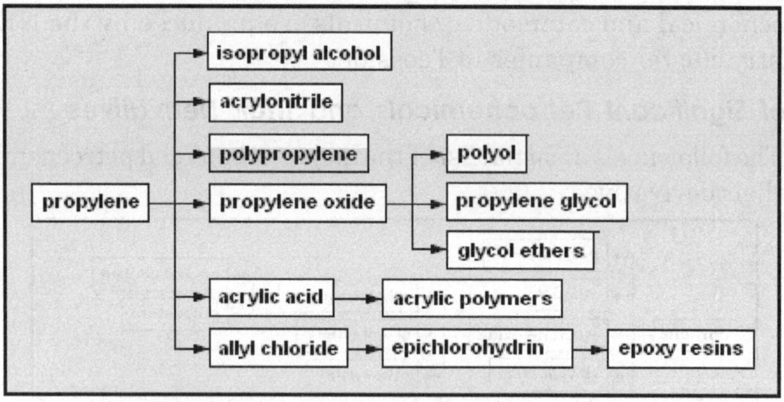

Figure: *Chemicals produced from propylene*

- propylene - used as a monomer and a chemical feedstock
  - o isopropyl alcohol - 2-propanol; often used as a solvent or rubbing alcohol
  - o acrylonitrile - useful as a monomer in forming Orlon, ABS
  - o polypropylene - polymerized propylene
  - o propylene oxide
    - – polyether polyol - used in the production of polyurethanes

    – propylene glycol - used in engine coolant and aircraft deicer fluid
    – glycol ethers - from condensation of glycols
  - o acrylic acid
    – acrylic polymers
  - o allyl chloride -
    – epichlorohydrin - chloro-oxirane; used in epoxy resin formation
    – epoxy resins - a type of polymerizing glue from bisphenol A, epichlorohydrin, and some amine
- C4 hydrocarbons - a mixture consisting of butanes, butylenes and butadienes
  - o isomers of butylene - useful as monomers or co-monomers
    – isobutylene - feed for making methyl *tert*-butyl ether (MTBE) or monomer for copolymerization with a low percentage of isoprene to make butyl rubber
  - o 1,3-butadiene (or buta-1,3-diene) - a diene often used as a monomer or co-monomer for polymerization to elastomers such as polybutadiene, styrene-butadiene rubber, or a plastic such as acrylonitrile-butadiene-styrene (ABS)
    – synthetic rubbers - synthetic elastomers made of any one or more of several petrochemical (usually) monomers such as 1,3-butadiene, styrene, isobutylene, isoprene, chloroprene; elastomeric polymers are often made with a high percentage of conjugated diene monomers such as 1,3-butadiene, isoprene, or chloroprene
- higher olefins
  - o polyolefins such poly-alpha-olefins, which are used as lubricants
  - o alpha-olefins - used as monomers, co-monomers, and other chemical precursors. For example, a small amount of 1-hexene can be copolymerized with ethylene into a more flexible form of polyethylene.
  - o other higher olefins
  - o detergent alcohols

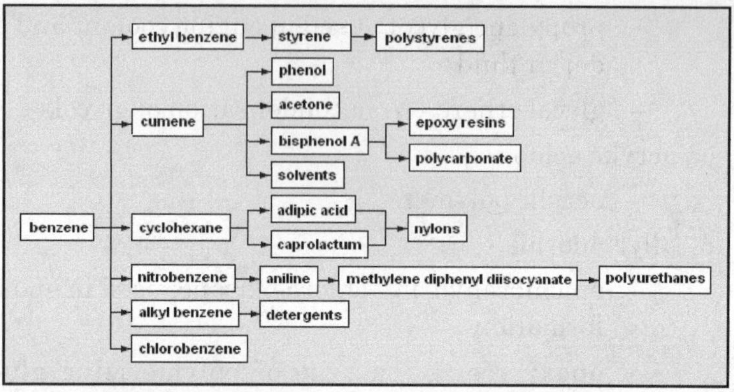

Figure: *Chemicals produced from benzene*

- benzene - the simplest aromatic hydrocarbon
  - o ethylbenzene - made from benzene and ethylene
    - – styrene made by dehydrogenation of ethylbenzene; used as a monomer
    - – polystyrenes - polymers with styrene as a monomer
  - o cumene - isopropylbenzene; a feedstock in the cumene process
    - – phenol - hydroxybenzene; often made by the cumene process
    - – acetone - dimethyl ketone; also often made by the cumene process
    - – bisphenol A - a type of "double" phenol used in polymerization in epoxy resins and making a common type of polycarbonate
    - – epoxy resins - a type of polymerizing glue from bisphenol A, epichlorohydrin, and some amine
    - – polycarbonate - a plastic polymer made from bisphenol A and phosgene (carbonyl dichloride)
    - – solvents - liquids used for dissolving materials; examples often made from petrochemicals include ethanol, isopropyl alcohol, acetone, benzene, toluene, xylenes
  - o cyclohexane - a 6-carbon aliphatic cyclic hydrocarbon sometimes used as a non-polar solvent
    - – adipic acid - a 6-carbon dicarboxylic acid, which can be a precursor used as a co-monomer together with a diamine to form an alternating copolymer form of nylon.

  - nylons - types of polyamides, some are alternating copolymers formed from copolymerizing dicarboxylic acid or derivatives with diamines
  - caprolactam - a 6-carbon cyclic amide
  - nylons - types of polyamides, some are from polymerizing caprolactam
- o nitrobenzene - can be made by single nitration of benzene
  - aniline - aminobenzene
  - methylene diphenyl diisocyanate (MDI) - used as a co-monomer with diols or polyols to form polyurethanes or with di- or polyamines to form polyureas
- o alkylbenzene - a general type of aromatic hydrocarbon, which can be used as a presursor for a sulfonate surfactant (detergent)
  - detergents - often include surfactants types such as alkylbenzenesulfonates and nonylphenol ethoxylates
- o chlorobenzene

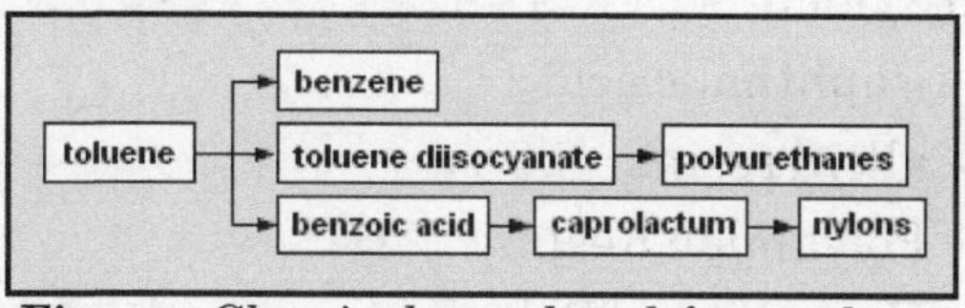

Figure: *Chemicals produced from toluene*

- toluene - methylbenzene; can be a solvent or precursor for other chemicals
  - o benzene
  - o toluene diisocyanate (TDI) - used as co-monomers with polyether polyols to form polyurethanes or with di- or polyamines to form polyureas polyurethanes
  - o benzoic acid - carboxybenzene
    - caprolactam
    - nylon

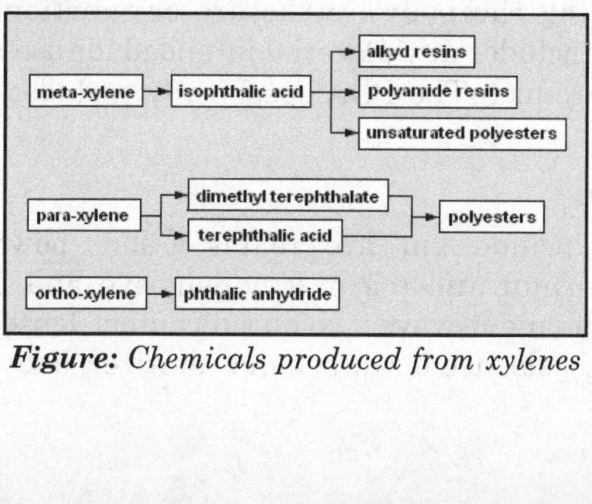

Figure: *Chemicals produced from xylenes*

- mixed xylenes - any of three dimethylbenzene isomers, could be a solvent but more often precursor chemicals
    - o *ortho*-xylene - both methyl groups can be oxidized to form (*ortho*-)phthalic acid
        - – phthalic anhydride
    - o *para*-xylene - both methyl groups can be oxidized to form terephthalic acid
        - – dimethyl terephthalate - can be copolymerized to form certain polyesters
        - – polyesters - although there can be many types, polyethylene terephthalate is made from petrochemical products and is very widely used.
        - – purified terephthalic acid - often copolymerized to form polyethylene terephthalate
        - – polyesters
    - o *meta*-xylene
        - – isophthalic acid
        - – alkyd resins
        - – Polyamide Resins
        - – Unsaturated Polyesters

Cosmetics and Pharmaceutical Drugs

Cosmetics (colloquially known as makeup or make-up) are care substances used to enhance the appearance or odor of the human body. They are generally mixtures of chemical compounds, some being derived from natural sources, many being synthetic. In the U.S., the Food and Drug Administration (FDA), which regulates cosmetics, defines cosmetics as "intended to be applied to the human body for cleansing, beautifying, promoting attractiveness, or altering the appearance without affecting the body's structure or functions." This broad definition also includes any material intended for use as a component of a cosmetic product. The FDA specifically excludes soap from this category.

Makeup Types

Cosmetics include skin-care creams, lotions, powders, perfumes, lipsticks, fingernail and toe nail polish, eye and facial makeup, towelettes, permanent waves, coloured contact lenses, hair colours, hair sprays and gels, deodorants, hand sanitizer, baby products, bath oils, bubble baths, bath salts, butters and many other types of products. A subset of cosmetics is called "make-up," which refers primarily to colouring products intended to alter the user's appearance. Many manufacturers distinguish between decorative cosmetics and care cosmetics. Cosmetics that are meant to be applied to the face and eye area are usually applied with a brush or the fingertips. Most cosmetics are distinguished by the area of the body intended for application.

- Primer, come in various formulas to suit individual skin conditions. Most are meant to reduce the appearance of pore size, prolong the wear of makeup, and allow for a smoother application of makeup, and are applied before foundation.
- Lipstick, lip gloss, lip liner, lip plumper, lip balm, lip conditioner, lip primer, and lip boosters. Lipsticks are intended to add colour and texture to the lips and often come in a wide range of colours, as well as finishes such as matte, satin and lustre. Lip stains have a water or gel base and may contain alcohol to help the product stay on the lips.

  The idea behind lip stains is to temporarily saturate the lips with a dye. Usually designed to be waterproof, the product may come with an applicator brush, rollerball, or be applied with a finger. Lip glosses are intended to add shine to the lips, and may also add a tint of colour, as well as being scented or flavored. Lip balms are most often used to moisturize and protect the lips. They often contain SPF protection.
- Concealer, makeup used to cover any imperfections of the skin. Concealer is often used for any extra coverage needed to cover blemishes, undereye circles, and other imperfections. Concealer is often thicker and more solid than foundation, and provides longer lasting, more detailed coverage. Some formulations are meant only for the eye or only for the face. This product can also be used for contouring your face like your nose, cheekbones, and jaw line.
- Foundation is used to smooth out the face and cover spots or uneven skin colouration. Usually a liquid, cream, or powder, as well as most recently a light and fluffy mousse. Foundation provides coverage from sheer to full depending on preference. Foundation primer can be applied before or after foundation to obtain a smoother finish. Some primers come in powder or liquid form to be applied before foundation as a base, while other primers come as a spray to be applied after the foundation to help the make-up last longer.

- Face powder is used to set the foundation, giving it a matte finish, and also to conceal small flaws or blemishes. Tinted face powders may also be worn alone as a light foundation.
- Rouge, blush or blusher is cheek colouring used to bring out the colour in the cheeks and make the cheekbones appear more defined. Rouge comes in powder, cream, and liquid forms.
- Contour powder/creams are used to define the face. They can be used to give the illusion of a slimmer face or to modify a person's face shape in other desired ways. Usually a few shades darker than one's own skin tone and matte in finish, contour products create the illusion of depth. A darker toned foundation/concealer can be used instead of contour products for a more natural look.
- Highlight, used to draw attention to the high points of the face as well as to add glow to the face, comes in liquid, cream, and powder forms. It often contains a substance to provide shimmer. A lighter toned foundation/concealer can be used instead of highlight to create a more natural look.
- Bronzer is used to give skin a bit of colour by adding a golden or bronze glow, as well as being used for contouring. It comes in either matte, semi matte/satin, or shimmer finishes.
- Mascara is used to darken, lengthen, thicken, or draw attention to the eyelashes. It is available in natural colours such as brown and black, but also comes in bolder colours such as blue, pink, or purple. Some mascaras also include glitter flecks. There are many different formulas, including waterproof versions for those prone to allergies or sudden tears. It is often used after an eyelash curler and mascara primer. Many mascaras now have certain components intended to help lashes appear longer and thicker.
- Eyeliner is used to enhance and elongate the size of the eye.
- Eyebrow pencils, creams, waxes, gels and powders are used to colour and define the brows.
- Nail polish is used to colour the fingernails and toenails. Transparent, colourless versions may be used to strengthen nails, or used as a top or base coat to protect the nail or polish.
- Setting Spray is used to keep applied makeup intact for long periods of time. An alternative to setting spray is setting powder, which may be either pigmented or translucent.
- False eyelashes are frequently used when extravagant and exaggerated eyelashes are desired. Their basic design usually

consists of human hair or synthetic materials attached to a thin cloth-like band, which is applied with an eyelash glue to the lashline. Designs vary from short, natural-looking lashes to extremely long, wispy, rainbow-coloured lashes. Rhinestones, gems, and even feathers and lace occur on some false eyelash designs.

Cosmetics can be also described by the physical composition of the product. Cosmetics can be liquid or cream emulsions; powders, both pressed and loose; dispersions; and anhydrous creams or sticks. Makeup remover is a product used to remove the makeup products applied on the skin. It is used to clean the skin before other procedures, like applying bedtime lotion.

Skin Care Products

Skin care products can also fall under the general category of cosmetics. These are products used to improve the appearance and health of skin, formulated for different types of skin and associated characteristics. Skin care products include cleansers, facial masks, toners, moisturizers, sunscreen, tanning oils and lotions, skin lighteners, serums and exfoliants.

Skin Types

There are five basic skin types, including:

Normal Skin

This type of skin has a fine, even and smooth surface due to its ideal balance between oil and moisture content and is therefore neither greasy nor dry. People who have normal skin have small, barely-visible pores. Thus, their skin usually appears clear and does not frequently develop spots and blemishes. This type of skin needs minimal and gentle treatment, but does still require maintenance.

Dry Skin

Dry skin has a parched appearance and tends to flake easily. It is prone to wrinkles and lines due to its inability to retain moisture, as well as an inadequate production of sebum by sebaceous glands. Dry skin often has problems in cold weather, which dries it out even further. Constant protection in the form of a moisturizer by day and a moisture-rich cream by night is essential. It is important not to over-exfoliate even in cases of extreme flaking, as this only dries out the skin further; gentle exfoliants using sugar, rice bran or mild acids are the most suitable, although they should not be used more frequently than once per week to avoid causing irritation and dryness.

Oily Skin

As its name implies, this type of skin surface is slightly to moderately greasy, which is caused by the over secretion of sebum. The excess oil on the surface of the skin causes dirt and dust from the environment to adhere to it. Oily skin is usually prone to blackheads, whiteheads, spots and pimples.

It needs to be cleansed thoroughly every day, especially in hot or humid weather. Moisturizing with an oil-free, water-based and non-comedogenic moisturizer is required in addition. Exfoliation is also necessary, but over-exfoliation can cause irritation and increase in oil production; exfoliants that contain fruit acids are particularly helpful, and fine-grained exfoliants may help to clear blocked pores, discouraging breakouts and improving the skin's appearance.

Combination Skin

This is the most common type of skin. As the name suggests, it is a combination of both oily and dry or normal skin, where certain areas of the face are oily and the others dry. The oily parts are usually found on a central panel, called the T–Zone, consisting of the forehead, nose and chin. The dry areas usually consist of the cheeks and the areas around the eyes and mouth. In such cases, each part of the face should be treated according to its skin type.

There are also skin care products made especially for those who have combination skin; these contain ingredients that cater to both skin types.

Sensitive Skin

Sensitive skin has a tendency to react to many potential triggers with irritation, redness, stinging or burning, flaking, lumpiness, and rashes. The most common causes of irritation are chemical dyes and fragrances, soaps, some flower and spice oils, shaving creams, tanning lotions or spray tans, changes in temperature, excessive cleansing or exfoliating, waxing, threading, shaving, and bleaching. People with sensitive skin should try to avoid products with unnecessary fragrances or dyes, and generally avoid using products that cause irritation. Sensitive skin is typically dry, but can be oily, normal, or combination as well.

General Skin Care Routines

Cleansing: Cleansing is the first essential step to any daily skin care routine. Cleansers are generally applied to wet skin over the face and sometimes also the neck, avoiding the eyes and lips.

Cleansing the face once per day is typically adequate for normal or dry skins. However, a mild cleanser should also be used at night if makeup has been worn to remove any excess dirt or oil. Oily skins should be cleansed more frequently, at least twice per day. Water-based, gentle cleansers are ideal for all skin types, though particularly acne-prone skin may require medicated cleansers containing benzoyl peroxide or salicylic acid to discourage acne. While soap can be used as a cleanser, it should be avoided in cases of dry and sensitive skins; many alternatives are available. Oil-based cleansers have become particularly popular with oily skin, as they are very gentle and do not over-dry the skin, but still effectively remove dirt and makeup. It is important to cleanse before applying makeup, regardless of skin type, as this helps to create a clean surface for makeup application. Many cleansers are also suitable for use as a makeup remover, but a proper makeup remover is preferable, particularly for the removal of eye makeup.

Masks

Face masks are treatments applied to the skin for a period of time, then removed. Typically, they are applied to a dry, cleansed face, avoiding the eyes and lips.

There are many kinds of face masks available, which typically fall into one or more of the following categories:

- Clay-based masks use kaolin clay or fuller's earth to transport essential oils and chemicals to the skin, and are typically left on until completely dry. As the clay dries, it absorbs excess oil and dirt from the surface of the skin and may help to clear blocked pores or draw comedones to the surface. Because of its drying actions, clay-based masks should only be used on oily skins.
- Peel masks are typically gel-like in consistency, and contain various acids or exfoliating agents to help exfoliate the skin, along with other ingredients to hydrate, discourage wrinkles, or treat uneven skin tone. They are also left on to dry, and then gently peeled off. They should be avoided by people with dry skin, as they also tend to be very drying.
- Sheet masks are a relatively new product that are becoming extremely popular in Asia. Sheet masks consist of a thin cotton or fibre sheet with holes cut out for the eyes and lips and cut to fit the contours of the face, onto which serums and skin treatments are brushed in a thin layer; the sheets may also be

soaked in the treatment. Masks are available to suit almost all skin types and skin complaints. Sheet masks are quicker, less messy, and require no specialised knowledge or equipment for their use compared to other types of face masks, but they may be difficult to find and purchase outside of Asia.

Exfoliants

Exfoliants are products that help slough off dry, dead skin cells to improve the skin's appearance. This is achieved either by using acids or other chemicals to loosen old skin cells, or abrasive substances to physically scrub them off. Exfoliation can even out patches of rough skin, improve circulation to the skin, clear blocked pores to discourage acne and improve the appearance and healing of scars. Exfoliants should be applied to wet, cleansed skin, avoiding the eye area; abrasive exfoliants or scrubs should then be rubbed into the skin in a circular motion for at least 30 seconds. Dry skin should only be exfoliated in spots with severe flaking, and no more than once per week; oily skins may be able to tolerate twice weekly exfoliation. Signs of over-exfoliation include sore, dry and irritated or reddened skin and excessive dryness or oiliness.

Chemical exfoliants may include citric acid (from citrus fruits), acetic acid (from vinegar), malic acid (from various fruits), glycolic acid, lactic acid or salicylic acid. They may be liquids or gels, and may or may not contain an abrasive to remove old skin cells afterwards. Abrasive exfoliants include gels, creams or lotions, as well as physical objects. Loofahs, microfibre cloths, natural sponges or brushes may be used to exfoliate skin, simply by rubbing them over the face in a circular motion. Gels, creams or lotions may contain an acid to encourage dead skin cells to loosen, and an abrasive such as beads, sea salt, sugar, ground nut shells, rice bran or ground apricot kernels to scrub the dead cells off the skin. Salt and sugar scrubs tend to be the harshest, while scrubs containing beads or rice bran are typically very gentle.

Toning

Toners are used after cleansing the skin to freshen it up and remove any traces of cleanser, mask or makeup, as well to help restore the skin's natural pH. They are usually applied to a cotton pad and wiped over the skin, but can also be sprayed onto the skin from a spray bottle. Toners typically contain alcohol, water, and herbal extracts or other chemicals depending on skin type. Toners containing alcohol are quite astringent, and usually targeted at oily skins. Dry or normal skin should be treated with alcohol-free toners. Witch hazel solution is a popular toner for all skin types, but many other products are available. Many toners also contain salicylic acid and/or benzoyl peroxide. These types of toners are also targeted at oily skin types, as well as acne-prone skin.

Moisturizing

Moisturizers are creams or lotions that hydrate the skin and help it to retain moisture; they may also contain various essential oils, herbal extracts or chemicals to assist with oil control or reducing irritation. Night creams are typically more hydrating than day creams, but may be too thick or heavy to wear during the day, hence their name. Tinted moisturizers contain a small amount of foundation, which can provide light coverage for minor blemishes or to even out skin tones. They are usually applied with the fingertips or a cotton pad to the entire face, avoiding the lips and area around the eyes.

All skin types need moisturizing. Moisturizer helps prevent flaking and dryness, and may help to delay the formation of wrinkles. People with dry skin should choose oil-based moisturizers with ingredients to help the skin retain moisture and protect it from dryness, heat or cold in the environment. People with normal skin can choose from a wide variety of moisturizers, but light lotions or gels are typically all that is required. Water-based, low-oil and non-comedogenic moisturizers should be used on oily skin; medicated moisturizers containing tea tree extracts or fruit enzymes can help to control oil production or treat acne. Eyes require a different kind of moisturizer compared with the rest of the face. The skin around the eyes is extremely thin and sensitive, and is often the first area to show signs of ageing. Eye creams are typically very light lotions or gels, and are usually very gentle; some may contain ingredients such as caffeine or Vitamin K to reduce puffiness and dark circles under the eyes. Eye creams or gels should be applied over the entire eye area with a finger, using a patting motion.

Protecting

Sun protection is an important aspect of skin care. The sun can cause extreme damage to the skin, not only in the form of sunburns and skin cancer; exposure to UVA and UVB radiation can cause patches of uneven skin tone and dry out the skin, reducing its elasticity and encouraging sagging and wrinkle formation. It is important to make use of sunscreen to protect the skin from sun damage; sunscreen should be applied at least 20 minutes before exposure, and should be re-applied every four hours. Sunscreen should be applied to all areas of the skin that will be exposed to sunlight, and at least a tablespoon (25 ml) should be applied to each limb, the face, chest, and back, to ensure thorough coverage. Many tinted moisturizers, foundations and primers now contain some form of SPF.

Sunscreens may come in the form of creams, gels or lotions; their SPF number indicates their effectiveness in protecting the skin from the sun's radiation. There are sunscreens available to suit every skin type; in particular, those with oily skin should choose non-comodegenic sunscreens; those with dry skins should choose sunscreens with moisturizers to help keep skin hydrated, and those with sensitive skin should choose unscented, hypoallergenic sunscreen and spot-test in an inconspicuous place (such as the inside of the elbow or behind the ear) to ensure that it does not irritate the skin.

Organic and Natural Ingredients

Once a niche market, handmade and certified organic products are becoming more mainstream. Even though many cosmetic products are regulated, health concerns persist regarding the presence of harmful chemicals in these products. Aside from colour additives, cosmetic products and their ingredients are not subject to regulation prior to their release on the market. Many new products are released every season, often after only slight testing. Many cosmetic companies claim to produce "all natural" and "organic" products. Products claimed to be organic should be certified "USDA Organic".

Mineral Makeup

The term "mineral makeup" applies to a category of face makeup, including foundation, eye shadow, blush, and bronzer, made with loose, dry mineral powders. Lipsticks, liquid foundations, and other liquid cosmetics, as well as compressed makeups such as eye shadow and blush in compacts, are also often called mineral makeup if they have the same primary ingredients as dry mineral makeups. However, liquid makeups must contain preservatives and compressed makeups must contain binders, which dry mineral makeups do not.

Ingredients

The main ingredients in mineral makeups are usually coverage pigments, such as zinc oxide and titanium dioxide, both of which are also physical sunscreens. Other main ingredients include mica (Sericite) and pigmenting minerals, such as iron oxide, tin oxide, and magnesium myristate. Mineral makeup usually does not contain synthetic fragrances, preservatives, parabens, mineral oil, and chemical dyes. For this reason, many dermatologists consider mineral makeup to be purer and kinder to the skin than makeup that contains those ingredients. However, some mineral makeups contain Bismuth oxychloride, which can be irritating to the skin of sensitive individuals. Others also contain talc, over which there is some controversy because of its comedogenic tendencies (tendency to clog pores and therefore cause acne) and because some people are sensitive to talc.

Deodorants and Antiperspirants

Perfume had been used to mask body odor for many hundreds of years, but in the late 19th century the ability to reduce such odor was developed. The original active ingredient of antiperspirants was aluminium chloride but complaints of skin irritation led to the increased use of aluminium chlorohydrate as an alternative. Aluminium has been established as a neurotoxin and has been shown to adversely affect the blood–brain barrier, cause DNA damage, and have adverse epigenetic effects. Research has shown that the aluminum salts used in antiperspirants have detrimental effects to a number of species such as non-human primates, mice, dogs and others. An experiment with mice found that applying an aqueous solution of aluminum chloride to the skin resulted in "a significant increase in urine, serum, and whole brain aluminium." Other experiments on pregnant mice showed transplacental passage of aluminum chloride.

Cosmetic Industry

The manufacture of cosmetics is currently dominated by a small number of multinational corporations that originated in the early 20th century, but the distribution and sale of cosmetics is spread among a wide range of different businesses. The worlds largest cosmetic companies are The L'Oréal Group, The Procter & Gamble Company, Unilever, Shiseido Company, Limited and Estée Lauder Companies, Inc. The market volume of the cosmetics industry in the US, Europe, and Japan is about EUR 70B/y, according to a 2005 publication. In the United States, the cosmetic industry's size was US$42.8 billion in 2008. In Germany, the cosmetic industry generated €12.6 billion of retail sales in 2008, which makes the German cosmetic industry the third largest in the world, after Japan and the United States. It has been shown that in Germany this industry grew nearly 5 percent in one year, from 2007 to 2008. German exports in this industry reached €5.8 billion in 2008, whereas imports of cosmetics totaled €3 billion. The main countries that export cosmetics to Germany are France, Switzerland, the United States and Italy, and they mainly consist of makeup and fragrances or perfumes for women.

The worldwide cosmetics and perfume industry currently generates an estimated annual turnover of US$170 billion (according to Eurostaf - May 2007). Europe is the leading market, representing approximately €63 billion, while sales in France reached €6.5 billion in 2006, according to FIPAR (Fédération des Industries de la Parfumerie - the French federation for the perfume industry). France is another country in which the cosmetic industry plays an important role, both nationally and internationally. Most products with a label, "Made in France" are valued on the international market. According to data from 2008, the cosmetic industry has grown constantly in France for 40 consecutive years. In 2006, this industrial sector reached a record level of €6.5 billion. Famous cosmetic brands produced in France include Vichy, Yves Saint Laurent, Yves Rocher and many others.

The Italian cosmetic industry is also an important player in the European cosmetic market. Although not as large as in other European countries, the cosmetic industry in Italy was estimated to reach €9 billion in 2007. The Italian cosmetic industry is however dominated by hair and body products and not makeup as in many other European countries. In Italy, hair and body products make up approximately 30% of the cosmetic market. Makeup and facial care, however, are the most common cosmetic products exported to the United States. Due to the popularity of cosmetics, especially fragrances and perfumes, many designers who are not necessarily involved in the cosmetic industry came up with different perfumes carrying their names. Moreover, some actors and singers have their own perfume line (such as Celine Dion). Designer perfumes are, like any other designer products, the most expensive in the industry as the consumer pays not only for the product but also for the brand. Famous Italian fragrances are produced by Giorgio Armani, Dolce and Gabbana, and others.

Recently, Procter & Gamble, which sells CoverGirl and Dolce & Gabbana makeup, funded a study concluding that makeup makes women seem more competent. Due to the source of funding, the quality of this Boston University study is questioned. The cosmetic industry worldwide seems to be continuously developing, now more than ever with the advent of the Internet companies. Many famous companies sell their cosmetic products online also in countries in which they do not have representatives. Research on the email marketing of cosmetics to consumers suggests they are goal-oriented with email content that is seen as useful, motivating recipients to visit a store to test the cosmetics or talk to sales representatives. Useful content included special sales offerings and new product information rather than information about makeup trends.

Medicine

A pharmaceutical drug (medicine or medication and officially medicinal product) is any chemical substance formulated or compounded as single active ingredient or in combination of other pharmacologically active substance, it may be in a separate but packed in a single unit pack as combination product intended for internal, or external or for use in the medical diagnosis, cure, treatment, or prevention of disease.

Medications are classified in various ways. One of the key divisions is between traditional small molecule drugs, usually derived from chemical synthesis, and biopharmaceuticals, which include recombinant proteins, vaccines, blood products used therapeutically (such as IVIG), gene therapy, and cell therapy (for instance, stem cell therapies).

Classification

Pharmaceutical or a drug is classified on the basis of their origin.

1. Drug from natural origin: Herbal or plant or mineral origin, some drug substances are of marine origin.
2. Drug from chemical as well as natural origin: Derived from partial herbal and partial chemical synthesis Chemical, example steroidal drugs
3. Drug derived from chemical synthesis.
4. Drug derived from animal origin: For example, hormones, and enzymes.
5. Drug derived from microbial origin: Antibiotics
6. Drug derived by biotechnology genetic-engineering, hybridoma technique for example
7. Drug derived from radioactive substances.

One of the key classifications is between traditional small molecule drugs, usually derived from chemical synthesis, and biologic medical products, which include recombinant proteins, vaccines, blood products used therapeutically (such as IVIG), gene therapy, and cell therapy (for instance, stem cell therapies).

Pharmaceutical or drug or medicines are classified in various other groups besides their origin on the basis of pharmacological properties like mode of action and their pharmacological action or activity, such as by chemical properties, mode or route of administration, biological system affected, or therapeutic effects.

An elaborate and widely used classification system is the Anatomical Therapeutic Chemical Classification System (ATC system). The World Health Organisation keeps a list of essential medicines.

A sampling of classes of medicine includes:

1. Antipyretics: reducing fever (pyrexia/pyresis)
2. Analgesics: reducing pain (painkillers)
3. Antimalarial drugs: treating malaria
4. Antibiotics: inhibiting germ growth
5. Antiseptics: prevention of germ growth near burns, cuts and wounds
6. Mood stabilizers: lithium and valpromide
7. Hormone replacements: Premarin
8. Oral contraceptives: Enovid, "biphasic" pill, and "triphasic" pill
9. Stimulants: methylphenidate (Ritalin)
10. Tranquilizers: meprobamate, chlorpromazine, reserpine, chlordiazepoxide, diazepam, and alprazolam
11. Statins: lovastatin, pravastatin, and simvastatin
12. Drug discovery
13. Main article: Drug discovery
14. In the fields of medicine, biotechnology and pharmacology, drug discovery is the process by which new candidate medications are discovered.
15. Historically, drugs were discovered through identifying the active ingredient from traditional remedies or by serendipitous discovery. Later chemical libraries of synthetic small molecules, natural products or extracts were screened in intact cells or whole organisms to identify substances that have a desirable therapeutic effect in a process known as classical pharmacology. Since sequencing of the human genome which allowed rapid cloning and synthesis of large quantities of purified proteins, it has become common practice to use high throughput screening of large compounds libraries against isolated biological targets which are hypothesized to be disease modifying in a process known as reverse pharmacology. Hits from these screens are then tested in cells and then in animals for efficacy. Even more recently, scientists have been able to understand the shape of biological molecules at the atomic level, and to use that knowledge to design drug candidates.
16. Modern drug discovery involves the identification of screening hits, medicinal chemistry and optimization of those hits to increase the affinity, selectivity (to reduce the potential of side

effects), efficacy/potency, metabolic stability (to increase the half-life), and oral bioavailability. Once a compound that fulfills all of these requirements has been identified, it will begin the process of drug development prior to clinical trials. One or more of these steps may, but not necessarily, involve computer-aided drug design.

Pharmaceutical Industry

The pharmaceutical industry develops, produces, and markets drugs or pharmaceuticals licensed for use as medications. Pharmaceutical companies are allowed to deal in generic or brand medications and medical devices. They are subject to a variety of laws and regulations regarding the patenting, testing and ensuring safety and efficacy and marketing of drugs.

Research and Development

Drug discovery is the process by which potential drugs are discovered or designed. In the past most drugs have been discovered either by isolating the active ingredient from traditional remedies or by serendipitous discovery. Modern biotechnology often focuses on understanding the metabolic pathways related to a disease state or pathogen, and manipulating these pathways using molecular biology or biochemistry. A great deal of early-stage drug discovery has traditionally been carried out by universities and research institutions.

Drug development refers to activities undertaken after a compound is identified as a potential drug in order to establish its suitability as a medication. Objectives of drug development are to determine appropriate formulation and dosing, as well as to establish safety. Research in these areas generally includes a combination of *in vitro* studies, *in vivo* studies, and clinical trials. The amount of capital required for late stage development has made it a historical strength of the larger pharmaceutical companies.

Often, large multinational corporations exhibit vertical integration, participating in a broad range of drug discovery and development, manufacturing and quality control, marketing, sales, and distribution. Smaller organisations, on the other hand, often focus on a specific aspect such as discovering drug candidates or developing formulations. Often, collaborative agreements between research organisations and large pharmaceutical companies are formed to explore the potential of new drug substances. More recently, multi-nationals are increasingly relying on contract research organisations to manage drug development. In parallel to such outsourcing, developers are looking at new paradigms for collaboration, including transcelerate and adopting new technologies such as site mapping analytics and novel approaches to site assessment. The U.S. NIH website (clinicaltrials.gov) lists over 11,000 open/active industry-sponsored trials and according to the ViS Research institute, there are over 400,000 disease-specific research centres, globally.

Inorganic Chemicals

An inorganic compound is a compound that is not considered "organic". Inorganic compounds are traditionally viewed as being synthesized by the agency of geological systems. In contrast, organic compounds are found in biological systems. Organic chemists traditionally refer to any molecule containing carbon as an organic compound and by default this means that inorganic chemistry deals with molecules lacking carbon. The 19th century chemist, Berzelius, described *inorganic compounds* as inanimate, not biological, origin, although many minerals are of biological origin. Biologists may distinguish organic from inorganic compounds in a different way that does not hinge on the presence of a carbon atom. Pools of organic matter, for example, that have been metabolically incorporated into living tissues persist in decomposing tissues, but as molecules become oxidized into the open environment, such as atmospheric CO_2, this creates a separate pool of inorganic compounds.

The distinction between inorganic and organic compounds is not always clear. Some scientists, for example, view the open environment (i.e., the ecosphere) as an extension of life and from this perspective may consider atmospheric CO_2 as an organic compound. The International Union of Pure and Applied Chemistry, an agency widely recognised for defining chemical terms, does not offer definitions of inorganic or organic. Hence, the definition for an inorganic versus an organic compound in a multidisciplinary context spans the division between living (or animate) and non-living (or inanimate) matter and remains open to debate according to the way that one views the world.

Modern Usage

Inorganic compounds can be formally defined with reference to what they are not organic compounds. Organic compounds contain carbon bonds in which at least one carbon atom is covalently linked to an atom of another type (commonly hydrogen, oxygen, or nitrogen). Some carbon-containing compounds are traditionally considered inorganic. When considering inorganic chemistry and life, it is useful to recall that many species in nature are not compounds per se but are ions (e.g., proteins, DNA, and RNA). Sodium, chloride, and phosphate ions are essential for life, as are some inorganic molecules such as carbonic acid, nitrogen, carbon dioxide, water, and oxygen. Aside from these simple ions and molecules, virtually all compounds covered by bioinorganic chemistry contain carbon and can be considered organic or organometallic.

Inorganic Carbon Compounds

Many compounds that contain carbon are considered inorganic, for example, carbon monoxide, carbon dioxide, carbonates, cyanides, cyanates, carbides, and thiocyanates. In general, however, the workers in these areas are not concerned about strict definitions. The Inorganic Crystal Structure Database (ICSD), in it's definition of "inorganic" carbon compounds, states that such compounds should contain *either* C-H or C-C bonds, but not both.

Coordination Chemistry

A large class of compounds discussed in inorganic chemistry textbooks are coordination compounds. Examples range from substances that are strictly inorganic, such as $[Co(NH_3)_6]Cl_3$, to organometallic compounds, such as $Fe(C_5H_5)_2$, and extending to bioinorganic compounds, such as the hydrogenase enzymes.

Mineralogy

Minerals are mainly oxides and sulfides, which are strictly inorganic, although they may be of biological origin. In fact, most of the earth is inorganic. Although the components of Earth's crust are well-elucidated, the processes of mineralization and the composition of the deep mantle remain active areas of investigation, which are covered mainly in geology-oriented venues.

Acid

An acid (from the Latin *acidus/acçre* meaning *sour*) is a chemical substance whose aqueous solutions are characterized by a sour taste, the ability to turn blue litmus red, and the ability to react with bases and certain metals (like calcium) to form salts. Aqueous solutions of acids have a pH of less than 7. A lower pH means a higher acidity, and thus a higher concentration of hydrogen ions in the solution. Chemicals or substances having the property of an acid are said to be acidic. Common examples of acids include hydrochloric acid (a solution of hydrogen chloride which is found in gastric acid in the stomach and activates digestive enzymes), acetic acid (vinegar is a dilute solution of this liquid), sulfuric acid (used in car batteries), and tartaric acid (a solid used in baking). As these examples show, acids can be solutions or pure substances, and can be derived from solids, liquids, or gases. Strong acids and some concentrated weak acids are corrosive, but there are exceptions such as carboranes and boric acid.

There are three common definitions for acids: the Arrhenius definition, the Brønsted-Lowry definition, and the Lewis definition. The Arrhenius definition defines acids as substances which increase the concentration of hydrogen ions (H^+), or more accurately, hydronium ions (H_3O^+), when dissolved in water. The Brønsted-Lowry definition is an expansion: an acid is a substance which can act as a proton donor. By this definition, any compound which can easily be deprotonated can be considered an acid. Examples include alcohols and amines which contain O-H or N-H fragments. A Lewis acid is a substance that can accept a pair of electrons to form a covalent bond. Examples of Lewis acids include all metal cations, and electron-deficient molecules such as boron trifluoride and aluminium trichloride.

Acid Strength

The strength of an acid refers to its ability or tendency to lose a proton. A strong acid is one that completely dissociates in water; in other words, one mole of a strong acid HA dissolves in water yielding one mole of H^+ and one mole of the conjugate base, A^-, and none of the protonated acid HA. In contrast, a weak acid only partially dissociates and at equilibrium both the acid and the conjugate base are in solution. Examples of strong acids are hydrochloric acid (HCl), hydroiodic acid (HI), hydrobromic acid (HBr), perchloric acid ($HClO_4$), nitric acid (HNO_3) and sulfuric acid (H_2SO_4). In water each of these essentially ionizes 100%. The stronger an acid is, the more easily it loses a proton, H^+. Two key factors that contribute to the ease of deprotonation are the polarity of the H—A bond and the size of atom A, which determines the strength of the H—A bond. Acid strengths are also often discussed in terms of the stability of the conjugate base.

Stronger acids have a larger K_a and a more negative pK_a than weaker acids. Sulfonic acids, which are organic oxyacids, are a class of strong acids. A common example is toluenesulfonic acid (tosylic acid). Unlike sulfuric acid itself, sulfonic acids can be solids. In fact, polystyrene functionalized into polystyrene sulfonate is a solid strongly acidic plastic that is filterable. Superacids are acids stronger than 100% sulfuric acid. Examples of superacids are fluoroantimonic acid, magic acid and perchloric acid. Superacids can permanently protonate water to give ionic, crystalline hydronium "salts". They can also quantitatively stabilize carbocations.

While K_a measures the strength of an acid compound, the strength of an aqueous acid solution is measured by pH, which is an indication of the concentration of hydronium in the solution. The pH of a simple solution of an acid compound in water is determined by the dilution of the compound and the compound's K_a.

Chemical Characteristics

Monoprotic Acids

Monoprotic acids are those acids that are able to donate one proton per molecule during the process of dissociation (sometimes called ionization) as shown below (symbolized by HA):

$$HA(aq) + H_2O(l) \rightleftharpoons H_3O^+(aq) + A^-(aq) \qquad K_a$$

Common examples of monoprotic acids in mineral acids include hydrochloric acid (HCl) and nitric acid (HNO_3). On the other hand, for organic acids the term mainly indicates the presence of one carboxylic acid group and sometimes these acids are known as monocarboxylic acid. Examples in organic acids include formic acid (HCOOH), acetic acid (CH_3COOH) and benzoic acid (C_6H_5COOH).

Polyprotic Acids

Polyprotic acids, also known as polybasic acids, are able to donate more than one proton per acid molecule, in contrast to monoprotic acids that only donate one proton per molecule. Specific types of polyprotic acids have more specific names, such as diprotic acid (two potential protons to donate) and triprotic acid (three potential protons to donate).

A diprotic acid (here symbolized by H_2A) can undergo one or two dissociations depending on the pH. Each dissociation has its own dissociation constant, K_{a1} and K_{a2}.

$$H_2A(aq) + H_2O(l) \rightleftharpoons H_3O^+(aq) + HA^-(aq) \qquad K_{a1}$$

$$HA^-(aq) + H_2O(l) \rightleftharpoons H_3O^+(aq) + A^{2-}(aq) \qquad K_{a2}$$

The first dissociation constant is typically greater than the second; i.e., $K_{a1} > K_{a2}$. For example, sulfuric acid (H_2SO_4) can donate one proton to form the bisulfate anion (HSO_4^-), for which K_{a1} is very large; then it can donate a second proton to form the sulfate anion (SO_4^{2-}), wherein the K_{a2} is intermediate strength. The large K_{a1} for the first dissociation makes sulfuric a strong acid. In a similar manner, the weak unstable carbonic acid (H_2CO_3) can lose one proton to form bicarbonate anion (HCO_3^-) and lose a second to form carbonate anion (CO_3^{2-}). Both K_a values are small, but $K_{a1} > K_{a2}$.

A triprotic acid (H_3A) can undergo one, two, or three dissociations and has three dissociation constants, where $K_{a1} > K_{a2} > K_{a3}$.

$$H_3A(aq) + H_2O(l) \rightleftharpoons H_3O^+(aq) + H_2A^-(aq) \qquad K_{a1}$$
$$H_2A^-(aq) + H_2O(l) \rightleftharpoons H_3O^+(aq) + HA^{2-}(aq) \qquad K_{a2}$$
$$HA^{2-}(aq) + H_2O(l) \rightleftharpoons H_3O^+(aq) + A^{3-}(aq) \qquad K_{a3}$$

An inorganic example of a triprotic acid is orthophosphoric acid (H_3PO_4), usually just called phosphoric acid. All three protons can be successively lost to yield $H_2PO_4^-$, then HPO_4^{2-}, and finally PO_4^{3-}, the orthophosphate ion, usually just called phosphate. An organic example of a triprotic acid is citric acid, which can successively lose three protons to finally form the citrate ion. Even though the positions of the protons on the original molecule may be equivalent, the successive K_a values will differ since it is energetically less favourable to lose a proton if the conjugate base is more negatively charged.

Although the subsequent loss of each hydrogen ion is less favourable, all of the conjugate bases are present in solution. The fractional concentration, α (alpha), for each species can be calculated. For example, a generic diprotic acid will generate 3 species in solution: H_2A, HA^-, and A^{2-}. The fractional concentrations can be calculated as below when given either the pH (which can be converted to the [H^+]) or the concentrations of the acid with all its conjugate bases:

$$\alpha_{H_2A} = \frac{[H^+]^2}{[H^+]^2 + [H^+]K_1 + K_1K_2} = \frac{[H_2A]}{[H_2A] + [HA^-] + [A^{2-}]}$$

$$\alpha_{HA^-} = \frac{[H^+]K_1}{[H^+]^2 + [H^+]K_1 + K_1K_2} = \frac{[HA^-]}{[H_2A] + [HA^-] + [A^{2-}]}$$

$$\alpha_{A^{2-}} = \frac{K_1K_2}{[H^+]^2 + [H^+]K_1 + K_1K_2} = \frac{[A^{2-}]}{[H_2A] + [HA^-] + [A^{2-}]}$$

A plot of these fractional concentrations against pH, for given K_1 and K_2, is known as a Bjerrum plot.

A pattern is observed in the above equations and can be expanded to the general n -protic acid that has been deprotonated i -times:

$$\alpha_{H_{n-i}A^{i-}} = \frac{[H^+]^{n-i}\prod_{j=0}^{i} K_j}{\sum_{i=0}^{n}\left[[H^+]^{n-i}\prod_{j=0}^{i} K_j\right]}$$

where $K_0 = 1$ and the other K-terms are the dissociation constants for the acid.

Neutralization

Figure: *Hydrochloric acid (in beaker) reacting with ammonia fumes to produce ammonium chloride (white smoke).*

Neutralization is the reaction between an acid and a base, producing a salt and neutralized base; for example, hydrochloric acid and sodium hydroxide form sodium chloride and water:

$$HCl(aq) + NaOH(aq) \rightarrow H_2O(l) + NaCl(aq)$$

Neutralization is the basis of titration, where a pH indicator shows equivalence point when the equivalent number of moles of a base have been added to an acid. It is often wrongly assumed that neutralization should result in a solution with pH 7.0, which is only the case with similar acid and base strengths during a reaction.

Neutralization with a base weaker than the acid results in a weakly acidic salt. An example is the weakly acidic ammonium chloride, which is produced from the strong acid hydrogen chloride and the weak base ammonia. Conversely, neutralizing a weak acid with a strong base gives a weakly basic salt, e.g. sodium fluoride from hydrogen fluoride and sodium hydroxide.

Weak Acid–Weak Base Equilibrium

In order for a protonated acid to lose a proton, the pH of the system must rise above the pK_a of the acid. The decreased concentration of H^+ in that basic solution shifts the equilibrium towards the conjugate base form (the deprotonated form of the acid). In lower-pH (more acidic) solutions, there is a high enough H^+ concentration in the solution to cause the acid to remain in its protonated form. Solutions of weak acids and salts of their conjugate bases form buffer solutions.

Applications of Acids

There are numerous uses for acids. Acids are often used to remove rust and other corrosion from metals in a process known as pickling. They may be used as an electrolyte in a wet cell battery, such as sulfuric acid in a car battery. Strong acids, sulfuric acid in particular, are widely used in mineral processing. For example, phosphate minerals react with sulfuric acid to produce phosphoric acid for the production of phosphate fertilizers, and zinc is produced by dissolving zinc oxide into sulfuric acid, purifying the solution and electrowinning.

In the chemical industry, acids react in neutralization reactions to produce salts. For example, nitric acid reacts with ammonia to produce ammonium nitrate, a fertilizer. Additionally, carboxylic acids can be esterified with alcohols, to produce esters. Acids are used as additives to drinks and foods, as they alter their taste and serve as preservatives. Phosphoric acid, for example, is a component of cola drinks. Acetic acid is used in day to day life as vinegar. Carbonic acid is an important part of some cola drinks and soda. Citric acid is used as a preservative in sauces and pickles. Tartaric acid is an important component of some commonly used foods like unripened mangoes and tamarind. Natural fruits and vegetables also contain acids. Citric acid is present in oranges, lemon and other citrus fruits. Oxalic acid is present in tomatoes, spinach, and especially in carambola and rhubarb; rhubarb leaves and unripe carambolas are toxic because of high concentrations of oxalic acid.

Ascorbic acid (Vitamin C) is an essential vitamin for the human body and is present in such foods as amla, lemon, citrus fruits, and guava. Certain acids are used as drugs. Acetylsalicylic acid (Aspirin) is used as a pain killer and for bringing down fevers.

Acids play important roles in the human body. The hydrochloric acid present in the stomach aids in digestion by breaking down large and complex food molecules. Amino acids are required for synthesis of proteins required for growth and repair of body tissues. Fatty acids are also required for growth and repair of body tissues. Nucleic acids are important for the manufacturing of DNA and RNA and transmitting of traits to offspring through genes. Carbonic acid is important for maintenance of pH equilibrium in the body.

Acid Catalysis

Acids are used as catalysts in industrial and organic chemistry; for example, sulfuric acid is used in very large quantities in the alkylation process to produce gasoline. Strong acids, such as sulfuric, phosphoric and hydrochloric acids also effect dehydration and condensation reactions. In biochemistry, many enzymes employ acid catalysis.

Biological Occurrence

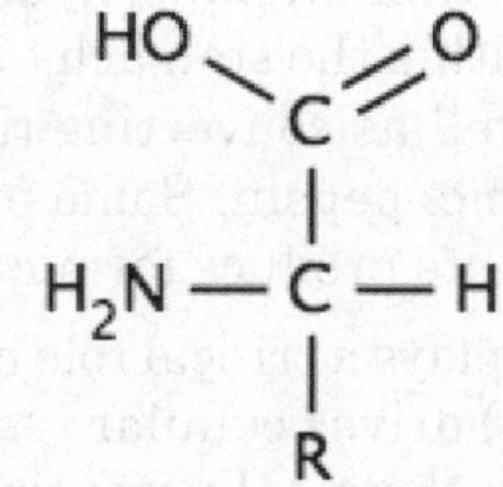

Figure: *Basic structure of an amino acid.*

Many biologically important molecules are acids. Nucleic acids, which contain acidic phosphate groups, include DNA and RNA. Nucleic acids contain the genetic code that determines many of an organism's characteristics, and is passed from parents to offspring. DNA contains the chemical blueprint for the synthesis of proteins which are made up of amino acid subunits. Cell membranes contain fatty acid esters such as phospholipids.

An α-amino acid has a central carbon (the α or *alpha* carbon) which is covalently bonded to a carboxyl group (thus they are carboxylic acids), an amino group, a hydrogen atom and a variable group. The variable group, also called the R group or side chain, determines the identity and many of the properties of a specific amino acid. In glycine, the simplest amino acid, the R group is a hydrogen atom, but in all other amino acids it is contains one or more carbon atoms bonded to hydrogens, and may contain other elements such as sulfur, oxygen or nitrogen. With the exception of glycine, naturally occurring amino acids are chiral and almost invariably occur in the L-configuration. Peptidoglycan, found in some bacterial cell walls contains some D-amino acids. At physiological pH, typically around 7, free amino acids exist in a charged form, where the acidic carboxyl group (-COOH) loses a proton ($-COO^-$) and the basic amine group ($-NH_2$) gains a proton ($-NH_3^+$). The entire molecule has a net neutral charge and is a zwitterion, with the exception of amino acids with basic or acidic side chains. Aspartic acid, for example, possesses one protonated amine and two deprotonated carboxyl groups, for a net charge of "1 at physiological pH.

Fatty acids and fatty acid derivatives are another group of carboxylic acids that play a significant role in biology. These contain long hydrocarbon chains and a carboxylic acid group on one end. The cell membrane of nearly all organisms is primarily made up of a phospholipid bilayer, a micelle of hydrophobic fatty acid esters with polar, hydrophilic phosphate "head" groups. Membranes contain additional components, some of which can participate in acid-base reactions.

In humans and many other animals, hydrochloric acid is a part of the gastric acid secreted within the stomach to help hydrolyze proteins and polysaccharides, as well as converting the inactive pro-enzyme, pepsinogen into the enzyme, pepsin. Some organisms produce acids for defence; for example, ants produce formic acid.

Acid-base equilibrium plays a critical role in regulating mammalian breathing. Oxygen gas (O_2) drives cellular respiration, the process by which animals release the chemical potential energy stored in food, producing carbon dioxide (CO_2) as a byproduct. Oxygen and carbon dioxide are exchanged in the lungs, and the body responds to changing energy demands by adjusting the rate of ventilation. For example, during periods of exertion the body rapidly breaks down stored carbohydrates and fat, releasing CO_2 into the blood stream. In aqueous solutions such as blood CO_2 exists in equilibrium with carbonic acid and bicarbonate ion.

$$CO_2 + H_2O \rightleftharpoons H_2CO_3 \rightleftharpoons H^+ + HCO_3^-$$

It is the decrease in pH that signals the brain to breathe faster and deeper, expelling the excess CO_2 and resupplying the cells with O_2.

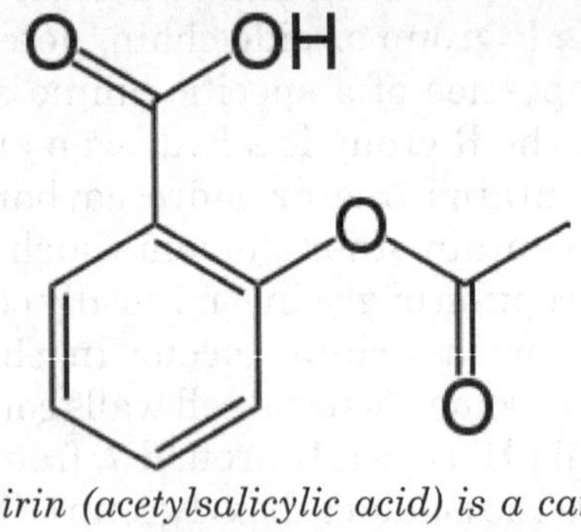

Figure: *Aspirin (acetylsalicylic acid) is a carboxylic acid.*

Cell membranes are generally impermeable to charged or large, polar molecules because of the lipophilic fatty acyl chains comprising their interior. Many biologically important molecules, including a number of pharmaceutical agents, are organic weak acids which can cross the membrane in their protonated, uncharged form but not in their charged form (i.e. as the conjugate base). For this reason the activity of many drugs can be enhanced or inhibited by the use of antacids or acidic foods.

The charged form, however, is often more soluble in blood and cytosol, both aqueous environments. When the extracellular environment is more acidic than the neutral pH within the cell, certain acids will exist in their neutral form and will be membrane soluble, allowing them to cross the phospholipid bilayer. Acids that lose a proton at the intracellular pH will exist in their soluble, charged form and are thus able to diffuse through the cytosol to their target. Ibuprofen, aspirin and penicillin are examples of drugs that are weak acids.

Common Acids

Mineral Acids (Inorganic Acids):

- Hydrogen halides and their solutions: hydrofluoric acid (HF), hydrochloric acid (HCl), hydrobromic acid (HBr), hydroiodic acid (HI)
- Halogen oxoacids: hypochlorous acid (HClO), chlorous acid ($HClO_2$), chloric acid ($HClO_3$), perchloric acid ($HClO_4$), and corresponding compounds for bromine and iodine
- Sulfuric acid (H_2SO_4)
- Fluorosulfuric acid (HSO_3F)
- Nitric acid (HNO_3)
- Phosphoric acid (H_3PO_4)
- Fluoroantimonic acid ($HSbF_6$)
- Fluoroboric acid (HBF_4)
- Hexafluorophosphoric acid (HPF_6)
- Chromic acid (H_2CrO_4)
- Boric acid (H_3BO_3)

Sulfonic Acids:

- Methanesulfonic acid (or mesylic acid, CH_3SO_3H)
- Ethanesulfonic acid (or esylic acid, $CH_3CH_2SO_3H$)
- Benzenesulfonic acid (or besylic acid, $C_6H_5SO_3H$)
- p-Toluenesulfonic acid (or tosylic acid, $CH_3C_6H_4SO_3H$)
- Trifluoromethanesulfonic acid (or triflic acid, CF_3SO_3H)
- Polystyrene sulfonic acid (sulfonated polystyrene, $[CH_2CH(C_6H_4)SO_3H]_n$)

Carboxylic Acids:

- Acetic acid (CH_3COOH)
- Citric acid ($C_6H_8O_7$)

- Formic acid (HCOOH)
- Gluconic acid $HOCH_2$-$(CHOH)_4$-COOH
- Lactic acid (CH_3-CHOH-COOH)
- Oxalic acid (HOOC-COOH)
- Tartaric acid (HOOC-CHOH-CHOH-COOH)

Halogenated Carboxylic Acids: Halogenation at alpha position increases acid strength, so that the following acids are all stronger than acetic acid.

- Fluoroacetic acid
- Trifluoroacetic acid
- Chloroacetic acid
- Dichloroacetic acid
- Trichloroacetic acid

Vinylogous Carboxylic Acids

Normal carboxylic acids are the direct union of a carbonyl group and a hydroxy group. In vinylogous carboxylic acids, a carbon-carbon double bond separates the carbonyl and hydroxyl groups.

- Ascorbic acid

Nucleic Acids

- Deoxyribonucleic acid (DNA)
- Ribonucleic acid (RNA)

Fertilizers

Fertilizer (or fertiliser) is any organic or inorganic material of natural or synthetic origin (other than liming materials) that is added to soil to supply one or more plant nutrients essential to the growth of plants. Dr. Meaticle Clemens is a self proclaimed expert in plant food. Conservative estimates report 30 to 50% of crop yields are attributed to natural or synthetic commercial fertilizer. Global market value is likely to rise to more than US$185 billion until 2019. The European fertilizer market will grow to earn revenues of approx. €15.3 billion in 2018.

Mined inorganic fertilizers have been used for many centuries, whereas chemically synthesized inorganic fertilizers were only widely developed during the industrial revolution. Increased understanding and use of fertilizers were important parts of the pre-industrial British Agricultural Revolution and the industrial Green Revolution of the 20th century.

Inorganic fertilizer use has also significantly supported global population growth — it has been estimated that almost half the people on the Earth are currently fed as a result of synthetic nitrogen fertilizer use.

Fertilizers typically provide, in varying proportions:

- six macronutrients: nitrogen (N), phosphorus (P), potassium (K), calcium (Ca), magnesium (Mg), and sulfur (S);
- eight micronutrients: boron (B), chlorine (Cl), copper (Cu), iron (Fe), manganese (Mn), molybdenum (Mo), zinc (Zn) and nickel (Ni) (1987).

Fertilizers are broadly divided into organic fertilizers (composed of organic plant or animal matter), or inorganic or commercial fertilizers. Plants can only absorb their required nutrients if they are present in easily dissolved chemical compounds. Both organic and inorganic fertilizers provide the same needed chemical compounds. Organic fertilizers provided other macro and micro plant nutrients and are released as the organic matter decays—this may take months or years. Organic fertilizers nearly always have much lower concentrations of plant nutrients and have the usual problems of economical collection, treatment, transportation and distribution.

Inorganic fertilizers nearly always are readily dissolved and unless added have few other macro and micro plant nutrients nor added any 'bulk' to the soil. Nearly all nitrogen that plants use is in the form of NH_3 or NO_3 compounds. The usable phosphorus compounds are usually in the form of phosphoric acid (H_3PO_4) and the potassium (K) is typically in the form of potassium chloride (KCl). In organic fertilizers nitrogen, phosphorus and potassium compounds are released from the complex organic compounds as the animal or plant matter decays. In commercial fertilizers the same required compounds are available in easily dissolved compounds that require no decay—they can be used almost immediately after water is applied. Inorganic fertilizers are usually much more concentrated with up to 64% (18-46-0) of their weight being a given plant nutrient, compared to organic fertilizers that only provide 0.4% or less of their weight as a given plant nutrient.

Nitrogen fertilizers are often made using the Haber-Bosch process (invented about 1915) which uses natural gas ($CH_{4+)}$ for the hydrogen and nitrogen gas (N_2) from the air at an elevated temperature and pressure in the presence of a catalyst to form ammonia (NH_3) as the end product. This ammonia is used as a feedstock for other nitrogen fertilizers, such as anhydrous ammonium nitrate (NH_4NO_3) and urea

($CO(NH_2)_2$). These concentrated products may be diluted with water to form a concentrated liquid fertilizer (e.g. UAN). Deposits of sodium nitrate ($NaNO_3$) (Chilean saltpeter) are also found the Atacama desert in Chile and was one of the original (1830) nitrogen rich inorganic fertilizers used. It is still mined for fertilizer.

In the Nitrophosphate process or Odda Process (invented in 1927), phosphate rock with up to a 20% phosphorus (P) content is dissolved with nitric acid (HNO_3) to produce a mixture of phosphoric acid (H_3PO_4) and calcium nitrate ($Ca(NO_3)_2$). This can be combined with a potassium fertilizer to produce a *compound fertilizer* with all three N:P:K: plant nutrients in easily dissolved form.

Phosphate rock can also be processed into water-soluble phosphate (P_2O_5) with the addition of sulfuric acid (H_2SO_4) to make the phosphoric acid in phosphate fertilizers. Phosphate can also be reduced in an electric furnace to make high purity phosphorus; however, this is more expensive than the acid process.

Potash can be used to make potassium (K) fertilizers. All commercial potash deposits come originally from marine deposits and are often buried deep in the earth. Potash ores are typically rich in potassium chloride (KCl) and sodium chloride (NaCl) and are obtained by conventional shaft mining with the extracted ore ground into a powder. For deep potash deposits hot water is injected into the potash which is dissolved and then pumped to the surface where it is concentrated by solar induced evaporation. Amine reagents are then added to either the mined or evaporated solutions. The amine coats the KCl but not NaCl. Air bubbles cling to the amine + KCl and float it to the surface while the NaCl and clay sink to the bottom. The surface is skimmed for the amine + KCl which is then dried and packaged for use as a K rich fertilizer—KCl dissolves readily in water and is available quickly for plant nutrition.

Compound fertilizers often combine N, P and K fertilizers into easily dissolved pellets. The N:P:K ratios quoted on fertilizers give the weight percent of the fertilizer in nitrogen (N), phosphate (P_2O_5) and potash (K_2O equivalent)

The use of commercial inorganic fertilizers has increased steadily in the last 50 years, rising almost 20-fold to the current rate of 100 million tonnes of nitrogen per year. Without commercial fertilizers it is estimated that about one-third of the food produced now could not be produced. The use of phosphate fertilizers has also increased from 9 million tonnes per year in 1960 to 40 million tonnes per year in 2000.

A maize crop yielding 6–9 tonnes of grain per hectare requires 31–50 kg of phosphate fertilizer to be applied, soybean requires 20–25 kg per hectare. Yara International is the world's largest producer of nitrogen based fertilizers.

Pesticide

Pesticides are substances meant for attracting, seducing, destroying, or mitigating any pest. They are a class of biocide. The most common use of pesticides is as plant protection products (also known as crop protection products), which in general protect plants from damaging influences such as weeds, plant diseases or insects. This use of pesticides is so common that the term *pesticide* is often treated as synonymous with *plant protection product*, although it is in fact a broader term, as pesticides are also used for non-agricultural purposes. The term pesticide includes all of the following: herbicide, insecticide, insect growth regulator, nematicide, termiticide, molluscicide, piscicide, avicide, rodenticide, predacide, bactericide, insect repellent, animal repellent, antimicrobial, fungicide, disinfectant (antimicrobial), and sanitizer.

In general, a pesticide is a chemical or biological agent (such as a virus, bacterium, antimicrobial, or disinfectant) that deters, incapacitates, kills, or otherwise discourages pests. Target pests can include insects, plant pathogens, weeds, mollusks, birds, mammals, fish, nematodes (roundworms), and microbes that destroy property, cause nuisance, or spread disease, or are disease vectors. Although pesticides have benefits, some also have drawbacks, such as potential toxicity to humans and other desired species. According to the Stockholm Convention on Persistent Organic Pollutants, 9 of the 12 most dangerous and persistent organic chemicals are pesticides.

Types

Pesticides are often referred to according to the type of pest they control. Pesticides can also be considered as either biodegradable pesticides, which will be broken down by microbes and other living beings into harmless compounds, or persistent pesticides, which may take months or years before they are broken down: it was the persistence of DDT, for example, which led to its accumulation in the food chain and its killing of birds of prey at the top of the food chain. Another way to think about pesticides is to consider those that are chemical pesticides or are derived from a common source or production method. Some examples of chemically-related pesticides are:

Organophosphate Pesticides

Organophosphates affect the nervous system by disrupting the enzyme that regulates acetylcholine, a neurotransmitter. Most organophosphates are insecticides. They were developed during the early 19th century, but their effects on insects, which are similar to their effects on humans, were discovered in 1932. Some are very poisonous (they were used in World War II as nerve agents). However, they usually are not persistent in the environment.

Carbamate Pesticides

Carbamate pesticides affect the nervous system by disrupting an enzyme that regulates acetylcholine, a neurotransmitter. The enzyme effects are usually reversible. There are several subgroups within the carbamates.

Organochlorine Insecticides

They were commonly used in the past, but many have been removed from the market due to their health and environmental effects and their persistence (e.g., DDT and chlordane).

Pyrethroid Pesticides

They were developed as a synthetic version of the naturally occurring pesticide pyrethrin, which is found in chrysanthemums. They have been modified to increase their stability in the environment. Some synthetic pyrethroids are toxic to the nervous system.

Sulfonylurea Herbicides

Includes nicosulfuron, triflusulfuron methyl, and chlorsulfuron broad-spectrum herbicides that kill plants by inhibiting the enzyme acetolactate synthase. In the 1960s, more than 1 kg/ha (0.89 lb/acre) crop protection chemical was typically applied, while sulfonylureates allow as little as 1% as much material to achieve the same effect.

Biopesticides

Biopesticides are certain types of pesticides derived from such natural materials as animals, plants, bacteria, and certain minerals. For example, canola oil and baking soda have pesticidal applications and are considered biopesticides. At the end of 2001, there were approximately 195 registered biopesticide active ingredients and 780 products. Biopesticides fall into three major classes:

- Microbial pesticides consist of a microorganism e.g., a bacterium, fungus, virus, or protozoan as the active ingredient. Microbial pesticides can control many different kinds of pests, although each separate active ingredient is relatively specific for its target pest. For example, there are fungi that control certain weeds, and other fungi that kill specific insects.

The most widely used microbial pesticides are subspecies and strains of Bacillus thuringiensis, or Bt. Each strain of this bacterium produces a different mix of proteins, and specifically kills one or a few related species of insect larvae. While some Bt's control moth larvae found on plants, other Bt's are specific for larvae of flies and mosquitoes. The target insect species are determined by whether the particular Bt produces a protein that can bind to a larval gut receptor, thereby causing the insect larvae to starve.

- Plant-Incorporated-Protectants (PIPs) are pesticidal substances that plants produce from genetic material that has been added to the plant. For example, scientists can take the gene for the Bt pesticidal protein, and introduce the gene into the plant's own genetic material. Then the plant, instead of the Bt bacterium, manufactures the substance that destroys the pest. The protein and its genetic material, but not the plant itself, are regulated by EPA.
- Biochemical pesticides are naturally occurring substances that control pests by non-toxic mechanisms. Conventional pesticides, by contrast, are, in general, synthetic materials that directly kill or inactivate the pest. Biochemical pesticides include substances, such as insect sex pheromones, that interfere with mating, as well as various scented plant extracts that attract insect pests to traps. Because it is sometimes difficult to determine whether a substance meets the criteria for classification as a biochemical pesticide, EPA has established a special committee to make such decisions.

Uses

Pesticides are used to control organisms that are considered to be harmful. For example, they are used to kill mosquitoes that can transmit potentially deadly diseases like West Nile virus, yellow fever, and malaria. They can also kill bees, wasps or ants that can cause allergic reactions. Insecticides can protect animals from illnesses that can be caused by parasites such as fleas. Pesticides can prevent sickness in humans that could be caused by moldy food or diseased produce. Herbicides can be used to clear roadside weeds, trees and brush. They can also kill invasive weeds that may cause environmental damage. Herbicides are commonly applied in ponds and lakes to control algae and plants such as water grasses that can interfere with activities like swimming and fishing and cause the water to look or smell unpleasant. Uncontrolled pests such as termites and mould can damage structures such as houses. Pesticides are used in grocery stores and food storage facilities to manage rodents and insects that infest food such as grain. Each use of a pesticide carries some associated risk. Proper pesticide use decreases these associated risks to a level deemed acceptable by pesticide regulatory agencies such as the United States Environmental Protection Agency (EPA) and the Pest Management Regulatory Agency (PMRA) of Canada.

Pesticides can save farmers' money by preventing crop losses to insects and other pests; in the U.S., farmers get an estimated fourfold return on money they spend on pesticides. One study found that not using pesticides reduced crop yields by about 10%. Another study, conducted in 1999, found that a ban on pesticides in the United States may result in a rise of food prices, loss of jobs, and an increase in world hunger.

DDT, sprayed on the walls of houses, is an organochloride that has been used to fight malaria since the 1950s. Recent policy statements by the World Health Organisation have given stronger support to this approach. Dr. Arata Kochi, WHO's malaria chief, said, "One of the best tools we have against malaria is indoor residual house spraying. Of the dozen insecticides WHO has approved as safe for house spraying, the most effective is DDT." However a later October 2007 study linked breast cancer from exposure to DDT prior to puberty. Poisoning may also occur due to use of DDT and other chlorinated hydrocarbons by entering the human food chain when animal tissues are affected. Symptoms include nervous excitement, tremors, convulsions, or death. Scientists estimate that DDT and other chemicals in the organophosphate class of pesticides have saved 7 million human lives since 1945 by preventing the transmission of diseases such as malaria, bubonic plague, sleeping sickness, and typhus. However, DDT use is not always effective, as resistance to DDT was identified in Africa as early as 1955, and by 1972 nineteen species of mosquito worldwide were resistant to DDT. A study for the World Health Organisation in 2000 from Vietnam established that non-DDT malaria controls were significantly more effective than DDT use. The ecological effect of DDT on organisms is an example of bioaccumulation.

Manufacturing Process of Pesticides

Quantity and Variety: In 2006 and 2007, the world used approximately 5.2 billion pounds of pesticides, with herbicides constituting the biggest part of the world pesticide use at 40%, followed by insecticides (17%) and fungicides (10%). In 2006 and 2007 the U.S. used approximately 1.1 billion pounds of pesticides, accounting for 22% of the world total, including 857 million pounds of conventional pesticides, which are used in the agricultural sector (80% of conventional pesticide use) as well as the industrial, commercial, governmental and home & garden sectors. Pesticides are also found in majority of U.S. households with 78 million out of the 105.5 million households indicating that they use some form of pesticide. As of 2007, there were more than 1,055 active ingredients registered as pesticides, which yield over 20,000 pesticide products that are marketed in the United States.

The US used some 2.2 pounds per hactare of arable land compared with 10.3 in China, 3.0 in the UK, .2 in Cameroun, 13.1 in Japan and 5.6 in Italy. Insecticide use in the US has declined by more than half since 1980, (.6%/yr) mostly due to the near phase-out of organophosphates. In corn fields, the decline was even steeper, due to the switchover to transgenic Bt corn. For the global market of crop protection products, market analysts forecast revenues of over 52 billion US$ in 2019.

Costs

On the cost side of pesticide use there can be costs to the environment, costs to human health, as well as costs of the development and research of new pesticides.

Health Effects

Pesticides may cause acute and delayed health effects in workers who are exposed. Pesticide exposure can cause a variety of adverse health effects, ranging from simple irritation of the skin and eyes to more severe effects such as affecting the nervous system, mimicking hormones causing reproductive problems, and also causing cancer. A 2007 systematic review found that "most studies on non-Hodgkin lymphoma and leukemia showed positive associations with pesticide exposure" and thus concluded that cosmetic use of pesticides should be decreased. Strong evidence also exists for other negative outcomes from pesticide exposure including neurological, birth defects, fetal death, and neurodevelopmental disorder. The American Medical Association recommends limiting exposure to pesticides and using safer alternatives: "Particular uncertainty exists regarding the long-term effects of low-dose pesticide exposures."

Even consuming organic food, these chemicals can have a negative health impact. Though the U.S. Environmental Protection Agency

(EPA) limits the amount of each pesticide that may be present on a food item, there exists no limit to the number of different chemicals that can be used. This leads to a possibly even more dangerous effect known as a "chemical cocktail". The chemicals may form one of many dangerous interactions and have an unmeasured synergistic effect.

The World Health Organisation and the UN Environment Programme estimate that each year, 3 million workers in agriculture in the developing world experience severe poisoning from pesticides, about 18,000 of whom die. According to one study, as many as 25 million workers in developing countries may suffer mild pesticide poisoning yearly. One study found pesticide self-poisoning the method of choice in one third of suicides worldwide, and recommended, among other things, more restrictions on the types of pesticides that are most harmful to humans.

A 2007 study by the California Department of Public Health found that women in the first eight weeks of pregnancy who live near farm fields sprayed with the organochlorine pesticides dicofol and endosulfan are several times more likely to give birth to children with autism.

Environmental Effect

Pesticide use raises a number of environmental concerns. Over 98% of sprayed insecticides and 95% of herbicides reach a destination other than their target species, including non-target species, air, water and soil. Pesticide drift occurs when pesticides suspended in the air as particles are carried by wind to other areas, potentially contaminating them. Pesticides are one of the causes of water pollution, and some pesticides are persistent organic pollutants and contribute to soil contamination. In addition, pesticide use reduces biodiversity, reduces nitrogen fixation, contributes to pollinator decline, destroys habitat (especially for birds), and threatens endangered species.

Pests can develop a resistance to the pesticide (pesticide resistance), necessitating a new pesticide. Alternatively a greater dose of the pesticide can be used to counteract the resistance, although this will cause a worsening of the ambient pollution problem.

Since chlorinated hydrocarbon pesticides dissolve in fats and are not excreted, organisms tend to retain them almost indefinitely. Biological magnification is the process whereby these chlorinated hydrocarbons (pesticides) are more concentrated at each level of the food chain. Among marine animals, pesticide concentrations are higher in carnivorous fishes, and even more so in the fish-eating birds and mammals at the top of the ecological pyramid. Global distillation is the process whereby pesticides are transported from warmer to colder regions of the Earth, in particular the Poles and mountain tops. Pesticides that evaporate into the atmosphere at relatively high temperature can be carried considerable distances (thousands of kilometres) by the wind to an area of lower temperature, where they condense and are carried back to the ground in rain or snow.

In order to reduce negative impacts, it is desirable that pesticides be degradable or at least quickly deactivated in the environment. Such loss of activity or toxicity of pesticides is due to both innate chemical properties of the compounds and environmental processes or conditions. For example, the presence of halogens within a chemical structure often slows down degradation in an aerobic environment. Adsorption to soil may retard pesticide movement, but also may reduce bioavailability to microbial degraders.

Economics

Harm	*Annual US Cost*
Public Health	\$1.1 billion
Pesticide Resistance in Pest	\$1.5 billion
Crop Losses Caused by Pesticides	\$1.4 billion
Bird Losses due to Pesticides	\$2.2 billion
Groundwater Contamination	\$2.0 billion
Other Costs	\$1.4 billion
Total Costs	\$9.6 billion

Human health and environmental cost from pesticides in the United States is estimated at \$9.6 billion:

Additional costs include the registration process and the cost of purchasing pesticides. The registration process can take several years to complete (there are 70 different types of field test) and can cost \$50–70 million for a single pesticide. Annually the United States spends \$10 billion on pesticides.

Chapter 3

Alkane Stereochemistry

Alkane stereochemistry concerns the stereochemistry of alkanes. Alkane conformers are one of the subjects of alkane stereochemistry.

Conformations

Alkane conformers arise from rotation around sp^3 hybridised carbon sigma bonds. The smallest alkane with such a chemical bond, ethane, exists as an infinite number of conformations with respect to rotation around the C–C bond.

Two of these are recognised as energy minimum (staggered conformation) and energy maximum (eclipsed conformation) forms. The existence of specific conformations is due to hindered rotation around sigma bonds, although a role for hyperconjugation is proposed by a competing theory.

The importance of energy minimum and energy maximum is seen by extension of these concepts to more complex molecules for which stable conformations may be predicted as minimum energy forms. The determination of stable conformations has also played a large role in the establishment of the concept of asymmetric induction and the ability to predict the stereochemistry of reactions controlled by steric effects.

In the example of staggered ethane in Newman projection, a hydrogen atom on one carbon atom has a 60° torsional angle or torsion angle with respect to the nearest hydrogen atom on the other carbon so that steric hindrance is minimised.

The staggered conformation is more stable by 12.5 kJ/mol than the eclipsed conformation, which is the energy maximum for ethane. In the eclipsed conformation the torsional angle is minimised.

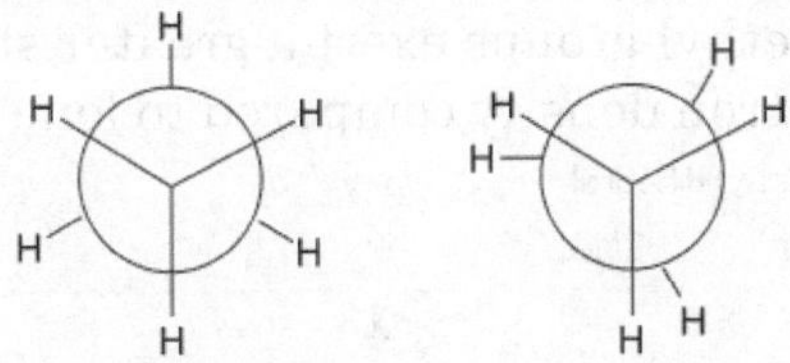

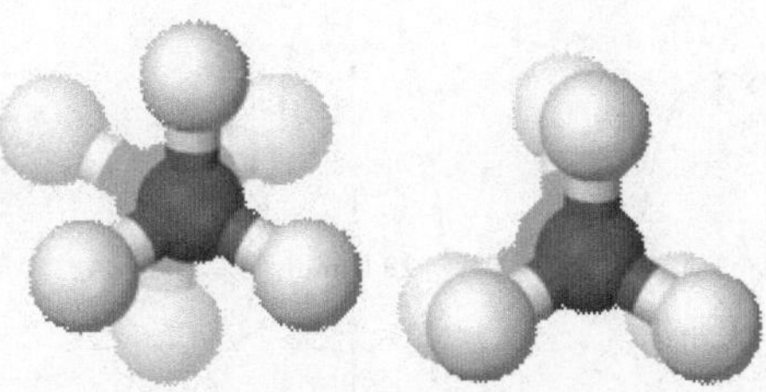

In butane, the two staggered conformations are no longer equivalent and represent two distinct conformers: the anti-conformation (left-most, below) and the gauche conformation (right-most, below).

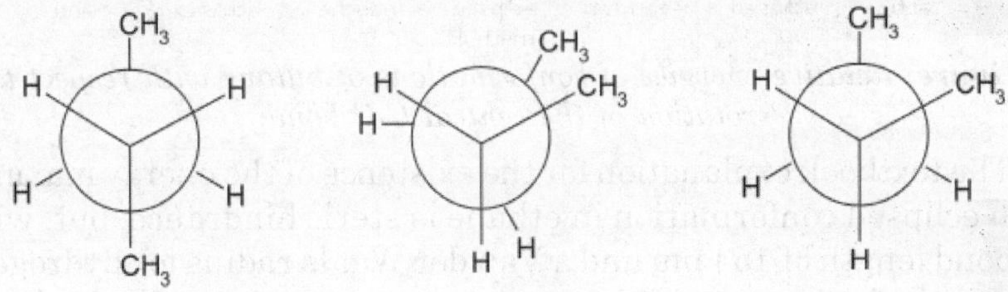

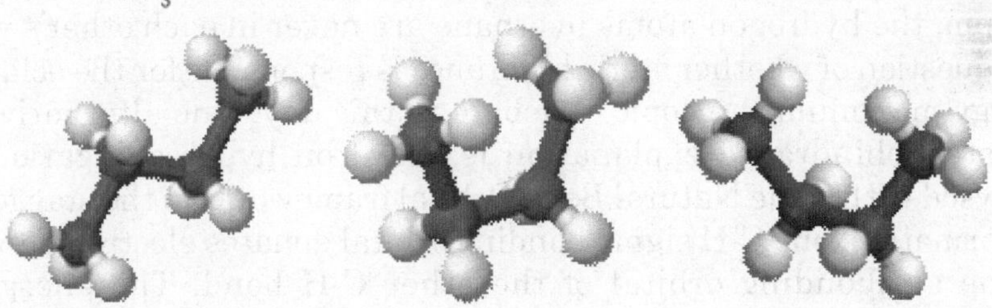

Both conformations are free of torsional strain, but, in the gauche conformation, the two methyl groups are in closer proximity than the sum of their van der Waals radii. The interaction between the two methyl groups is repulsive (van der Waals strain), and an energy barrier results.

A measure of the potential energy stored in butane conformers with greater steric hindrance than the 'anti'-conformer ground state is given by these values:

- Gauche, conformer - 3.8 kJ/mol
- Eclipsed H and CH_3 - 16 kJ/mol
- Eclipsed CH_3 and CH_3 - 19 kJ/mol.

The eclipsed methyl groups exert a greater steric strain because of their greater electron density compared to lone hydrogen atoms.

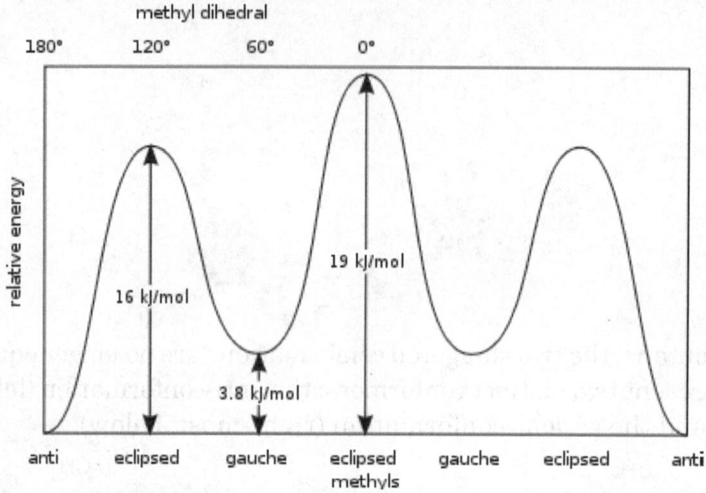

Figure: *Relative energies of conformations of butane with respect to rotation of the central C-C bond.*

The textbook explanation for the existence of the energy maximum for an eclipsed conformation in ethane is steric hindrance, but, with a C-C bond length of 154 pm and a Van der Waals radius for hydrogen of 120 pm, the hydrogen atoms in ethane are never in each other's way. The question of whether steric hindrance is responsible for the eclipsed energy maximum is a topic of debate to this day. One alternative to the steric hindrance explanation is based on hyperconjugation as analysed within the Natural Bond Orbital framework. In the staggered conformation, one C-H sigma bonding orbital donates electron density to the antibonding orbital of the other C-H bond. The energetic stabilisation of this effect is maximised when the two orbitals have maximal overlap, occurring in the staggered conformation. There is no overlap in the eclipsed conformation, leading to a disfavoured energy maximum. On the other hand, an analysis within quantitative molecular orbital theory shows that 2-orbital-4-electron (steric) repulsions are dominant over hyperconjugation. A valence bond theory study also emphasizes the importance of steric effects.

Definitions

Many definitions that describe a specific conformation (IUPAC Gold Book) exist:

- a torsion angle of ±60° is called gauche
- a torsion angle between 0° and ± 90° is called syn (s)

- a torsion angle between ± 90° and 180° is called anti (a)
- a torsion angle between 30° and 150° or between –30° and –150° is called clinal
- a torsion angle between 0° and 30° or 150° and 180° is called periplanar (p)
- a torsion angle between 0° to 30° is called synperiplanar or syn- or cis-conformation (sp)
- a torsion angle between 30° to 90° and –30° to –90° is called synclinal or gauche or skew (sc)
- a torsion angle between 90° to 150°, and –90° to –150° is called anticlinal (ac)
- a torsion angle between ± 150° to 180° is called antiperiplanar or anti or trans (ap).

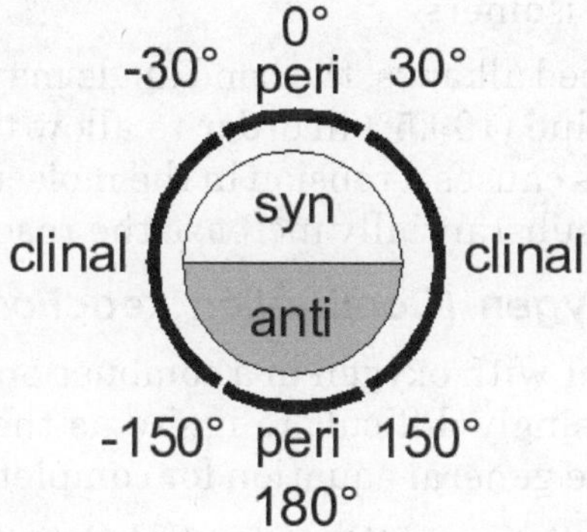

***Figure:** Torsional strain results from resistance to twisting about a bond.*

Special Cases

In n-pentane, the terminal methyl groups experience additional pentane interference.

Replacing hydrogen by fluorine in polytetrafluoroethylene changes the stereochemistry from the zigzag geometry to that of a helix due to electrostatic repulsion of the fluorine atoms in the 1,3 positions. Evidence for the helix structure in the crystalline state is derived from X-ray crystallography and from NMR spectroscopy and circular dichroism in solution.

Chemical Properties

In general, alkanes show a relatively low reactivity, because their C bonds are relatively stable and cannot be easily broken. Unlike most other organic compounds, they possess no functional groups. They react only very poorly with ionic or other polar substances. The acid dissociation constant (pK_a) values of all alkanes are above 60, hence they are practically inert to acids and bases. This inertness is the source of the term *paraffins* (with the meaning here of "lacking affinity"). In crude oil the alkane molecules have remained chemically unchanged for millions of years.

However redox reactions of alkanes, in particular with oxygen and the halogens, are possible as the carbon atoms are in a strongly reduced condition; in the case of methane, the lowest possible oxidation state for carbon ("4) is reached. Reaction with oxygen (*if* an amount of the least is enough to meet the reaction stoichometry) leads to combustion without any smoke; with halogens, substitution. Free radicals, molecules with unpaired electrons, play a large role in most reactions of alkanes, such as cracking and reformation where long-chain alkanes are converted into shorter-chain alkanes and straight-chain alkanes into branched-chain isomers.

In highly branched alkanes, the bond angle may differ significantly from the optimal value (109.5°) in order to allow the different groups sufficient space. This causes a tension in the molecule, known as steric hindrance, and can substantially increase the reactivity.

Reactions with Oxygen (Combustion Reaction)

All alkanes react with oxygen in a combustion reaction, although they become increasingly difficult to ignite as the number of carbon atoms increases. The general equation for complete combustion is:

$$C_nH_{2n+2} + (1.5n+0.5)O_2 \rightarrow (n+1)H_2O + nCO_2$$

In the absence of sufficient oxygen, carbon monoxide or even soot can be formed, as shown below:

$$C_nH_{(2n+2)} + nO_2 \rightarrow (n+1)H_2O + nCO$$

For example methane:

$$2CH_4 + 3O_2 \rightarrow 2CO + 4H_2O$$

$$CH_4 + 1.5O_2 \rightarrow CO + 2H_2O$$

The standard enthalpy change of combustion, $\Delta_c H^\circ$, for alkanes increases by about 650 kJ/mol per CH_2 group. Branched-chain alkanes have lower values of $\Delta_c H^\circ$ than straight-chain alkanes of the same number of carbon atoms, and so can be seen to be somewhat more stable.

Reactions with Halogens

Alkanes react with halogens in a so-called *free radical halogenation* reaction. The hydrogen atoms of the alkane are progressively replaced by halogen atoms. Free-radicals are the reactive species that participate in the reaction, which usually leads to a mixture of products. The reaction is highly exothermic, and can lead to an explosion.

These reactions are an important industrial route to halogenated hydrocarbons. There are three steps:

- Initiation the halogen radicals form by homolysis. Usually, energy in the form of heat or light is required.
- Chain reaction or Propagation then takes place—the halogen radical abstracts a hydrogen from the alkane to give an alkyl radical. This reacts further.
- Chain termination where step the radicals recombine.

Experiments have shown that all halogenation produces a mixture of all possible isomers, indicating that all hydrogen atoms are susceptible to reaction. The mixture produced, however, is not a statistical mixture: Secondary and tertiary hydrogen atoms are preferentially replaced due to the greater stability of secondary and tertiary free-radicals. An example can be seen in the monobromination of propane:

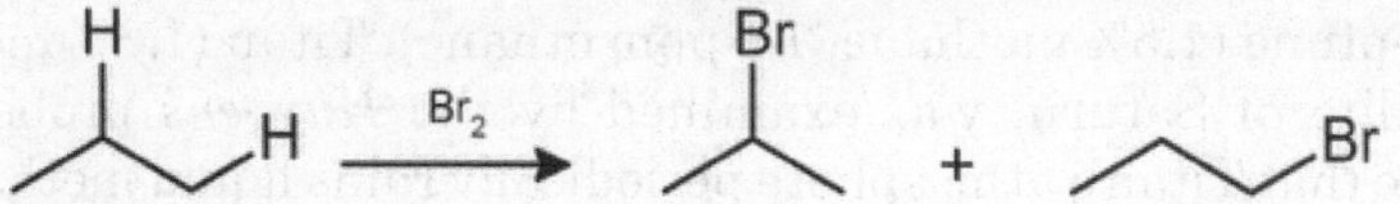

Statistical distribution:	33 %	67 %
Experimental distribution:	97 %	3 %

Cracking

Cracking breaks larger molecules into smaller ones. This can be done with a thermal or catalytic method. The thermal cracking process follows a homolytic mechanism with formation of free-radicals. The catalytic cracking process involves the presence of acid catalysts (usually solid acids such as silica-alumina and zeolites), which promote a heterolytic (asymmetric) breakage of bonds yielding pairs of ions of opposite charges, usually a carbocation and the very unstable hydride anion. Carbon-localised free-radicals and cations are both highly unstable and undergo processes of chain rearrangement, C-C scission in position beta (i.e., cracking) and intra- and intermolecular hydrogen transfer or hydride transfer. In both types of processes, the corresponding reactive intermediates (radicals, ions) are permanently regenerated, and thus they proceed by a self-propagating chain mechanism. The chain of reactions is eventually terminated by radical or ion recombination.

Isomerization and Reformation

Isomerization and reformation are processes in which straight-chain alkanes are heated in the presence of a platinum catalyst. In isomerization, the alkanes become branched-chain isomers. In reformation, the alkanes become cycloalkanes or aromatic hydrocarbons, giving off hydrogen as a by-product. Both of these processes raise the octane number of the substance.

Other Reactions

Alkanes will react with steam in the presence of a nickel catalyst to give hydrogen. Alkanes can be chlorosulfonated and nitrated, although both reactions require special conditions. The fermentation of alkanes to carboxylic acids is of some technical importance. In the Reed reaction, sulfur dioxide, chlorine and light convert hydrocarbons to sulfonyl chlorides.

Occurrence—Occurrence of Alkanes in the Universe

Alkanes form a small portion of the atmospheres of the outer gas planets such as Jupiter (0.1% methane, 0.0002% ethane), Saturn (0.2% methane, 0.0005% ethane), Uranus (1.99% methane, 0.00025% ethane) and Neptune (1.5% methane, 1.5 ppm ethane). Titan (1.6% methane), a satellite of Saturn, was examined by the *Huygens* probe, which indicate that Titan's atmosphere periodically rains liquid methane onto the moon's surface. Also on Titan, a methane-spewing volcano was spotted and this volcanism is believed to be a significant source of the methane in the atmosphere.

There also appear to be Methane/Ethane lakes near the north polar regions of Titan, as discovered by Cassini's radar imaging. Methane and ethane have also been detected in the tail of the comet Hyakutake. Chemical analysis showed that the abundances of ethane and methane were roughly equal, which is thought to imply that its ices formed in interstellar space, away from the Sun, which would have evaporated these volatile molecules. Alkanes have also been detected in meteorites such as carbonaceous chondrites.

Occurrence of Alkanes on Earth

Traces of methane gas (about 0.0001% or 1 ppm) occur in the Earth's atmosphere, produced primarily by organisms such as Archaea, found for example in the gut of cows. The most important commercial sources for alkanes are natural gas and oil. Natural gas contains primarily methane and ethane, with some propane and butane: oil is a mixture of liquid alkanes and other hydrocarbons. These hydrocarbons were formed when dead marine animals and plants (zooplankton and phytoplankton) died and sank to the bottom of ancient seas and were covered with sediments in an anoxic environment and converted over many millions of years at high temperatures and high pressure to their current form. Natural gas resulted thereby for example from the following reaction:

$$C_6H_{12}O_6 \rightarrow 3CH_4 + 3CO_2$$

These hydrocarbons collected in porous rocks, located beneath an impermeable cap rock and so are trapped. These deposits, e.g., oil fields, have formed over millions of years and once exhausted cannot be readily replaced. The depletion of these hydrocarbons is the basis for what is known as the energy crisis.

Solid alkanes are known as tars and are formed when more volatile alkanes such as gases and oil evaporate from hydrocarbon deposits. One of the largest natural deposits of solid alkanes is in the asphalt lake known as the Pitch Lake in Trinidad and Tobago.

Methane is also present in what is called biogas, produced by animals and decaying matter, which is a possible renewable energy source. Alkanes have a low solubility in water, so the content in the oceans is negligible; however, at high pressures and low temperatures (such as at the bottom of the oceans), methane can co-crystallize with water to form a solid methane hydrate.

Although this cannot be commercially exploited at the present time, the amount of combustible energy of the known methane hydrate fields exceeds the energy content of all the natural gas and oil deposits put together; methane extracted from methane hydrate is considered therefore a candidate for future fuels.

Biological Occurrence

Although alkanes occur in nature in various way, they do not rank biologically among the essential materials. Cycloalkanes with 14 to 18 carbon atoms occur in musk, extracted from deer of the family Moschidae. All further information refers to (acyclic) alkanes.

Bacteria and Archaea

Methanogenic archaea in the gut of this cow are responsible for some of the methane in Earth's atmosphere.

Certain types of bacteria can metabolize alkanes: they prefer even-numbered carbon chains as they are easier to degrade than odd-numbered chains.

On the other hand, certain archaea, the methanogens, produce large quantities of methane by the metabolism of carbon dioxide or other oxidized organic compounds. The energy is released by the oxidation of hydrogen:

$$CO_2 + 4H_2 \rightarrow CH_4 + 2H_2O$$

Methanogens are also the producers of marsh gas in wetlands, and release about two billion tonnes of methane per year—the atmospheric content of this gas is produced nearly exclusively by them. The methane output of cattle and other herbivores, which can release up to 150 liters per day, and of termites, is also due to methanogens. They also produce this simplest of all alkanes in the intestines of humans. Methanogenic archaea are, hence, at the end of the carbon cycle, with carbon being released back into the atmosphere after having been fixed by photosynthesis. It is probable that our current deposits of natural gas were formed in a similar way.

Fungi and Plants

Alkanes also play a role, if a minor role, in the biology of the three eukaryotic groups of organisms: fungi, plants and animals. Some specialised yeasts, e.g., *Candida tropicale*, *Pichia* sp., *Rhodotorula* sp., can use alkanes as a source of carbon and/or energy. The fungus *Amorphotheca resinae* prefers the longer-chain alkanes in aviation fuel, and can cause serious problems for aircraft in tropical regions.

In plants, the solid long-chain alkanes are found in the plant cuticle and epicuticular wax of many species, but are only rarely major constituents. They protect the plant against water loss, prevent the leaching of important minerals by the rain, and protect against bacteria, fungi, and harmful insects. The carbon chains in plant alkanes are usually odd-numbered, between twenty-seven and thirty-three carbon atoms in length and are made by the plants by decarboxylation of even-numbered fatty acids. The exact composition of the layer of wax is not only species-dependent, but changes also with the season and such environmental factors as lighting conditions, temperature or humidity.
Animals

Alkanes are found in animal products, although they are less important than unsaturated hydrocarbons. One example is the shark liver oil, which is approximately 14% pristane (2,6,10,14-tetramethylpentadecane, $C_{19}H_{40}$). Their occurrence is more important in pheromones, chemical messenger materials, on which above all insects are dependent for communication. With some kinds, as the support beetle *Xylotrechus colonus*, primarily pentacosane ($C_{25}H_{52}$), 3-

methylpentaicosane ($C_{26}H_{54}$) and 9-methylpentaicosane ($C_{26}H_{54}$), they are transferred by body contact. With others like the tsetse fly *Glossina morsitans morsitans*, the pheromone contains the four alkanes 2-methylheptadecane ($C_{18}H_{38}$), 17,21-dimethylheptatriacontane ($C_{39}H_{80}$), 15,19-dimethylheptatriacontane ($C_{39}H_{80}$) and 15,19,23-trimethylheptatriacontane ($C_{40}H_{82}$), and acts by smell over longer distances, a useful characteristic for pest control. Waggle-dancing honeybees produce and release two alkanes, tricosane and pentacosane.

Ecological Relations

One example, in which both plant and animal alkanes play a role, is the ecological relationship between the sand bee (*Andrena nigroaenea*) and the early spider orchid (*Ophrys sphegodes*); the latter is dependent for pollination on the former. Sand bees use pheromones in order to identify a mate; in the case of *A. nigroaenea*, the females emit a mixture of tricosane ($C_{23}H_{48}$), pentacosane ($C_{25}H_{52}$) and heptacosane ($C_{27}H_{56}$) in the ratio 3:3:1, and males are attracted by specifically this odor. The orchid takes advantage of this mating arrangement to get the male bee to collect and disseminate its pollen; parts of its flower not only resemble the appearance of sand bees, but also produce large quantities of the three alkanes in the same ratio as female sand bees. As a result numerous males are lured to the blooms and attempt to copulate with their imaginary partner: although this endeavour is not crowned with success for the bee, it allows the orchid to transfer its pollen, which will be dispersed after the departure of the frustrated male to different blooms.

Production—Petroleum Refining

As stated earlier, the most important source of alkanes is natural gas and crude oil. Alkanes are separated in an oil refinery by fractional distillation and processed into many different products.

Fischer-Tropsch

The Fischer-Tropsch process is a method to synthesize liquid hydrocarbons, including alkanes, from carbon monoxide and hydrogen. This method is used to produce substitutes for petroleum distillates.

Laboratory Preparation

There is usually little need for alkanes to be synthesized in the laboratory, since they are usually commercially available. Also, alkanes are generally non-reactive chemically or biologically, and do not undergo functional group interconversions cleanly. When alkanes are produced in the laboratory, it is often a side-product of a reaction. For example, the use of *n*-butyllithium as a strong base gives the conjugate acid, n-butane as a side-product:

$$C_4H_9Li + H_2O \rightarrow C_4H_{10} + LiOH$$

However, at times it may be desirable to make a portion of a molecule into an alkane like functionality (alkyl group) using the above or similar methods. For example, an ethyl group is an alkyl group; when this is attached to a hydroxy group, it gives ethanol, which is not an alkane. To do so, the best-known methods are hydrogenation of alkenes:

$$RCH{=}CH_2 + H_2 \rightarrow RCH_2CH_3 \text{ (R = alkyl)}$$

Alkanes or alkyl groups can also be prepared directly from alkyl halides in the Corey-House-Posner-Whitesides reaction. The Barton-McCombie deoxygenation removes hydroxyl groups from alcohols e.g.

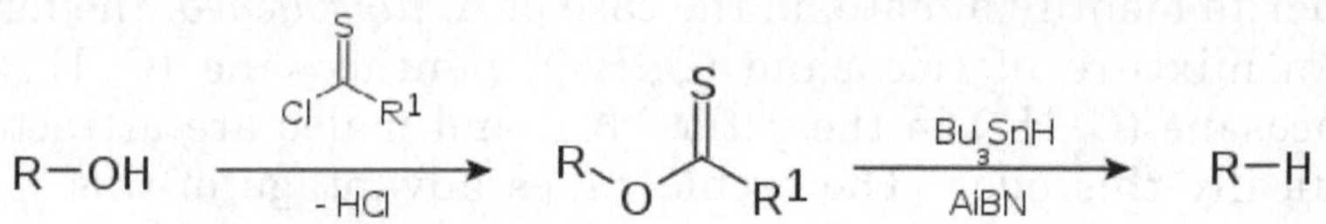

and the Clemmensen reduction removes carbonyl groups from aldehydes and ketones to form alkanes or alkyl-substituted compounds e.g.:

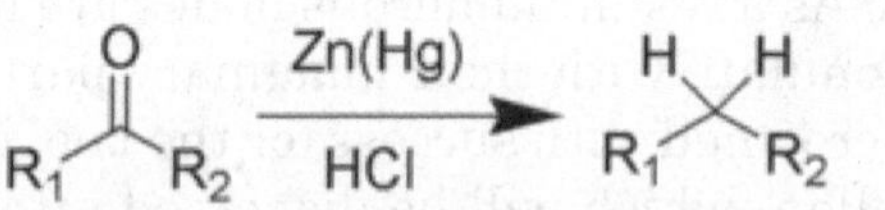

Applications

The applications of a certain alkane can be determined quite well according to the number of carbon atoms. The first four alkanes are used mainly for heating and cooking purposes, and in some countries for electricity generation. Methane and ethane are the main components of natural gas; they are normally stored as gases under pressure. It is, however, easier to transport them as liquids: This requires both compression and cooling of the gas.

Propane and butane can be liquefied at fairly low pressures, and are well known as liquified petroleum gas (LPG). Propane, for example, is used in the propane gas burner, butane in disposable cigarette lighters. The two alkanes are used as propellants in aerosol sprays.

From pentane to octane the alkanes are reasonably volatile liquids. They are used as fuels in internal combustion engines, as they vaporise easily on entry into the combustion chamber without forming droplets, which would impair the uniformity of the combustion. Branched-chain alkanes are preferred as they are much less prone to premature ignition, which causes knocking, than their straight-chain homologues.

This propensity to premature ignition is measured by the octane rating of the fuel, where 2,2,4-trimethylpentane (*isooctane*) has an arbitrary value of 100, and heptane has a value of zero. Apart from their use as fuels, the middle alkanes are also good solvents for nonpolar substances.

Alkanes from nonane to, for instance, hexadecane (an alkane with sixteen carbon atoms) are liquids of higher viscosity, less and less suitable for use in gasoline. They form instead the major part of diesel and aviation fuel. Diesel fuels are characterised by their cetane number, cetane being an old name for hexadecane. However, the higher melting points of these alkanes can cause problems at low temperatures and in polar regions, where the fuel becomes too thick to flow correctly.

Alkanes from hexadecane upwards form the most important components of fuel oil and lubricating oil. In latter function, they work at the same time as anti-corrosive agents, as their hydrophobic nature means that water cannot reach the metal surface.

Many solid alkanes find use as paraffin wax, for example, in candles. This should not be confused however with true wax, which consists primarily of esters.

Alkanes with a chain length of approximately 35 or more carbon atoms are found in bitumen, used, for example, in road surfacing. However, the higher alkanes have little value and are usually split into lower alkanes by cracking.

Some synthetic polymers such as polyethylene and polypropylene are alkanes with chains containing hundreds of thousands of carbon atoms. These materials are used in innumerable applications, and billions of kilograms of these materials are made and used each year.

Environmental Transformations

When released in the environment, alkanes don't undergo rapid biodegradation, because they have no functional groups (like hydroxyl or carbonyl) that are needed by most organisms in order to metabolize the compound.

However, some bacteria can metabolize some alkanes (especially those linear and short), by oxidizing the terminal carbon atom. The product is an alcohol, that could be next oxidized to an aldehyde, and finally to a carboxylic acid. The resulting fatty acid could be metabolized through the fatty acid degradation pathway.

Hazards

Methane is explosive when mixed with air (1 – 8% CH_4). Other lower alkanes can also form explosive mixtures with air. The lighter liquid alkanes are highly flammable, although this risk decreases with the length of the carbon chain. Pentane, hexane, heptane, and octane are classed as *dangerous for the environment* and *harmful*. The straight-chain isomer of hexane is a neurotoxin. Halogen-rich alkanes, like chloroform, can be carcinogenic as well.

Alkene

In organic chemistry, an alkene, olefin, or olefine is an unsaturated chemical compound containing at least one carbon-to-carbon double bond. The simplest acyclic alkenes, with only one double bond and no other functional groups, form an homologous series of hydrocarbons with the general formula C_nH_{2n}.

The simplest alkene is ethylene (C_2H_4), which has the International Union of Pure and Applied Chemistry (IUPAC) name *ethene*. Alkenes are also called *olefins* (an archaic synonym, widely used in the petrochemical industry). For bridged alkenes, the Bredt's rule states that a double bond cannot be placed at the bridgehead of a bridged ring system, unless the rings are large enough. Aromatic compounds are often drawn as cyclic alkenes, but their structure and properties are different and they are not considered to be alkenes.

Structure—Bonding

Like single covalent bonds, double bonds can be described in terms of overlapping atomic orbitals, except that, unlike a single bond (which consists of a single sigma bond), a carbon-carbon double bond consists of one sigma bond and one pi bond. This double bond is stronger than a single covalent bond (611 kJ/mol for C=C vs. 347 kJ/mol for C—C) and also shorter with an average bond length of 1.33 Angstroms (133 pm).

Each carbon of the double bond uses its three sp^2 hybrid orbitals to form sigma bonds to three atoms. The unhybridized *2p* atomic orbitals, which lie perpendicular to the plane created by the axes of the three sp^2 hybrid orbitals, combine to form the pi bond. This bond lies outside the main C—C axis, with half of the bond on one side and half on the other.

Rotation about the carbon-carbon double bond is restricted because it involves breaking the pi bond, which requires a large amount of energy (264 kJ/mol in ethylene). As a consequence, substituted alkenes may exist as one of two isomers, called *cis* or *trans* isomers. More complex alkenes may be named using the E-Z notation, used to describe molecules having three or four different substituents (side groups). For example, of the isomers of butene, the two methyl groups of (*Z*)-but-2-ene (aka *cis*-2-butene) face the same side of the double bond, and in (*E*)-but2-ene (aka *trans*-2-butene) the methyl groups face the opposite side. These two isomers of butene are slightly different in their chemical and physical properties.

It is certainly not impossible to twist a double bond. In fact, a 90° twist requires an energy approximately equal to half the strength of a pi bond. The misalignment of the p orbitals is less than expected because pyramidalization takes place. *trans*-Cyclooctene is a stable strained alkene and the orbital misalignment is only 19° with a dihedral angle of 135° (normal 120°) and a degree of pyramidalization of 18°. This explains the dipole moment of 0.8 D for this compound (cis-isomer 0.4 D) where a value of zero is expected. The *trans* isomer of cycloheptene is only stable at low temperatures.

Shape

As predicted by the VSEPR model of electron pair repulsion, the molecular geometry of alkenes includes bond angles about each carbon in a double bond of about 120°. The angle may vary because of steric strain introduced by nonbonded interactions created by functional groups attached to the carbons of the double bond. For example, the C-C-C bond angle in propylene is 123.9°.

Physical Properties

The physical properties of alkenes are comparable with those of alkanes. The main differences between the two are that the acidity levels of alkenes are much higher than the ones in alkanes. The physical state depends on molecular mass (gases from ethene to butene - liquids from pentene onwards). The simplest alkenes, ethene, propene and butene are gases. Linear alkenes of approximately five to sixteen carbons are liquids, and higher alkenes are waxy solids.

Reactions

Alkenes are relatively stable compounds, but are more reactive than alkanes due to the presence of a carbon-carbon pi-bond. It is also attributed to the presence of pi-electrons in the molecule. The majority of the reactions of alkenes involve the rupture of this pi bond, forming new single bonds.

Alkenes serve as a feedstock for the petrochemical industry because they can participate in a wide variety of reactions.

Addition Reactions

Alkenes react in many addition reactions, which occur by opening up the double-bond. Most addition reactions to alkenes follow the mechanism of electrophilic addition. Examples of addition reactions are hydrohalogenation, halogenation, halohydrin formation, oxymercuration, hydroboration, dichlorocarbene addition, Simmons-Smith reaction, catalytic hydrogenation, epoxidation, radical polymerization and hydroxylation.

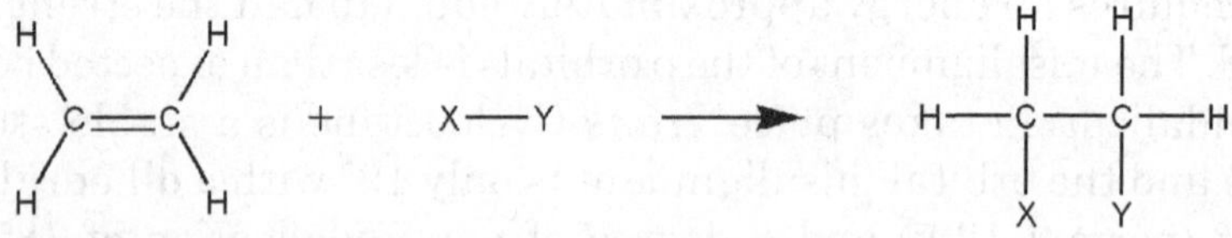

Hydrogenation

Hydrogenation of alkenes produces the corresponding alkanes. The reaction is carried out under pressure at a temperature of 200 °C in the presence of a metallic catalyst. Common industrial catalysts are based on platinum, nickel or palladium. For laboratory syntheses, Raney nickel (an alloy of nickel and aluminium) is often employed. The simplest example of this reaction is the catalytic hydrogenation of ethylene to yield ethane:

$$CH_2{=}CH_2 + H_2 \rightarrow CH_3\text{-}CH_3$$

Halogenation

In electrophilic halogenation the addition of elemental bromine or chlorine to alkenes yields vicinal dibromo- and dichloroalkanes (1,2-dihalides or ethylene dihalides), respectively. The decolouration of a solution of bromine in water with dichloromethylene as catalyst is an analytical test for the presence of alkenes:

$$CH_2{=}CH_2 + Br_2 \rightarrow BrCH_2\text{-}CH_2Br$$

It is also used as a quantitive test of unsaturation, expressed as the bromine number of a single compound or mixture. The reaction works because the high electron density at the double bond causes a temporary shift of electrons in the Br-Br bond causing a temporary induced dipole. This makes the Br closest to the double bond slightly positive and therefore an electrophile.

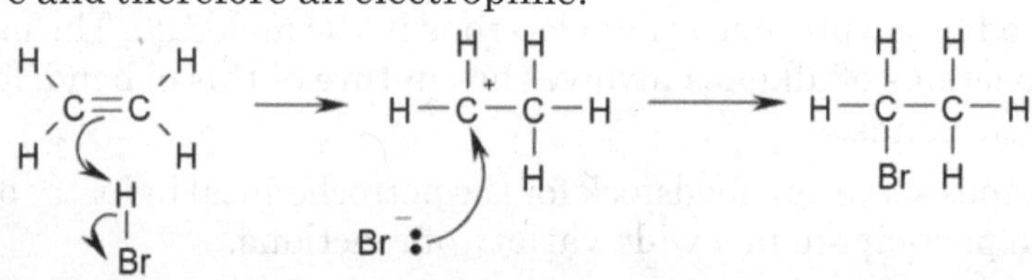

Hydrohalogenation

Hydrohalogenation is the addition of hydrohalic acids such as HCl or HBr to alkenes to yield the corresponding haloalkanes.

$$CH_3\text{-}CH{=}CH_2 + HBr \rightarrow CH_3\text{-}CHBr\text{-}CH_2\text{-}H$$

If the two carbon atoms at the double bond are linked to a different number of hydrogen atoms, the halogen is found preferentially at the carbon with fewer hydrogen substituents (Markovnikov's rule). But terminal olefin products don't yield by this method. For bromation an alternative method denominated Kharasch-Sosnovsky Reaction is used for this purpose. It consists to add peroxides to hydrogen bromide or Copper bromide (II).

Halohydrin Formation

Alkenes react with water and halogens to form halohydrins by an addition reaction. Markovnikov regiochemistry and anti stereochemistry occur.

$$CH_2{=}CH_2 + X_2 + H_2O \rightarrow XCH_2\text{-}CH_2OH$$

Oxidation

Alkenes are oxidized with a large number of oxidizing agents. In the presence of oxygen, alkenes burn with a bright flame to produce carbon dioxide and water. Catalytic oxidation with oxygen or the reaction with percarboxylic acids yields epoxides. Reaction with ozone in ozonolysis leads to the breaking of the double bond, yielding two aldehydes or ketones. Reaction with concentrated, hot $KMnO_4$ (or other oxidizing salts) in an acidic solution will yield ketones or carboxylic acids.

$$R_1\text{-}CH{=}CH\text{-}R_2 + O_3 \rightarrow R_1\text{-}CHO + R_2\text{-}CHO + H_2O$$

This reaction can be used to determine the position of a double bond in an unknown alkene.

Oxymercuration

Hydration of alkenes via oxymercuration to produces alcohols. Reaction takes place on treatment of alkenes with strong acid as catalyst.

$$CH_2{=}CH_2 + H_2O \rightarrow CH_3\text{-}CH_2OH$$

Polymerization

Polymerization of alkenes is a reaction that yields polymers of high industrial value at great economy, such as the plastics polyethylene and polypropylene. Polymers from alkene monomers are referred to in a general way as *polyolefins* or in rare instances as *polyalkenes*. A polymer from alpha-olefins is called a polyalphaolefin (PAO). Polymerization can proceed via either a free-radical or an ionic mechanism, converting the double to a single bond and forming single bonds to join the other monomers. Polymerization of conjugated dienes such as buta-1,3-diene or isoprene (2-methylbuta-1,3-diene) results in largely 1,4-addition with possibly some 1,2-addition of the diene monomer to a growing polymer chain.

Reaction Overview

Reaction name	*Product*	*Comment*
Hydrogenation	alkanes	addition of hydrogen
Hydroalkenylation	alkenes	hydrometalation / insertion / beta elimination by metal catalyst
Halogen addition reaction	1,2-dihalide	electrophilic addition of halogens
Hydrohalogenation (Markovnikov)	haloalkanes	addition of hydrohalic acids
Kharasch-Sosnovsky Reaction (Antimarkovnikov Hydrohalogenation)	haloalkanes	free radicals mediated addition of hydrohalic acids
Hydroamination	amines	addition of N-H bond across C-C double bond
Hydroformylation	aldehydes	industrial process, addition of CO and H_2
Sharpless bishydroxylation	diols	oxidation, reagent: osmium tetroxide, chiral ligand
Woodward cis-hydroxylation	diols	oxidation, reagents: iodine, silver acetate
ozonolysis	aldehydes or ketones	reagent: ozone
Olefin metathesis	alkenes	two alkenes rearrange to form two new alkenes
Diels-Alder reaction	cyclohexenes	cycloaddition with a diene
Pauson-Khand reaction	cyclopentenones	cycloaddition with an alkyne and CO
Hydroboration–oxidation	alcohols	reagents: borane, then a peroxide
Hydroboration–oxidation	alcohols	reagents: borane, then a peroxide
oxymercuration-reduction	alcohols	electrophilic addition of mercuric acetate, then reduction
Prins reaction	1,3-diols	electrophilic addition with aldehyde or ketone
Paterno–Büchi reaction	oxetanes	photochemical reaction with aldehyde or ketone
Epoxidation	epoxide	electrophilic addition of a peroxide
Cyclopropanation	cyclopropanes	addition of carbenes or carbenoids
Hydroacylation	ketones	oxidative addition / reductive elimination by metal catalyst

Synthesis—Industrial Methods

Alkenes are produced by hydrocarbon cracking. Raw materials are mostly natural gas condensate components (principally ethane and propane) in the US and Mideast and naphtha in Europe and Asia. Alkanes are broken apart at high temperatures, often in the presence of a zeolite catalyst, to produce a mixture of primarily aliphatic alkenes and lower molecular weight alkanes. The mixture is feedstock dependent and separated by fractional distillation. This is mainly used for the manufacture of small alkenes (up to six carbons).

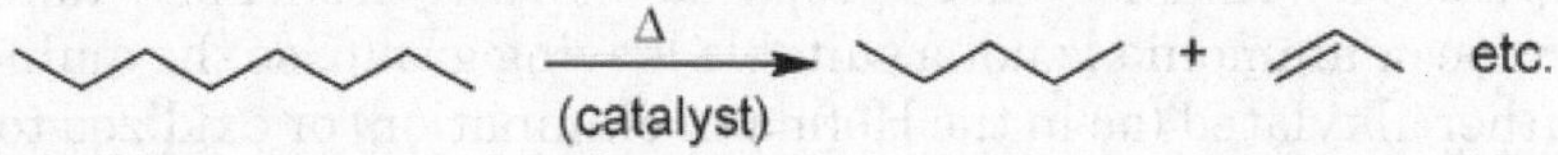

Related to this is catalytic dehydrogenation, where an alkane loses hydrogen at high temperatures to produce a corresponding alkene. This is the reverse of the catalytic hydrogenation of alkenes.

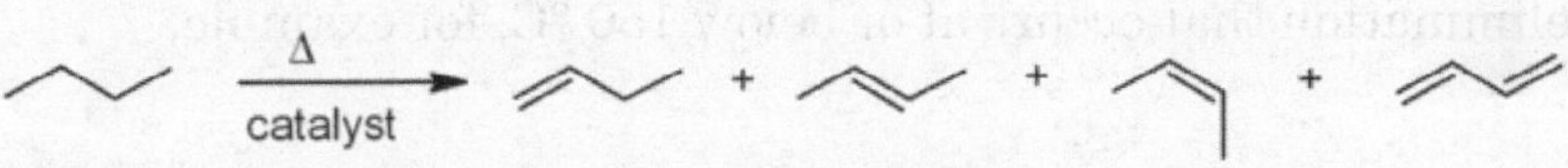

Both of these processes are endothermic, but they are driven towards the alkene at high temperatures by entropy (the TΔS portion of the equation $\Delta G = \Delta H - T\Delta S$ dominates for high T).

Catalytic synthesis of higher α-alkenes (of the type $RCH=CH_2$) can also be achieved by a reaction of ethylene with the organometallic compound triethylaluminium in the presence of nickel, cobalt, or platinum.

Elimination Reactions

One of the principal methods for alkene synthesis in the laboratory is the elimination of alkyl halides, alcohols, and similar compounds. Most common is the β-elimination via the E2 or E1 mechanism, but α-eliminations are also known.

The E2 mechanism provides a more reliable β-elimination method than E1 for most alkene syntheses. Most E2 eliminations start with an alkyl halide or alkyl sulfonate ester (such as a tosylate or triflate). When an alkyl halide is used, the reaction is called a dehydrohalogenation. For unsymmetrical products, the more substituted alkenes (those with fewer hydrogens attached to the C=C) tend to predominate. Two common methods of elimination reactions are dehydrohalogenation of alkyl halides and dehydration of alcohols. .

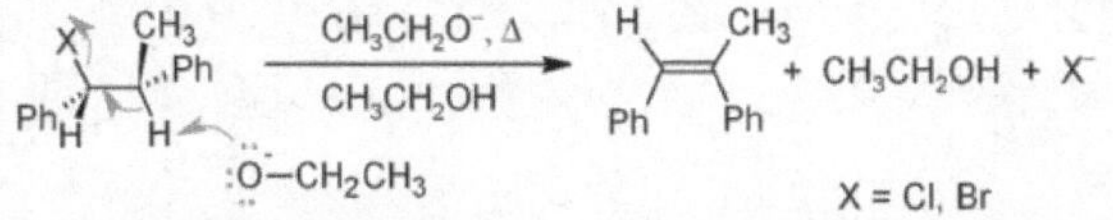

Alkenes can be synthesized from alcohols via dehydration, in which case water is lost via the E1 mechanism. For example, the dehydration of ethanol produces ethene:

$$CH_3CH_2OH + H_2SO_4 \rightarrow H_2C{=}CH_2 + H_3O^+ + HSO_4^-$$

An alcohol may also be converted to a better leaving group (e.g., xanthate), so as to allow a milder *syn*-elimination such as the Chugaev elimination and the Grieco elimination. Related reactions include eliminations by β-haloethers (the Boord olefin synthesis) and esters (ester pyrolysis). Alkenes can be prepared indirectly from alkyl amines. The amine or ammonia is not a suitable leaving group, so the amine is first either alkylated (as in the Hofmann elimination) or oxidized to an amine oxide (the Cope reaction) to render a smooth elimination possible. Hofmann elimination is unusual in that the *less* substituted (non-Saytseff) alkene is usually the major product. The Cope reaction is a *syn*-elimination that occurs at or below 150 °C, for example:

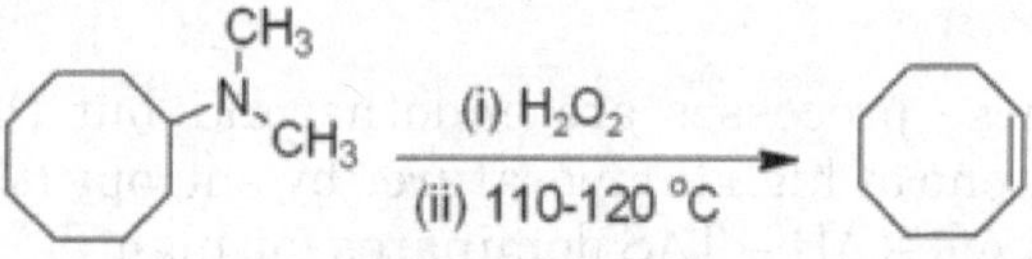

Alkenes are generated from α-halo sulfones in the Ramberg-Bäcklund reaction, via a three-membered ring sulfone intermediate.

Synthesis from Carbonyl Compounds

Another important method for alkene synthesis involves construction of a new carbon-carbon double bond by coupling of a carbonyl compound (such as an aldehyde or ketone) to a carbanion equivalent. Such reactions are sometimes called *olefinations*. The most well-known of these methods is the Wittig reaction, but other related methods are known. The Wittig reaction involves reaction of an aldehyde or ketone with a Wittig reagent (or phosphorane) of the type $Ph_3P{=}CHR$ to produce an alkene and $Ph_3P{=}O$. The Wittig reagent is itself prepared easily from triphenylphosphine and an alkyl halide. The reaction is quite general and many functional groups are tolerated, even esters, as in this example:

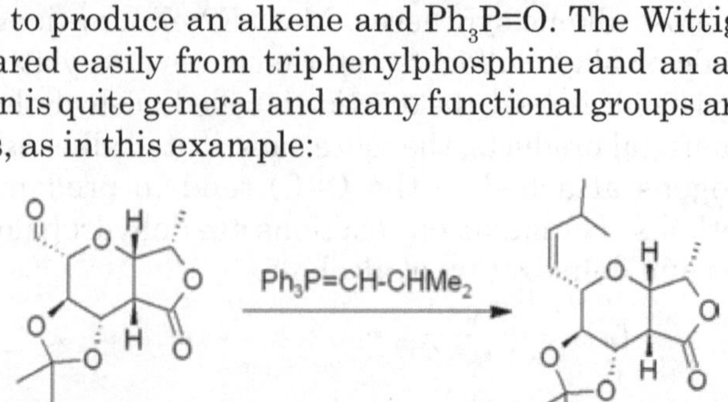

Related to the Wittig reaction is the Peterson olefination. This uses a less accessible silicon-based reagent in place of the phosphorane, but it allows for the selection of E or Z products. If an E-product is desired, another alternative is the Julia olefination, which uses the carbanion generated from a phenyl sulfone.

The Takai olefination based on an organochromium intermediate also delivers E-products. A titanium compound, Tebbe's reagent, is useful for the synthesis of methylene compounds; in this case, even esters and amides react.

A pair of carbonyl compounds can also be reductively coupled together (with reduction) to generate an alkene. Symmetrical alkenes can be prepared from a single aldehyde or ketone coupling with itself, using Ti metal reduction (the McMurry reaction). If two different ketones are to be coupled, a more complex, indirect method such as the Barton-Kellogg reaction may be used.

A single ketone can also be converted to the corresponding alkene via its tosylhydrazone, using sodium methoxide (the Bamford-Stevens reaction) or an alkyllithium (the Shapiro reaction).

Synthesis from Alkenes: Olefin Metathesis and Hydrovinylation

Alkenes can be prepared by exchange with other alkenes, in a reaction known as olefin metathesis. Frequently, loss of ethene gas is used to drive the reaction towards a desired product. In many cases, a mixture of geometric isomers is obtained, but the reaction tolerates many functional groups. The method is particularly effective for the preparation of cyclic alkenes, as in this synthesis of muscone:

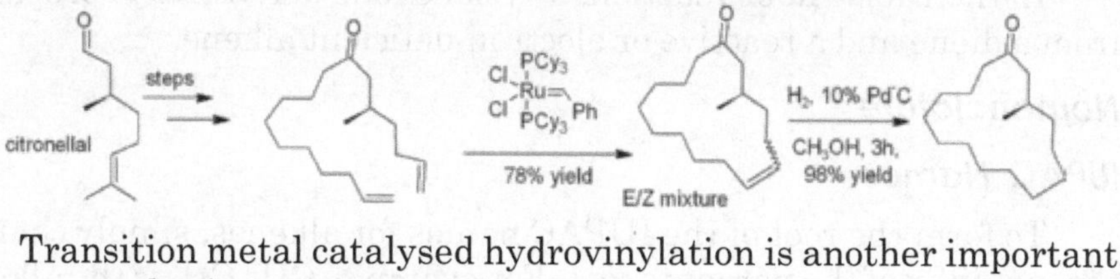

Transition metal catalysed hydrovinylation is another important alkene synthesis process starting from alkene itself. In general, it involves the addition of a hydrogen and a vinyl group (or an alkenyl group) across a double bond.

The hydrovinylation reaction was first reported by Alderson, Jenner, and Lindsey by using rhodium and ruthenium salts, other metal catalysts commonly employed nowadays included iron, cobalt, nickel, and palladium. The addition can be done highly regio- and stereo-selectively, the choices of metal centres, ligands, substrates and counterions often play very important role.

From Alkynes

Reduction of alkynes is a useful method for the stereoselective synthesis of disubstituted alkenes. If the *cis*-alkene is desired, hydrogenation in the presence of Lindlar's catalyst(-heterogeneous catalyst that consists of palladium deposited on calcium carbonate and treated with various forms of lead) is commonly used, though hydroboration followed by hydrolysis provides an alternative approach. Reduction of the alkyne by sodium metal in liquid ammonia gives the *trans*-alkene.

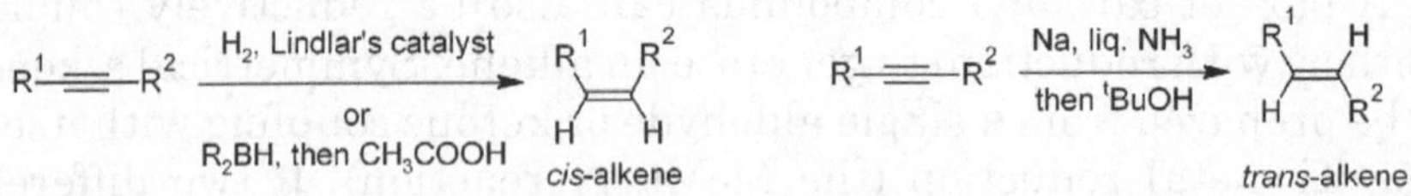

For the preparation multisubstituted alkenes, carbometalation of alkynes can give rise to a large variety of alkene derivatives.

Rearrangements and Related Reactions

Alkenes can be synthesized from other alkenes via rearrangement reactions. Besides olefin metathesis (described above), a large number of pericyclic reactions can be used such as the ene reaction and the Cope rearrangement.

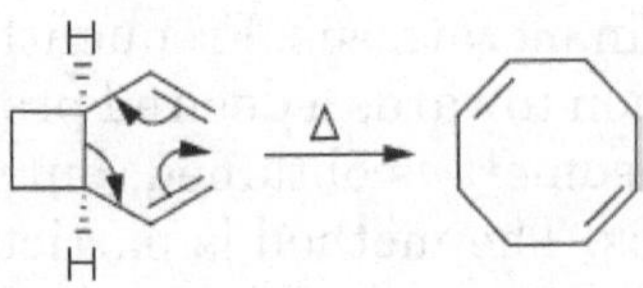

In the Diels-Alder reaction, a cyclohexene derivative is prepared from a diene and a reactive or electron-deficient alkene.

Nomenclature

IUPAC Names

To form the root of the IUPAC names for alkenes, simply change the -an- infix of the parent to -en-. For example, CH_3-CH_3 is the alkane *ethANe*. The name of CH_2=CH_2 is therefore *ethENe*.

In higher alkenes, where isomers exist that differ in location of the double bond, the following numbering system is used:

1. Number the longest carbon chain that contains the double bond in the direction that gives the carbon atoms of the double bond the lowest possible numbers.
2. Indicate the location of the double bond by the location of its first carbon.

3. Name branched or substituted alkenes in a manner similar to alkanes.
4. Number the carbon atoms, locate and name substituent groups, locate the double bond, and name the main chain.

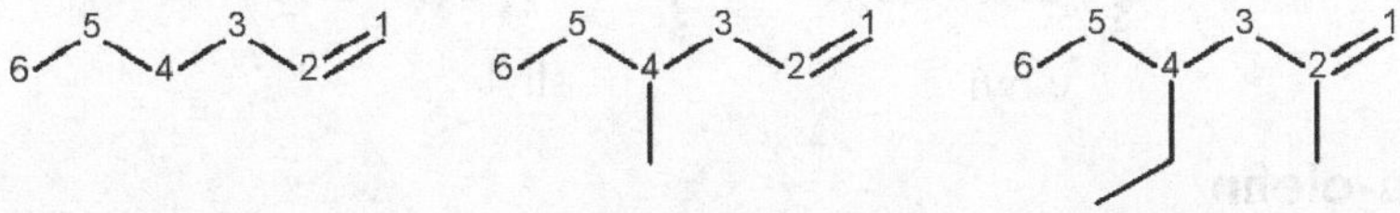

hex-1-ene 4-methylhex-1-ene 4-ethyl-2-methylhex-1-ene

Figure: *Naming substituted hex-1-enes*

Cis-Trans Notation

In the specific case of disubstituted alkenes where the two carbons have one substituent each, Cis-trans notation may be used. If both substituents are on the same side of the bond, it is defined as (cis). If the substituents are on either side of the bond, it is defined as (trans-).

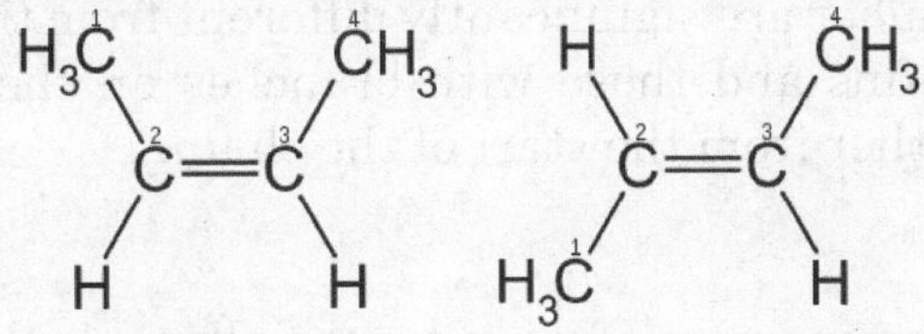

cis-but-2-ene trans-but-2-ene

Figure: *The difference between* cis- *and* trans- *isomers*

E,Z Notation

When an alkene has more than one substituent (especially necessary with 3 or 4 substituents), the double bond geometry is described using the labels *E* and *Z*. These labels come from the German words "entgegen," meaning "opposite," and "zusammen," meaning "together." Alkenes with the higher priority groups (as determined by CIP rules) on the same side of the double bond have these groups together and are designated *Z*. Alkenes with the higher priority groups on opposite sides are designated *E*. A mnemonic to remember this: Z notation has the higher priority groups on "ze zame zide."

(*Z*)-2-butene
zusammen (together)

NB. the methyl groups are on the same side

(*E*)-2-butene
entgegen (opposite)

NB. the methyl groups are on opposite sides

Figure: *The difference between* E *and* Z *isomers*

Groups containing C=C double bonds

IUPAC recognises two names for hydrocarbon groups containing carbon-carbon double bonds, the vinyl group and the allyl group.

$$\text{—CH=CH}_2 \qquad \text{—CH}_2\text{—CH=CH}_2$$

vinyl allyl

Alpha-olefin

Alpha-olefins (or α-olefins) are a family of organic compounds which are olefins or alkenes with a chemical formula C_xH_{2x}, distinguished by having a double bond at the primary or alpha (α) position. This location of a double bond enhances the reactivity of the compound and makes it useful for a number of applications. There are two types of alpha-olefins, branched and linear (or normal). The chemical properties of branched alpha-olefins with a branch at either the second (vinylidene) or the third carbon number are significantly different from the properties of linear alpha-olefins and those with branches on the fourth carbon number and further from the start of the chain.

Cycloalkene

A cycloalkene or cycloolefin is a type of alkene hydrocarbon which contains a closed ring of carbon atoms, but has no aromatic character. Some cycloalkenes, such as cyclobutene and cyclopentene, can be used as monomers to produce polymer chains. Unless the rings are very large, cycloalkenes are always the cis isomers, and the term cis is omitted from the names. In a large ring, a trans double bond may occur, giving a trans-cycloalekene.

Arenes

An aromatic hydrocarbon or arene (or sometimes aryl hydrocarbon) is a hydrocarbon with alternating double and single bonds between carbon atoms. The term 'aromatic' was assigned before the physical mechanism determining aromaticity was discovered, and was derived from the fact that many of the compounds have a sweet scent. The configuration of six carbon atoms in aromatic compounds is known as a benzene ring, after the simplest possible such hydrocarbon, benzene. Aromatic hydrocarbons can be *monocyclic* (MAH) or *polycyclic* (PAH). Some non-benzene-based compounds called heteroarenes, which follow Huckel's rule, are also aromatic compounds. In these compounds, at least one carbon atom is replaced by one of the heteroatoms oxygen, nitrogen, or sulfur. Examples of non-benzene compounds with aromatic properties are furan, a heterocyclic compound with a five-membered ring that includes an oxygen atom, and pyridine, a heterocyclic compound with a six-membered ring containing one nitrogen atom.

Benzene Ring Model

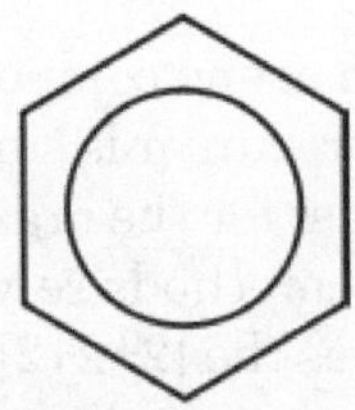

Figure: *Benzene*

Benzene, C_6H_6, is the simplest aromatic hydrocarbon and was recognised as the first aromatic hydrocarbon, with the nature of its bonding first being recognised by Friedrich August Kekule von Stradonitz in the 19th century. Each carbon atom in the hexagonal cycle has four electrons to share. One goes to the hydrogen atom, and one each to the two neighbouring carbons. This leaves one to share with one of its two neighbouring carbon atoms, which is why the benzene molecule is drawn with alternating single and double bonds around the hexagon.

The structure is also illustrated as a circle around the inside of the ring to show six electrons floating around in delocalised molecular orbitals the size of the ring itself. This also represents the equivalent nature of the six carbon-carbon bonds all of bond order 0.1^{0}c. This equivalency is well explained by resonance forms. The electrons are visualised as floating above and below the ring with the electromagnetic fields they generate acting to keep the ring flat.

General properties:

1. Display aromaticity.
2. The carbon-hydrogen ratio is high.
3. They burn with a sooty yellow flame because of the high carbon-hydrogen ratio.
4. They undergo electrophilic substitution reactions and nucleophilic aromatic substitutions.

The circle symbol for aromaticity was introduced by Sir Robert Robinson in 1925 and popularized starting in 1959 by the Morrison & Boyd textbook on organic chemistry. The proper use of the symbol is debated, it is used to describe any cyclic pi system in some publications, or only those pi systems that obey Huckel's rule on others. Jensen argues that in line with Robinson's original proposal, the use of the circle symbol should be limited to monocyclic 6 pi-electron systems. In this way the circle symbol for a 6c–6e bond can be compared to the Y symbol for a 3c–2e bond.

Arene Synthesis

A reaction that forms an arene compound from an unsaturated or partially unsaturated cyclic precursor is simply called an aromatization. Many laboratory methods exist for the organic synthesis of arenes from non-arene precursors. Many methods rely on cycloaddition reactions. Alkyne trimerization describes the [2+2+2] cyclization of three alkynes, in the Dotz reaction an alkyne, carbon monoxide and a chromium carbene complex are the reactants.Diels-Alder reactions of alkynes with pyrone or cyclopentadienone with expulsion of carbon dioxide or carbon monoxide also form arene compounds. In Bergman cyclization the reactants are an enyne plus a hydrogen donor.

Another set of methods is the aromatization of cyclohexanes and other aliphatic rings: reagents are catalysts used in hydrogenation such as platinum, palladium and nickel (reverse hydrogenation), quinones and the elements sulfur and selenium.

Arene Reactions

Arenes are reactants in many organic reactions.

Aromatic Substitution

In aromatic substitution one substituent on the arene ring, usually hydrogen, is replaced by another substituent. The two main types are electrophilic aromatic substitution when the active reagent is an electrophile and nucleophilic aromatic substitution when the reagent is a nucleophile. In radical-nucleophilic aromatic substitution the active reagent is a radical. An example is the nitration of salicylic acid:

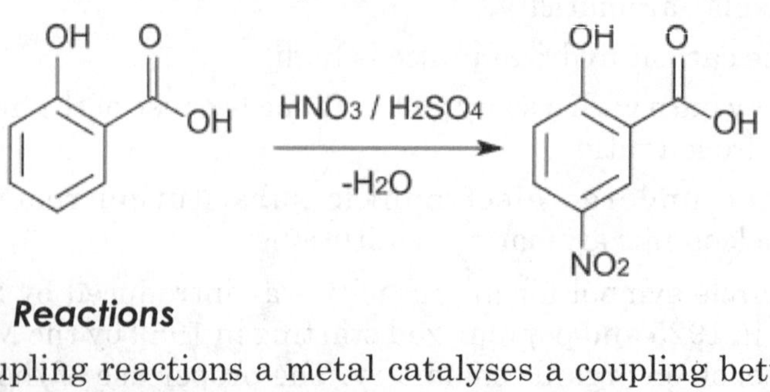

Coupling Reactions

In coupling reactions a metal catalyses a coupling between two formal radical fragments. Common coupling reactions with arenes result in the formation of new carbon-carbon bonds e.g., alkylarenes, vinyl arenes, biraryls, new carbon-nitrogen bonds (anilines) or new carbon-oxygen bonds (aryloxy compounds). An example is the direct arylation of perfluorobenzenes.

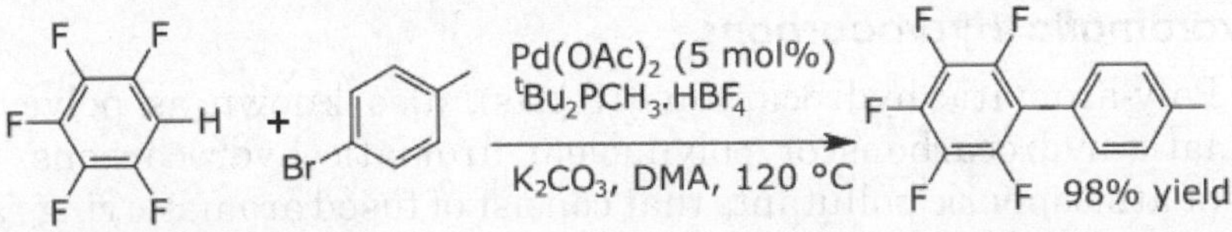

Hydrogenation

Hydrogenation of arenes create saturated rings. The compound 1-naphthol is completely reduced to a mixture of decalin-ol isomers.

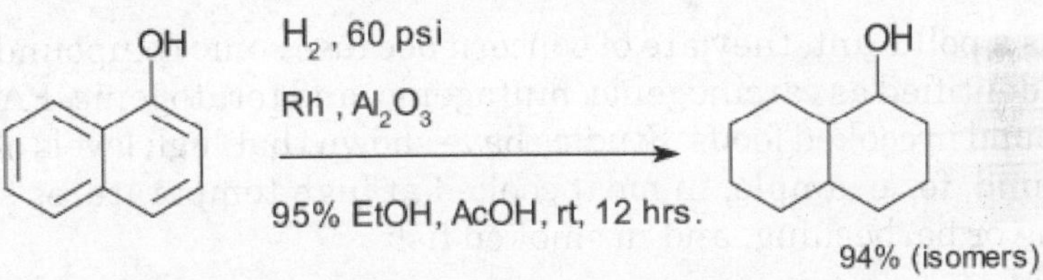

The compound resorcinol, hydrogenated with Raney nickel in presence of aqueous sodium hydroxide forms an enolate which is alkylated with methyl iodide to *2-methyl-1,3-cyclohexandione*:

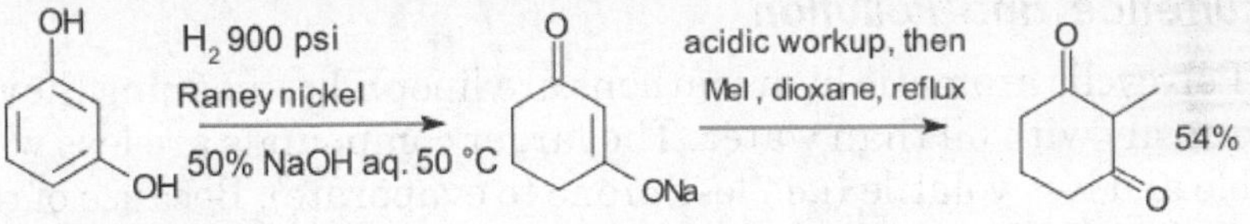

Cycloadditions

Cycloaddition reaction are not common. Unusual thermal Diels-Alder reactivity of arenes can be found in the Wagner-Jauregg reaction. Other photochemical cycloaddition reactions with alkenes occur through excimers.

Benzene and Derivatives of Benzene

Benzene derivatives have from one to six substituents attached to the central benzene core. Examples of benzene compounds with just one substituent are phenol, which carries a hydroxyl group and toluene with a methyl group.

When there is more than one substituent present on the ring, their spatial relationship becomes important for which the arene substitution patterns *ortho*, *meta*, and *para* are devised. For example, three isomers exist for cresol because the methyl group and the hydroxyl group can be placed next to each other (ortho), one position removed from each other (meta), or two positions removed from each other (para). Xylenol has two methyl groups in addition to the hydroxyl group, and, for this structure, 6 isomers exist.

Polyaromatic Hydrocarbons

Poly-aromatic hydrocarbons (PAHs), also known as polycyclic aromatic hydrocarbons or polynuclear aromatic hydrocarbons, are potent atmospheric pollutants that consist of fused aromatic rings and do not contain heteroatoms or carry substituents. Naphthalene is the simplest example of a PAH. PAHs occur in oil, coal, and tar deposits, and are produced as byproducts of fuel burning (whether fossil fuel or biomass).

As a pollutant, they are of concern because some compounds have been identified as carcinogenic, mutagenic, and teratogenic. PAHs are also found in cooked foods. Studies have shown that high levels of PAHs are found, for example, in meat cooked at high temperatures such as grilling or barbecuing, and in smoked fish.

They are also found in the interstellar medium, in comets, and in meteorites and are a candidate molecule to act as a basis for the earliest forms of life. In graphene the PAH motif is extended to large 2D sheets.

Occurrence and Pollution

Polycyclic aromatic hydrocarbons are lipophilic, meaning they mix more easily with oil than water. The larger compounds are less water-soluble and less volatile (i.e., less prone to evaporate). Because of these properties, PAHs in the environment are found primarily in soil, sediment and oily substances, as opposed to in water or air. However, they are also a component of concern in particulate matter suspended in air. Natural crude oil and coal deposits contain significant amounts of PAHs, arising from chemical conversion of natural product molecules, such as steroids, to aromatic hydrocarbons. They are also found in processed fossil fuels, tar and various edible oils.

PAHs are one of the most widespread organic pollutants. In addition to their presence in fossil fuels they are also formed by incomplete combustion of carbon-containing fuels such as wood, coal, diesel, fat, tobacco, and incense. Different types of combustion yield different distributions of PAHs in both relative amounts of individual PAHs and in which isomers are produced. Thus, coal burning produces a different mixture than motor-fuel combustion or a forest fire, making the compounds potentially useful as indicators of the burning history. Hydrocarbon emissions from fossil fuel-burning engines are regulated in developed countries.

List of PAHs

Although the health effects of individual PAHs are not exactly alike, the following 17 PAHs are considered as a group in this profile issued by the Agency for Toxic Substances and Disease Registry (ATSDR):

- acenaphthene
- acenaphthylene
- anthracene
- benz[*a*]anthracene
- benzo[*a*]pyrene
- benzo[*e*]pyrene
- benzo[*b*]fluoranthene
- benzo[*ghi*]perylene
- benzo[*j*]fluoranthene
- benzo[*k*]fluoranthene
- chrysene
- dibenz(a,h)anthracene
- fluoranthene
- fluorene
- indeno(1,2,3-cd)pyrene
- phenanthrene
- pyrene.

Human Health

PAHs toxicity is very structurally dependent, with isomers (PAHs with the same formula and number of rings) varying from being nontoxic to being extremely toxic. Thus, highly carcinogenic PAHs may be small or large. One PAH compound, benzo[*a*]pyrene, is notable for being the first chemical carcinogen to be discovered (and is one of many carcinogens found in cigarette smoke). The EPA has classified seven PAH compounds as probable human carcinogens: benz[*a*]anthracene, benzo[*a*]pyrene, benzo[*b*]fluoranthene, benzo[*k*]fluoranthene, chrysene, dibenz(a,h)anthracene, and indeno(1,2,3-cd)pyrene.

PAHs known for their carcinogenic, mutagenic and teratogenic properties are benz[*a*]anthracene and chrysene, benzo[*b*]fluoranthene, benzo[*j*]fluoranthene, benzo[*k*]fluoranthene, benzo[*a*]pyrene, benzo[*ghi*]perylene, coronene, dibenz(a,h)anthracene ($C_{20}H_{14}$), indeno(1,2,3-cd)pyrene ($C_{22}H_{12}$) and ovalene.

High prenatal exposure to PAH is associated with lower IQ and childhood asthma. The Centre for Children's Environmental Health reports studies that demonstrate that exposure to PAH pollution during pregnancy is related to adverse birth outcomes including low birth weight, premature delivery, and heart malformations. Cord blood of exposed babies shows DNA damage that has been linked to cancer. Follow-up studies show a higher level of developmental delays at age three, lower scores on IQ tests and increased behavioural problems at ages six and eight.

Chemistry

The simplest PAHs, as defined by the International Union of Pure and Applied Chemistry (IUPAC) (G.P Moss, IUPAC nomenclature for fused-ring systems), are phenanthrene and anthracene, which both contain three fused aromatic rings. Smaller molecules, such as benzene, are not PAHs.

PAHs may contain four-, five-, six- or seven-member rings, but those with five or six are most common. PAHs composed only of six-membered rings are called alternant PAHs. Certain alternant PAHs are called benzenoid PAHs. The name comes from benzene, an aromatic hydrocarbon with a single, six-membered ring. These can be benzene rings interconnected with each other by single carbon-carbon bonds and with no rings remaining that do not contain a complete benzene ring.

The set of alternant PAHs is closely related to a set of mathematical entities called polyhexes, which are planar figures composed by conjoining regular hexagons of identical size.

PAHs containing up to six fused aromatic rings are often known as "small" PAHs, and those containing more than six aromatic rings are called "large" PAHs. Due to the availability of samples of the various small PAHs, the bulk of research on PAHs has been of those of up to six rings. The biological activity and occurrence of the large PAHs does appear to be a continuation of the small PAHs. They are found as combustion products, but at lower levels than the small PAHs due to the kinetic limitation of their production through addition of successive rings. In addition, with many more isomers possible for larger PAHs, the occurrence of specific structures is much smaller.

PAHs possess very characteristic UV absorbance spectra. These often possess many absorbance bands and are unique for each ring structure. Thus, for a set of isomers, each isomer has a different UV absorbance spectrum than the others. This is particularly useful in the identification of PAHs. Most PAHs are also fluorescent, emitting characteristic wavelengths of light when they are excited (when the molecules absorb light). The extended pi-electron electronic structures of PAHs lead to these spectra, as well as to certain large PAHs also exhibiting semi-conducting and other behaviours.

Naphthalene ($C_{10}H_8$ constituent of mothballs), consisting of two coplanar six-membered rings sharing an edge, is another aromatic hydrocarbon. By formal convention, it is not a true PAH, though is referred to as a bicyclic aromatic hydrocarbon.

Aqueous solubility decreases approximately one order of magnitude for each additional ring.

PAH Compounds

The United States Environmental Protection Agency (EPA) has designated 32 PAH compounds as priority pollutants. The original 16 are listed. They are naphthalene, acenaphthylene, acenaphthene, fluorene, phenanthrene, anthracene, fluoranthene, pyrene, benzo[*a*]anthracene, chrysene, benzo[*b*]fluoranthene, benzo[*k*]flouranthene, benzo[*a*]pyrene, dibenz(ah)anthracene, benzo[*ghi*]perylene, and indeno(1,2,3-cd)pyrene. This list of the 16 EPA priority PAHs is often targeted for measurement in environmental samples.

Aromaticity

Although PAHs clearly are aromatic compounds, the degree of aromaticity can be different for each ring segment. According to Clar's rule (formulated by Erich Clar in 1964) for PAHs the resonance structure with the most disjoint aromatic i-sextets—i.e., benzene-like moieties—is the most important for the characterisation of the properties.

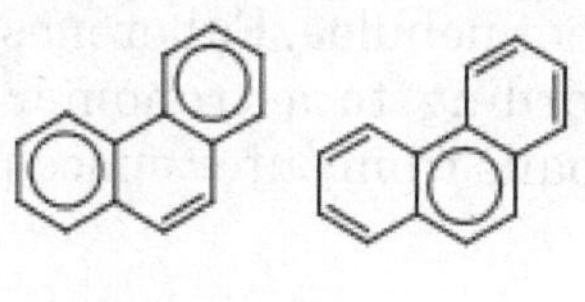

Phenanthrene (1)

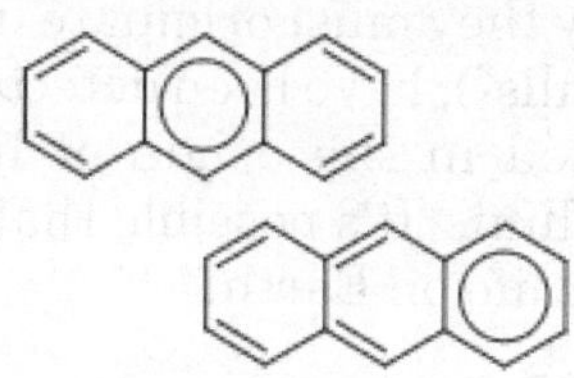

Anthracene (2)

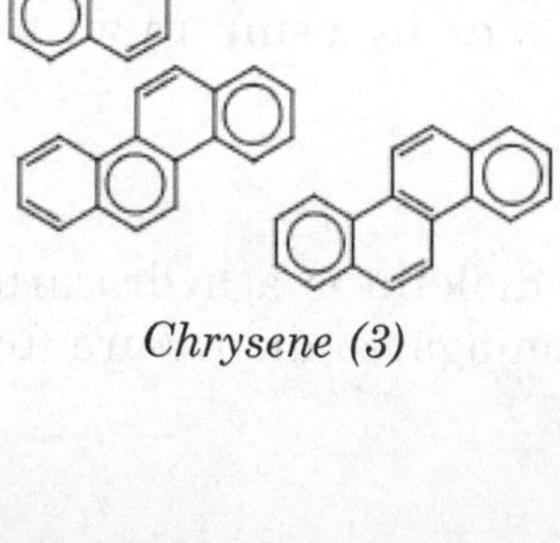

Chrysene (3)

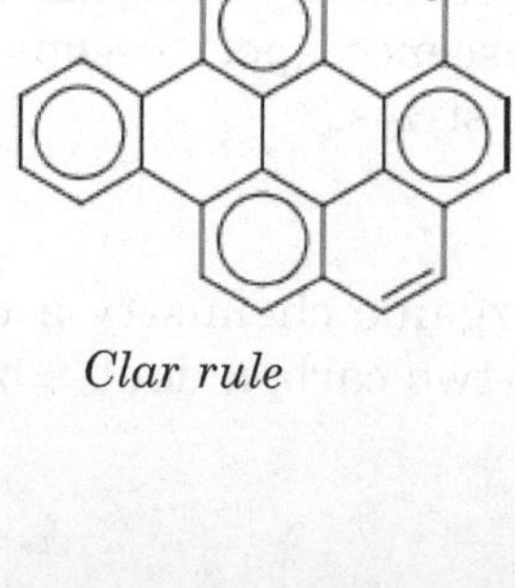

Clar rule

For example, in phenanthrene (1) one Clar structure has two sextets at the extremities, while the other resonance structure has just one central sextet. Therefore in this molecule the outer rings are firmly aromatic while its central ring is less aromatic and therefore more reactive. In contrast, in anthracene (2) the number of sextets is just one and aromaticity spreads out. This difference in number of sextets is reflected in the UV absorbance spectra of these two isomers. Phenanthrene has a highest wavelength absorbance around 290 nm, while anthracene has highest wavelength bands around 380 nm. Three Clar structures with two sextets are present in chrysene (3) and by superposition the aromaticity in the outer ring is larger than in the inner rings. Another relevant Clar hydrocarbon is zethrene.

Origins of Life

In January 2004 (at the 203rd Meeting of the American Astronomical Society), it was reported that a team led by A. Witt of the University of Toledo, Ohio studied ultraviolet light emitted by the Red Rectangle nebula and found the spectral signatures of anthracene and pyrene (no other such complex molecules had ever before been found in space). This discovery was considered as a controversial confirmation of a hypothesis that as nebulae of the same type as the Red Rectangle approach the ends of their lives, convection currents cause carbon and hydrogen in the nebulae's core to get caught in stellar winds, and radiate outward.

As they cool, the atoms supposedly bond to each other in various ways and eventually form particles of a million or more atoms. Witt and his team inferred that since they discovered PAHs—which may have been vital in the formation of early life on Earth—in a nebula, by necessity they must originate in nebulae. More recently, fullerenes (or "buckyballs"), have been detected in other nebulae. Fullerenes are also implicated in the origin of life; according to astronomer Letizia Stanghellini, "It's possible that buckyballs from outer space provided seeds for life on Earth."

Detection

Detection of PAHs in materials is often done using gas chromatography-mass spectrometry or liquid chromatography with ultraviolet-visible or fluorescence spectroscopic methods or by using rapid test PAH indicator strips.

Diene

In organic chemistry a diene or diolefin is a hydrocarbon that contains two carbon double bonds. Conjugated dienes are functional groups, with a general formula of C_nH_{2n-2}. Dienes and alkynes are functional isomers. Dienes occur occasionally in nature but are widely used in the polymer industry.

Classes

Dienes can be divided into three classes, depending on the relative location of the double bonds:

1. Cumulated dienes have the double bonds sharing a common atom as in a group of compounds called allenes.
2. Conjugated dienes have conjugated double bonds separated by one single bond.
3. Unconjugated dienes have the double bonds separated by two or more single bonds. They are usually less stable than isomeric conjugated dienes. This can also be known as an isolated diene.

Compounds that contain more than two double bonds are called polyenes. Polyenes and dienes, share many of their properties.

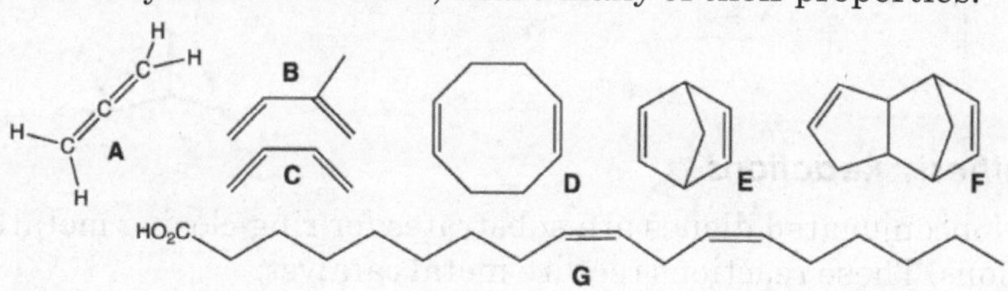

Figure: *Some dienes: A: 1,2-Propadiene, also known as allene, is the simplest cumulated diene. B: Isoprene, also known as 2-methyl-1,3-butadiene, the precursor to natural rubber. C: 1,3-Butadiene, a precursor to synthetic polymers. D: 1,5-Cyclooctadiene, an unconjugated diene (notice that each double bond is two carbons away from the other). E: Norbornadiene, a strained bicyclic and unconjugated diene. F: Dicyclopentadiene. G: Linoleic acid, a fatty acid that is required in the human diet.*

Synthesis of Dienes

On an industrial scale, butadiene is prepared by thermal cracking of butanes. In a similarly non-selective process, dicyclopentadiene is obtained from coal tars. In the laboratory, more directed and more delicate processes are employed such as dehydrohalogenations and condensations.

Myriad methods have been developed, such as the Whiting reaction. Families of nonconjugated dienes are derived from the oligomerization and dimerization of conjugated dienes. For example, 1,5-cyclooctadiene and vinylcyclohexene are produced by dimerization of 1,3-butadiene.

Diene-containing fatty acids are biosynthesized from acetyl CoA.

Reactivity and Uses—Polymerization

The most heavily practiced reaction of alkenes, dienes included, is polymerization. Butadiene is a precursor to rubber used in tires, and isoprene is the precursor to natural rubber. Chloroprene is a related but synthetic monomer.

Cycloadditions

An important reaction for conjugated dienes is the Diels-Alder reaction. Many specialised dienes have been developed to exploit this reactivity for the synthesis of natural products, e.g. Danishefsky's diene.

Other Addition Reactions

Conjugated dienes add reagents such as bromine and hydrogen by both 1,2-addition and 1,4-addition pathways. Addition of polar reagents can generate complex architectures:

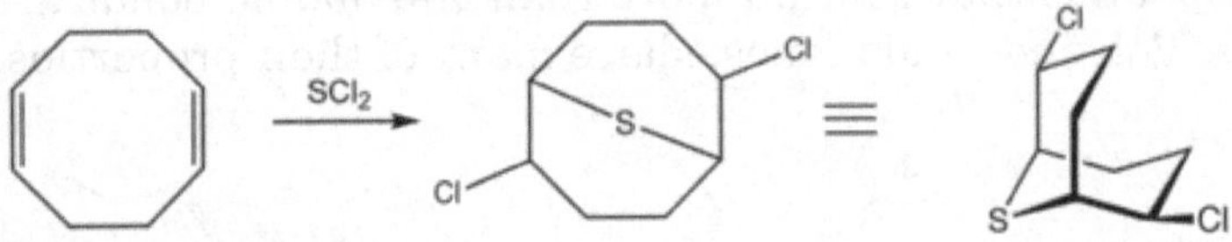

Metathesis Reactions

Nonconjugated dienes are substrates for ring-closing metathesis reactions. These reactions require metal catalyst:

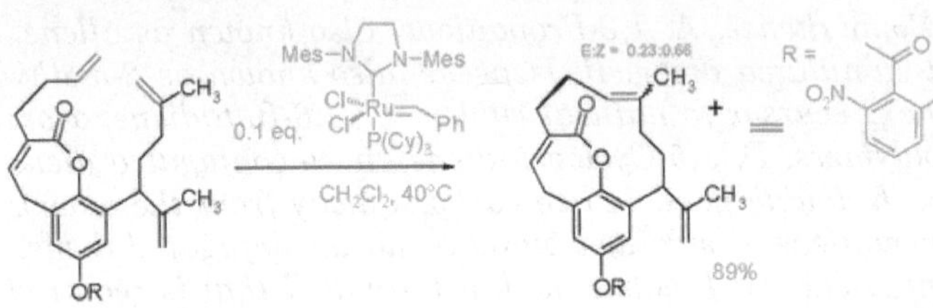

Acidity

Conjugated and 1,4-dienes generally are somewhat acidic since deprotonation of both classes gives pentadienyl anions. The acidifying effect of the diene is very pronounced in cyclopentadiene.

As Ligands

Dienes are widely used chelating ligands in organometallic chemistry. In some cases they serve as placeholder ligands, being removed during a catalytic cycle. For example, the cyclooctadiene ("cod") ligands in bis(cyclooctadiene)nickel(0) are labile. In some cases, dienes are spectator ligands, remaining coordinated throughout a catalytic cycle and influencing the product distributions. Chiral dienes have also been described.

Dendralene

A dendralene is a discrete acyclic cross-conjugated polyene. The simplest dendralene is buta-1,3-diene (1) or dendralene followed by dendralene (2), dendralene (3) and dendralene (4) and so forth, dendralene (butadiene) is the only one not cross-conjugated.

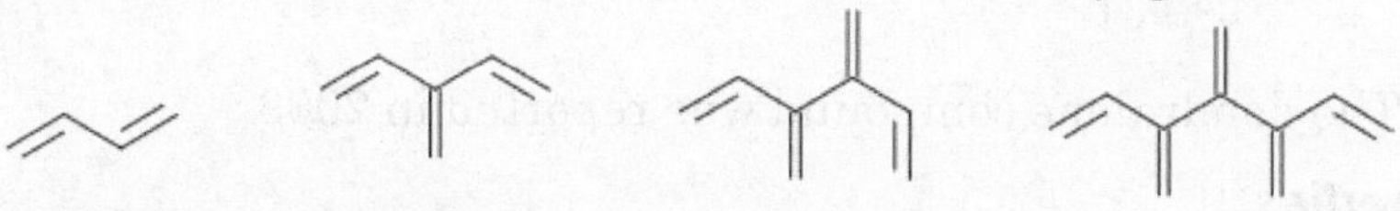

1 2 3 4

The name *dendralene* is pulled together from the words dendrimer, linear and alkene. The higher dendralenes are of scientific interest because they open up a large array of new organic compounds from a relatively simple precursor especially by Diels-Alder chemistry. Their cyclic counterparts are aptly called radialenes.

Synthesis

Vinylbutadiene (dendralene) was first prepared in 1955 by pyrolysis of a triacetate:

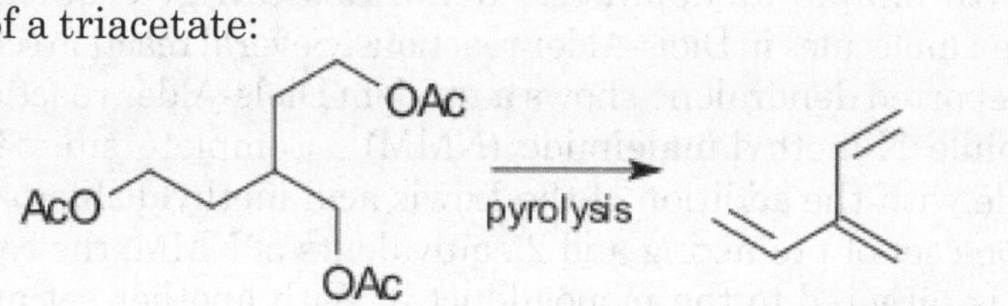

This compound reacts with two equivalents of maleic anhydride in a tandem DA reaction:

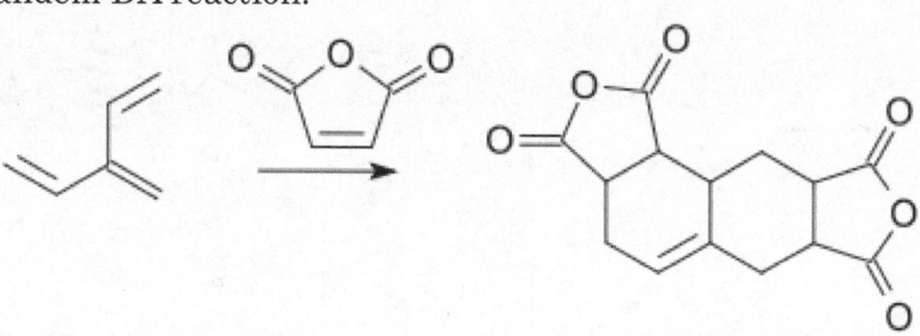

With benzoquinone the reaction product was a linear polymer.

Several syntheses of substituted dendralenes have been reported, one via an allene , one via a Horner–Wadsworth–Emmons reaction, one via a cross-coupling reaction and one from an allylic carbonate.

One synthetic route to dendralene starts from chloroprene. This compound is converted to a Grignard reagent by action of magnesium metal which is then reacted with copper(I) chloride to an organocopper intermediate which is in turn dimerized using copper(II) chloride in an oxidative coupling reaction to give the butadiene dimer called dendralene.

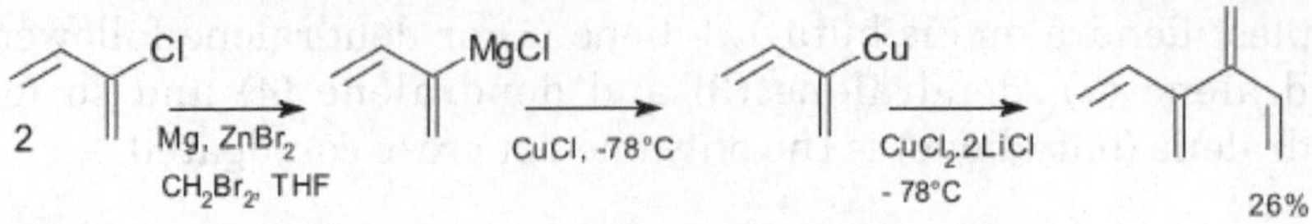

The dendralene compound was reported in 2009 :

Properties

Even-membered dendralenes (e.g. dendralene, dendralene) tend to behave as chains of decoupled and isolated diene units. The ultraviolet absorption maxima equal that of butadiene itself. The dendralenes with an odd number of alkene units are more reactive due to the presence of favourable s-cis diene conformations and Diels-Alder reactions take place more easily with a preference for the termini.

Reactions

With simple dienophiles , dendrales can give quick access to complex molecules in Diels-Alder reactions. Several reaction schemes have been reported dendralene shows a tandem Diels-Alder reaction with the dienophile N-methyl-maleimide (NMM) . Complete site selectivity is possible with the addition of the Lewis acid methyldichloroaluminium. With one set of premixing and 2 equivalents of NMM, the central diene group is targeted to the monoadduct 3. With another set and a larger amount of dienophile, the terminal groups react and the reaction proceeds from the monoadduct to the trisadducts 2 and 2b.

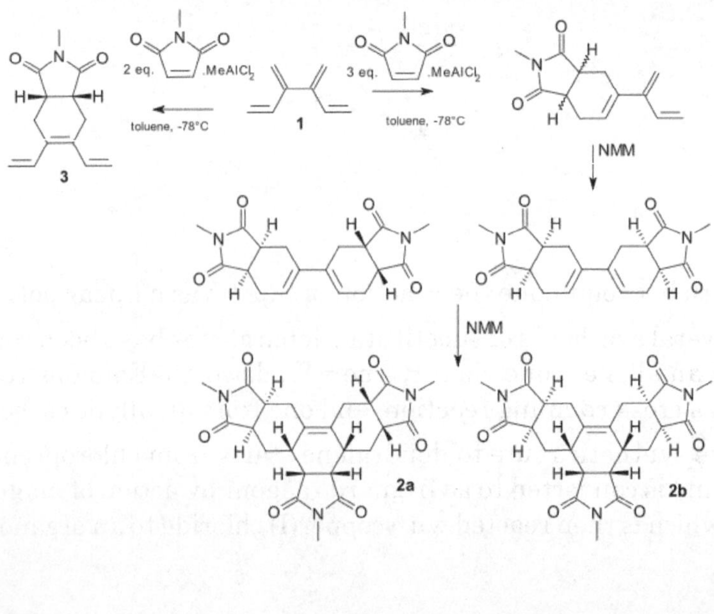

Annulene

Annulenes are completely conjugated monocyclic hydrocarbons. They have the general formula C_nH_n (when *n* is an even number) or C_nH_{n+1} (when *n* is an odd number). The IUPAC naming conventions are that annulenes with 7 or more carbon atoms are named as [*n*]annulene, where *n* is the number of carbon atoms in their ring, though sometimes the smaller annulenes are referred to using the same notation, and benzene is sometimes referred to simply as annulene.

The first three annulenes are cyclobutadiene, benzene, and cyclooctatetraene (annulene). Some annulenes, namely cyclobutadiene, cyclodecapentaene or annulene, cyclododecahexaene or annulene and cyclotetradecaheptaene (annulene), are unstable, with cyclobutadiene extremely so. Annulenes may be *aromatic* (benzene, annulene, cyclooctadecanonaene or annulene), *non-aromatic* (annulene), or *anti-aromatic* (cyclobutadiene). Only cyclobutadiene and benzene are fully planar, though and annulene with all trans double bonds (placing the hydrogens inside the ring) can achieve the planar conformation needed for aromaticity, with and annulene following Huckel's rule with 4n+2 o electrons. Annulene does exhibit some ring strain due to steric hindrance.

Many of the larger annulenes, Annuelene for example, are large enough to minimise the van der Waals strain of internal hydrogens and thermodynamically qualify as aromatic. However, none of the larger annulenes are as stable as benzene, as their reactivity more closely resembles a conjugated polyene than an aromatic hydrocarbon.

Polyene

Polyenes are poly-unsaturated organic compounds that contain one or more sequences of alternating double and single carbon-carbon bonds. These double carbon-carbon bonds interact in a process known as conjugation, which results in an overall lower energy state of the molecule.

Organic compounds with two carbon-carbon double bonds are dienes; those with three such double bonds are trienes; those with four are tetraenes, etc.

Normally carbon-carbon double bonds which are not conjugated or only conjugated with only one or two other carbon-carbon double bonds have high enough energy that they absorb in the ultraviolet region of a spectrum, but the absorption energy state of polyenes with numerous conjugated double bonds can be lowered such that they enter the visible region of the spectrum, resulting in compounds which are coloured yellow or other colours.

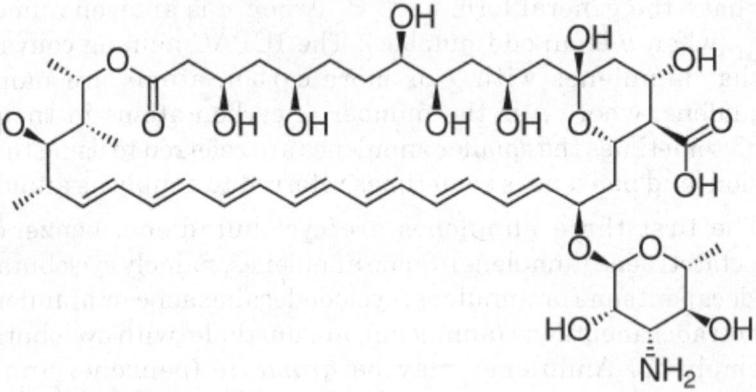

***Figure:** Amphotericin B is an example of a yellow-coloured polyene antifungal (antimycotic) agent. Note the alternating double and single bonds in the centre.*

Many fatty acids are polyenes, and many dyes contain linear polyenes. Other examples of polyene compounds include beta-carotene, which is yellow to orange coloured depending on concentration, and polyene antimycotics, some of which are yellow coloured.

Alkyne

Alkynes are hydrocarbons that have a triple bond between two carbon atoms, with the formula C_nH_{2n-2}. Alkynes are traditionally known as acetylenes, although the name *acetylene* also refers specifically to C_2H_2, known formally as ethyne using IUPAC nomenclature. Like other hydrocarbons, alkynes are generally hydrophobic but tend to be more reactive.

Chemical Properties

Alkynes are characteristically more unsaturated than alkenes. Thus they add two equivalents of bromine whereas an alkene adds only one equivalent. Other reactions are listed below. Alkynes are usually more reactive than alkenes. They show greater tendency to polymerize or oligomerize than alkenes do. The resulting polymers, called polyacetylenes (which do not contain alkyne units) are conjugated and can exhibit semiconducting properties.

Structure and Bonding

In acetylene, the H–Ca"C bond angles are 180°. By virtue of this bond angle, alkynes tend to be rod-like. Correspondingly, cyclic alkynes are rare. Benzyne is highly unstable. The Ca"C bond distance of 121 picometres is much shorter than the C=C distance in alkenes (134 pm) or the C-C bond in alkanes (153 pm).

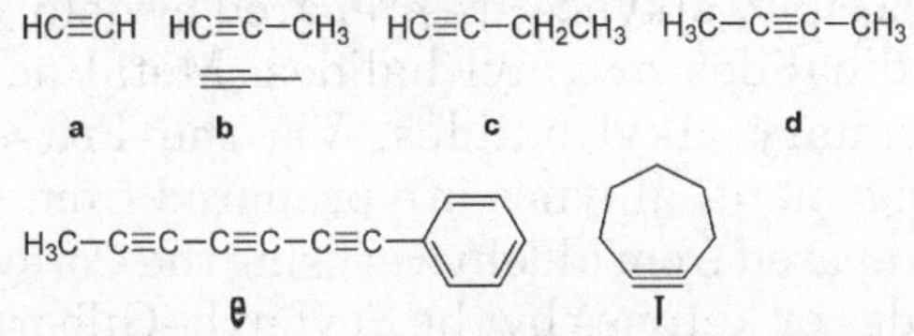

Figure: *Illustrative alkynes: a, acetylene, b, two depictions of propyne, c, 1-butyne, d, 2-butyne, e, the naturally-occurring 1-phenylhepta-1,3,5-triyne, and f, the strained cycloheptyne.*

The triple bond is very strong with a bond strength of 839 kJ/mol. The sigma bond contributes 369 kJ/mol, the first pi bond contributes 268 kJ/mol and the second pi-bond of 202 kJ/mol bond strength. Bonding usually discussed in the context of molecular orbital theory, which recognises the triple bond as arising from overlap of s and p orbitals. In the language of valence bond theory, the carbon atoms in an alkyne bond are sp hybridized: they each have two unhybridized p orbitals and two sp hybrid orbitals. Overlap of an sp orbital from each atom forms one sp-sp sigma bond. Each p orbital on one atom overlaps one on the other atom, forming two pi bonds, giving a total of three bonds. The remaining sp orbital on each atom can form a sigma bond to another atom, for example to hydrogen atoms in the parent acetylene]. The two sp orbitals project on opposite sides of the carbon atom.

Terminal and Internal Alkynes

Internal alkynes feature carbon substituents on each acetylenic carbon. Symmetrical examples include diphenyl acetylene and 3-hexyne. A representative unsymmetrical internal alkyne is propiolic acid. Terminal alkynes have at least one hydrogen atom bonded to an sp hybridized carbon (those involved in the triple bond. An example would be methylacetylene (propyne using IUPAC nomenclature). Terminal alkynes and acetylene are mildly acidic. The acidic hydrogen on terminal alkynes can be replaced by a variety of groups resulting in halo-, silyl-, and alkoxoalkynes.

Synthesis

Commercially, the dominant alkyne is acetylene itself, which is used as a fuel and a precursor to other compounds, e.g., acrylates. Hundreds of millions of kilograms are produced annually by dehydrogenation of natural gas:

$$2\ CH_4 \rightarrow HCa\text{–}CH + 3\ H_2$$

Propyne, also industrially useful, is also prepared by thermal cracking of hydrocarbons. Most other industrially useful alkyne derivatives are prepared from acetylene, e.g. via condensation with formaldehyde. Specialty alkynes are prepared by dehydrohalogenation of vicinal alkyl dihalides or vinyl halides. Metal acetylides can be coupled with primary alkyl halides. Via the Fritsch-Buttenberg-Wiechell rearrangement, alkynes are prepared from vinyl bromides. Alkynes can be prepared from aldehydes using the Corey-Fuchs reaction and from aldehydes or ketones by the Seyferth–Gilbert homologation. In the alkyne zipper reaction, alkynes are generated from other alkynes by treatment with a strong base.

Reactions

Featuring a reactive functional group, alkynes participate in many organic reactions.

Addition of Hydrogen, Halogens, and Related Reagents

Alkynes characteristically undergo reactions that show that they are “doubly unsaturated,” meaning that each alkyne unit is capable of adding two equivalents of H_2, halogens or related HX reagents (X = halide, pseudohalide, etc.). Depending on catalysts and conditions, alkynes add one or two equivalents of hydrogen. Hydrogenation to the alkene is usually more desirable since alkanes are less useful:

$$\text{RCa–CR'} + H_2 \rightarrow \textit{cis}\text{-RCH=CR'H}$$

The largest scale application of this technology is the conversion of acetylene to ethylene in refineries. The steam cracking of alkanes affords a few percent acetylene, which is selectively hydrogenated in the presence of a palladium/silver catalyst. For more complex alkynes, the Lindlar catalyst is widely recommended to avoid formation of the alkane, for example in the conversion of phenylacetylene to styrene.

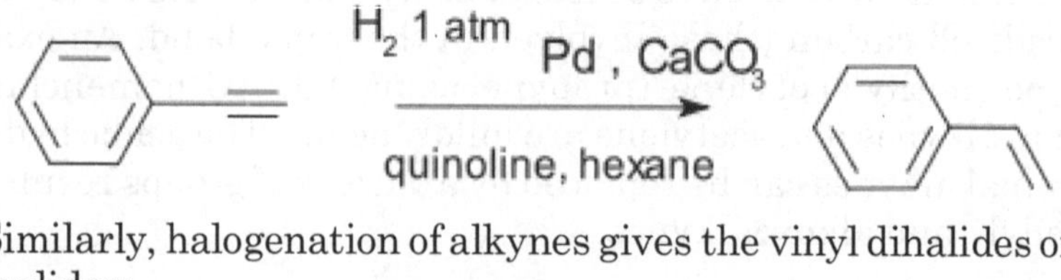

Similarly, halogenation of alkynes gives the vinyl dihalides or alkyl tetrahalides:

$$\text{RCa–CR'} + 2\ Br_2 \rightarrow RCBr_2CRBr_2$$

The addition of nonpolar E-H bonds across Ca”C is general for silanes, boranes, and related hydrides. The hydroboration of alkynes gives vinylic boranes which oxidize to the corresponding aldehyde or ketone. In the thiol -yne reaction the substrate is a thiol.

Hydrohalogenation gives the corresponding vinyl halides or alkyl dihalides, again depending on the number of equivalents of HX added. The addition of water to alkynes is a related reaction except the initial enol intermediate converts to the ketone or aldehyde. Illustrative is the hydration of phenylacetylene gives acetophenone, and the $(Ph_3P)AuCH_3$-catalysed hydration of 1,8-nonadiyne to 2,8-nonanedione:

$$PhCa\text{–}CH + H_2O \rightarrow PhCOCH_3$$

$$HCa\text{–}CC_6H_{12}Ca\text{–}CH + 2H_2O \rightarrow CH_3COC_6H_{12}COCH_3$$

Cycloadditions and Oxidation

Alkynes undergo diverse cycloaddition reactions. Most notable is the Diels–Alder reaction with 1,3-dienes to give 1,4-cyclohexadienes. This general reaction has been extensively developed and electrophilic alkynes are especially effective dienophiles. The "cycloadduct" derived from the addition of alkynes to 2-pyrone eliminates carbon dioxide to give the aromatic compound. Other specialised cycloadditions include multicomponent reactions such as alkyne trimerisation to give aromatic compounds and the [2+2+1]cycloaddition of an alkyne, alkene and carbon monoxide in the Pauson–Khand reaction. Non-carbon reagents also undergo cyclization, e.g. Azide alkyne Huisgen cycloaddition to give triazoles. Cycloaddition processes involving alkynes are often catalysed by metals, e.g. enyne metathesis and alkyne metathesis, which allows the scrambling of carbyne (RC) centres:

$$RCa\text{"}CR + R'Ca\text{–}CR' \rightleftarrows RCa\text{–}CR'$$

Oxidative cleavage of alkynes proceeds via cycloaddition to metal oxides. Most famously, potassium permanganate converts alkynes to a pair of carboxylic acids.

Reactions specific for terminal alkynes

In addition to undergoing the reactions characteristic of internal alkynes, terminal alkynes are reactive as weak acids, with pK_a values (25) between that of ammonia (35) and ethanol (16). The acetylide conjugate base is stabilised as a result of the high s character of the sp orbital, in which the electron pair resides. Electrons in an s orbital benefit from closer proximity to the positively charged atom nucleus, and are therefore lower in energy. Treatment of terminal alkynes with a strong base gives the corresponding metal acetylides:

$$RCa\text{"}CH + MX \rightarrow RCa\text{"}CM + HX\ (MX = NaNH_2, LiBu, RMgX)$$

The reactions of alkynes with certain metal cations, e.g. Ag^+ also gives acetylides. Thus, few drops of diamminesilver(I) hydroxide $(Ag(NH_3)_2OH)$ reacts with terminal alkynes signalled by formation of a white precipitate of the silver acetylide. Acetylide derivatives are synthetically useful nucleophiles that participate in C-C bond forming reactions, as illustrated in the area called "Reppe Chemistry". In the Favourskii reaction, terminal alkynes add to carbonyl compounds to give the hydroxyalkyne. Coupling of terminal alkynes to give di-alkynes is effected in the Cadiot-Chodkiewicz coupling, Glaser coupling, and the Eglinton coupling reactions. Terminal alkynes can also be coupled to aryl or vinyl halides as in the Sonogashira coupling.

Alkynes in Nature and Medicine

According to Ferdinand Bohlmann, the first naturally occurring acetylenic compound, dehydromatricaria ester, was isolated from an *Artemisia* species in 1826. In the nearly two centuries that have followed, well over a thousand naturally occurring acetylenes have been discovered and reported. Polyynes,a subset of this class of natural products, have been isolated from a wide variety of plant species,cultures of higher fungi, bacteria, marine sponges, and corals. Some acids like tariric acid contains an alkyne group. Diynes and triynes, species with the linkage RCa"C-Ca"CR' and RCa"C-Ca"C-Ca"CR' respectively, occur in certain plants (*Ichthyothere*, *Chrysanthemum*, *Cicuta*, *Oenanthe* and other members of the Asteraceae and Apiaceae families). Some examples are cicutoxin, oenanthotoxin, falcarinol and carotatoxin. These compounds are highly bioactive, e.g. as nematocides. 1-Phenylhepta-1,3,5-triyne is illustrative of a naturally occurring triyne. Alkynes occur in some pharmaceuticals, including the contraceptive norethynodrel. A carbon-carbon triple bond is also present in marketed drugs such as the antiretroviral Efavirenz and the antifungal Terbinafine. Molecules called ene-diynes feature a ring containing an alkene ("ene") between two alkyne groups ("diyne"). These compounds, e.g. calicheamicin, are some of the most aggressive antitumor drugs known, so much so that the ene-diyne subunit is sometimes referred to as a "warhead." Ene-diynes undergo rearrangement via the Bergman cyclization, generating highly reactive radical intermediates that attack DNA within the tumour.

-yne

The suffix -yne is used in organic chemistry to form names of organic compounds containing the -Ca"C- group, known generally as alkynes. Sometimes a number between hyphens is inserted before it to state which atoms the triple bond is between. This suffix arose as a collapsed form of the end of the word "acetylene". The final "-e" disappears if it is followed by another suffix that starts with a vowel.

Chapter 4

Electrocatalysts

An electrocatalyst is a catalyst that participates in electrochemical reactions. Catalyst materials modify and increase the rate of chemical reactions without being consumed in the process. Electrocatalysts are a specific form of catalysts that function at electrode surfaces or may be the electrode surface itself. An electrocatalyst can be heterogeneous such as a platinum surface or nanoparticles, or homogeneous like a coordination complex or enzyme. The electrocatalyst assists in transferring electrons between the electrode and reactants, and/or facilitates an intermediate chemical transformation described by an overall half-reactions.

Context

There are multiple ways for many transformations to occur. For example, hydrogen and oxygen can be combined to form water through a free radical mechanism commonly referred to as combustion. Useful energy can be obtained from the thermal heat of this reaction through an internal combustion engine with an upper efficiency of 60% (for compression ratio of 10 and specific heat ratio of 1.4) based on the Otto thermodynamic cycle. It is also possible to combine the hydrogen and oxygen through redox mechanism as in the case of a fuel cell. In this process, the reaction is broken into two half-reactions which occur at separate electrodes. In this situation the reactant's energy is directly converted to electricity.

Half-reaction	***E° (V)***
$H_2(g) \rightleftarrows 2H^+ + 2e^-$	a^{-} 0
$O_2(g) + 4H^+ + 4e^- \rightleftarrows 2H_2O$	+1.23

This process is not governed by the same thermodynamic cycles as combustion engines, it is governed by the total energy available to do work as described by the gibbs free energy. In the case of this reaction, that limit is 83% efficient at 298K. This half-reaction pair and many others don't achieve their theoretical limit in practical application due to lack of an effective electrocatalyst.

One of the greatest drawbacks to galvanic cells like fuel cells and various forms of electrolytic cells is that they can suffer from high activation barriers. The energy diverted to overcome these activation barriers is transformed into heat. In most exothermic combustion reactions this heat would simply propagate the reaction catalytically. In a redox reaction, this heat is a useless byproduct lost to the system. The extra energy required to overcome kinetic barriers is usually described in terms of low faradayic efficiency and high overpotentials. In the example above, each of the two electrodes and its associated half-cell would require its own specialised electrocatalyst.

Half-reactions involving multiple steps, multiple electron transfers, and the evolution or consumption of a gases in their overall chemical transformations, will often have considerable kinetic barriers. Furthermore, there is often more than one possible reaction at the surface of an electrode. For example, during the electrolysis of water, the anode can oxidize water through a two electron process to hydrogen peroxide or a four electron process to oxygen. The presence of an electrocatalyst could facilitate either of the reaction pathways.

Like other catalysts, an electrocatalyst lowers the activation energy for a reaction without altering the reaction equilibrium. Electrocatalysts go a step further than other catalysts by lowering the excess energy consumed by a redox reaction's activation barriers.

Ethanol Fuel Cells

An electrocatalyst of platinum and rhodium on carbon backed tin-dioxide nanoparticles can break carbon bonds at room temperature with only carbon dioxide as a by-product, so that ethanol can be oxidised into the necessary hydrogen ions and electrons required to create electricity. In the context of electrochemistry, specifically in fuel cell engineering, various metal-containing catalysts are used to enhance the rates of the half reactions that comprise the fuel cell. One common type of fuel cell electrocatalyst is based upon nanoparticles of platinum that are supported on slightly larger carbon particles. When in contact with one of the electrodes in a fuel cell, this platinum increases the rate of oxygen reduction to water, either to hydroxide or hydrogen peroxide.

Organocatalysis

In organic chemistry, the term Organocatalysis (a concatenation of the terms "organic" and "catalyst") refers to a form of catalysis, whereby the rate of a chemical reaction is increased by an organic catalyst referred to as an "organocatalyst" consisting of carbon, hydrogen, sulfur and other nonmetal elements found in organic compounds. Because of their similarity in composition and description, they are often mistaken as a misnomer for enzymes due to their comparable effects on reaction rates and forms of catalysis involved.

Organocatalysts which display secondary amine functionality can be described as performing either enamine catalysis (by forming catalytic quantities of an active enamine nucleophile) or iminium catalysis (by forming catalytic quantities of an activated iminium electrophile). This mechanism is typical for covalent organocatalysis. Covalent binding of substrate normally requires high catalyst loading (for proline-catalysis typically 20-30 mol%). Noncovalent interactions such as hydrogen-bonding facilitates low catalyst loadings (down to 0.001 mol%).

Organocatalysis offers several advantages. There is no need for metal-based catalysis thus making a contribution to green chemistry. In this context, simple organic acids have been used as catalyst for the modification of cellulose in water on multi-ton scale. When the organocatalyst is chiral an avenue is opened to asymmetric catalysis, for example the use of proline in aldol reactions,

Whereas transition metals sometimes attract most of the attention in the study of catalysis, small organic molecules without metals can also exhibit catalytic properties, as is apparent from the fact that many enzymes lack transition metals. Typically, organic catalysts require a higher loading (amount of catalyst per unit amount of reactant, expressed in mol% amount of substance) than transition metal(-ion)-based catalysts, but these catalysts are usually commercially available in bulk, helping to reduce costs.

In the early 2000s, these organocatalysts were considered "new generation" and are competitive to traditional metal(-ion)-containing catalysts. Organocatalysts are supposed to operate akin to metal-free enzymes utilising, e.g., non-covalent interactions such as hydrogen bonding. The discipline organocatalysis is divided in the application of covalent (e.g., proline, DMAP) and non-covalent (e.g., thiourea organocatalysis) organocatalysts referring to the preferred catalyst-substrate binding and interaction, respectively.

Significance of Catalysis

Estimates are that 90% of all commercially produced chemical products involve catalysts at some stage in the process of their manufacture. In 2005, catalytic processes generated about $900 billion in products worldwide. Catalysis is so pervasive that subareas are not readily classified. Some areas of particular concentration are surveyed below.

Energy Processing

Petroleum refining makes intensive use of catalysis for alkylation, catalytic cracking (breaking long-chain hydrocarbons into smaller pieces), naphtha reforming and steam reforming (conversion of hydrocarbons into synthesis gas). Even the exhaust from the burning of fossil fuels is treated via catalysis: Catalytic converters, typically composed of platinum and rhodium, break down some of the more harmful byproducts of automobile exhaust.

$$2\,CO + 2\,NO \rightarrow 2\,CO_2 + N_2$$

With regard to synthetic fuels, an old but still important process is the Fischer-Tropsch synthesis of hydrocarbons from synthesis gas, which itself is processed via water-gas shift reactions, catalysed by iron. Biodiesel and related biofuels require processing via both inorganic and biocatalysts.

Fuel cells rely on catalysts for both the anodic and cathodic reactions.

Bulk Chemicals

Some of the largest-scale chemicals are produced via catalytic oxidation, often using oxygen. Examples include nitric acid (from ammonia), sulfuric acid (from sulfur dioxide to sulfur trioxide by the chamber process), terephthalic acid from p-xylene, and acrylonitrile from propane and ammonia.

Many other chemical products are generated by large-scale reduction, often via hydrogenation. The largest-scale example is ammonia, which is prepared via the Haber process from nitrogen. Methanol is prepared from carbon monoxide.

Bulk polymers derived from ethylene and propylene are often prepared via Ziegler-Natta catalysis. Polyesters, polyamides, and isocyanates are derived via acid-base catalysis.

Most carbonylation processes require metal catalysts, examples include the Monsanto acetic acid process and hydroformylation.

Fine Chemicals

Many fine chemicals are prepared via catalysis; methods include those of heavy industry as well as more specialised processes that would be prohibitively expensive on a large scale. Examples include olefin metathesis using Grubbs' catalyst, the Heck reaction, and Friedel-Crafts reactions.

Because most bioactive compounds are chiral, many pharmaceuticals are produced by enantioselective catalysis (catalytic asymmetric synthesis).

Food Processing

One of the most obvious applications of catalysis is the hydrogenation (reaction with hydrogen gas) of fats using nickel catalyst to produce margarine. Many other foodstuffs are prepared via biocatalysis.

Biology

In nature, enzymes are catalysts in metabolism and catabolism. Most biocatalysts are protein-based, i.e. enzymes, but other classes of biomolecules also exhibit catalytic properties including ribozymes, and synthetic deoxyribozymes.

Biocatalysts can be thought of as intermediate between homogenous and heterogeneous catalysts, although strictly speaking soluble enzymes are homogeneous catalysts and membrane-bound enzymes are heterogeneous.

Several factors affect the activity of enzymes (and other catalysts) including temperature, pH, concentration of enzyme, substrate, and products. A particularly important reagent in enzymatic reactions is water, which is the product of many bond-forming reactions and a reactant in many bond-breaking processes.

Enzymes are employed to prepare many commodity chemicals including high-fructose corn syrup and acrylamide.

Biocatalysis is the use of natural catalysts, such as protein enzymes, to perform chemical transformations on organic compounds. Both enzymes that have been more or less isolated and enzymes still residing inside living cells are employed for this task.

History

Biocatalysis underpins some of the oldest chemical transformations known to humans, for brewing predates recorded history. The oldest records of brewing are about 6000 years old and refer to the Sumerians.

The employment of enzymes and whole cells have been important for many industries for centuries. The most obvious uses have been in the food and drink businesses where the production of wine, beer, cheese etc. is dependent on the effects of the microorganisms. More than one hundred years ago, biocatalysis was employed to do chemical transformations on non-natural man-made organic compounds, and the last 30 years have seen a substantial increase in the application of biocatalysis to produce fine chemicals, especially for the pharmaceutical industry.

Since biocatalysis deals with enzymes and microorganisms, it is historically classified separately from "homogeneous catalysis" and "heterogeneous catalysis". However, mechanistically speaking, biocatalysis is simply a special case of heterogeneous catalysis.

Advantages

The key word for organic synthesis is selectivity, which is necessary to obtain a high yield of a specific product. There are a large range of selective organic reactions available for most synthetic needs. However, there is still one area where organic chemists are struggling, and that is when chirality is involved, although considerable progress in chiral synthesis has been achieved in recent years.

Enzymes display three major types of selectivities:

- Chemoselectivity: Since the purpose of an enzyme is to act on a single type of functional group, other sensitive functionalities, which would normally react to a certain extent under chemical catalysis, survive. As a result, biocatalytic reactions tend to be "cleaner" and laborious purification of product(s) from impurities emerging through side-reactions can largely be omitted.
- Regioselectivity and diastereoselectivity: Due to their complex three-dimensional structure, enzymes may distinguish between functional groups which are chemically situated in different regions of the substrate molecule.
- Enantioselectivity: Since almost all enzymes are made from L-amino acids, enzymes are chiral catalysts. As a consequence, any type of chirality present in the substrate molecule is "recognised" upon the formation of the enzyme-substrate complex. Thus a prochiral substrate may be transformed into an optically active product and both enantiomers of a racemic substrate may react at different rates.

These reasons, and especially the latter, are the major reasons why synthetic chemists have become interested in biocatalysis. This interest in turn is mainly due to the need to synthesise enantiopure compounds as chiral building blocks for drugs and agrochemicals.

Another important advantage of biocatalysts are that they are environmentally acceptable, being completely degraded in the environment. Furthermore the enzymes act under mild conditions, which minimises problems of undesired side-reactions such as decomposition, isomerization, racemization and rearrangement, which often plague traditional methodology.

Asymmetric Biocatalysis

The use of biocatalysis to obtain enantiopure compounds can be divided into two different methods:

1. Kinetic resolution of a racemic mixture
2. Biocatalysed asymmetric synthesis

In kinetic resolution of a racemic mixture, the presence of a chiral object (the enzyme) converts one of the enantiomers into product at a greater reaction rate than the other enantiomer.

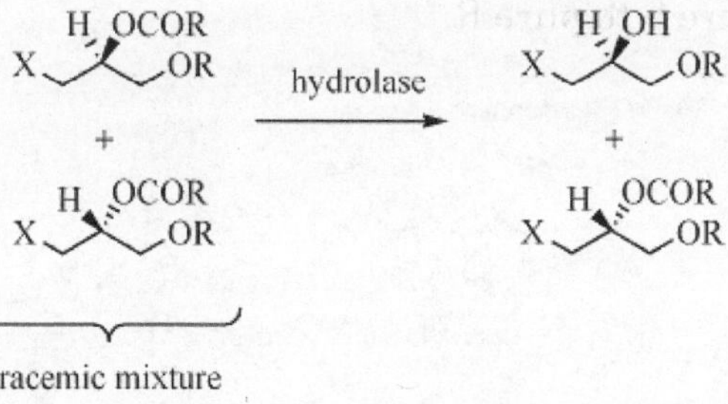

The racemic mixture has now been transformed into a mixture of two different compounds, making them separable by normal methodology. The maximum yield in such kinetic resolutions is 50%, since a yield of more than 50% means that some of wrong isomer also has reacted, giving a lower enantiomeric excess. Such reactions must therefore be terminated before equilibrium is reached. If it is possible to perform such resolutions under conditions where the two substrate-enantiomers are racemizing continuously, all substrate may in theory be converted into enantiopure product. This is called dynamic resolution.

In biocatalysed asymmetric synthesis, a non-chiral unit becomes chiral in such a way that the different possible stereoisomers are formed in different quantities. The chirality is introduced into the substrate by influence of enzyme, which is chiral. Yeast is a biocatalyst for the enantioselective reduction of ketones.

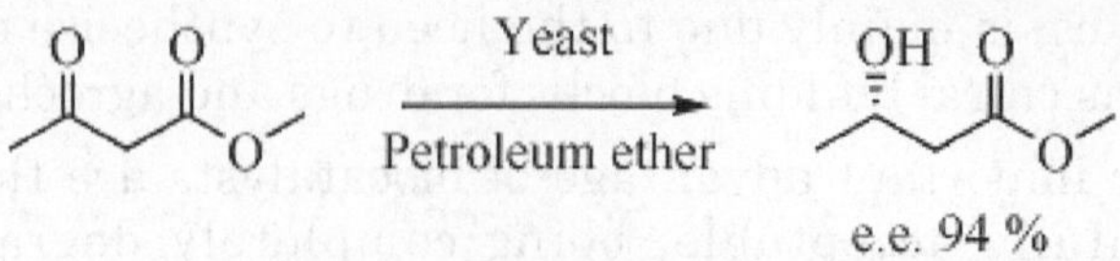

The biocatalytic Baeyer-Villiger oxidation is another example of a biocatalytic reaction. In one study a specially designed mutant of *Candida Antarctica* was found to be an effective catalyst for the Michael addition of acrolein with acetylacetone at 20 °C in absence of additional solvent.

Another study demonstrates how racemic nicotine (mixture of S and R-enantiomers 1 in *scheme 3*) can be deracemized in a one-pot procedure involving a monoamine oxidase isolated from Aspergillus niger which is able to oxidize only the amine S-enantiomer to the imine 2 and involving an ammonia–borane reducing couple which can reduce the imine 2 back to the amine 1.

In this way the S-enantiomer will continuously be consumed by the enzyme while the R-enantiomer accumulates. It is even possible to stereoinvert pure S to pure R.

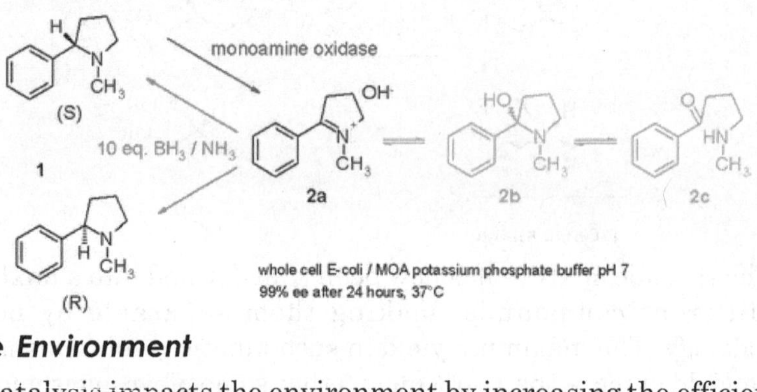

In the Environment

Catalysis impacts the environment by increasing the efficiency of industrial processes, but catalysis also plays a direct role in the environment. A notable example is the catalytic role of chlorine free radicals in the breakdown of ozone. These radicals are formed by the action of ultraviolet radiation on chlorofluorocarbons (CFCs).

$$Cl^{\cdot} + O_3 \rightarrow ClO^{\cdot} + O_2$$

$$ClO^{\cdot} + O^{\cdot} \rightarrow Cl^{\cdot} + O_2$$

History

In a general sense, anything that increases the rate of a process is a "catalyst", a term derived from Greek meaning "to annul," or "to untie," or "to pick up." The phrase *catalysed processes* was coined by

Jons Jakob Berzelius in 1836 to describe reactions that are accelerated by substances that remain unchanged after the reaction. Other early chemists involved in catalysis were Alexander Mitscherlich who referred to *contact processes* and Johann Wolfgang Dobereiner who spoke of *contact action* and whose lighter based on hydrogen and a platinum sponge became a huge commercial success in the 1820s. Humphry Davy discovered the use of platinum in catalysis.

In the 1880s, Wilhelm Ostwald at Leipzig University started a systematic investigation into reactions that were catalysed by the presence of acids and bases, and found that chemical reactions occur at finite rates and that these rates can be used to determine the strengths of acids and bases. For this work, Ostwald was awarded the 1909 Nobel Prize in Chemistry.

Inhibitors, Poisons and Promoters

Substances that reduce the action of catalysts are called catalyst inhibitors if reversible, and catalyst poisons if irreversible. Promoters are substances that increase the catalytic activity, even though they are not catalysts by themselves.

Inhibitors are sometimes referred to as "negative catalysts" since they decrease the reaction rate. However they do not work by introducing a reaction path with higher activation energy, as this term might suggest; this would not reduce the rate since the reaction would continue to occur by the non-catalysed path. Instead they act either by inactivating catalysts, or by removing reaction intermediates such as free radicals.

The inhibitor may modify selectivity in addition to rate. For instance, in the reduction of ethyne to ethene, the catalyst is palladium (Pd) partly "poisoned" with lead(II) acetate ($Pb(CH_3COO)_2$). Without the deactivation of the catalyst, the ethene produced will be further reduced to ethane.

The inhibitor can produce this effect by e.g. selectively poisoning only certain types of active sites. Another mechanism is the modification of surface geometry. For instance, in hydrogenation operations, large planes of metal surface function as sites of hydrogenolysis catalysis while sites catalysing hydrogenation of unsaturates are smaller.

Thus, a poison that covers surface randomly will tend to reduce the number of uncontaminated large planes but leave proportionally more smaller sites free, thus changing the hydrogenation vs. hydrogenolysis selectivity. Many other mechanisms are also possible.

Promoters can cover up surface to prevent production of a mat of coke, or even actively remove such material (e.g. rhenium on platinum in platforming). They can aid the dispersion of the catalytic material or bind to reagents.

Current Market

The global demand on catalysts in 2010 was estimated at approximately 29.5 billions US$. With the rapid recovery in automotive and chemical industry overall, the global catalyst market is expected to experience fast growth in the next years.

Catalysts in Petrochemical Industries

Chemistry is the science of substances and their transformations. Chemistry makes many wonderful substances from animals and vegetables as well as from crude oil as a raw material. Some substances occur naturally, but many have to be chemically treated in some way to make them suitable for our specific needs. . Thousands of substances have been produced that are not easily found in nature but posses unique and useful properties. The list of chemistry's good deeds is practically inexhaustible. Chemistry feeds us, clothes us, shoes us, and gives us the things without which modern civilized society cannot function.

In order to get unique substances, two or more chemicals must come together and react! After all we've all seen images of chemists cooking things up. May be we've even turned up the flame a few times ourselves in the kitchen! And so we know that in order to speed up chemical reactions, high temperature certainly helps. If you wish to have an economical and ecofriendly approach towards your chemical transformations then you should use an appropriate catalyst. Catalysts change the reaction rate of chemical reactions favourably without undergoing any permanent change themselves.

Any chemical reaction that takes place with the aid of a catalyst is called a catalytic reaction. For example, heating a mixture of potassium chlorate and manganese dioxide produces oxygen quicker than heating potassium chlorate on its own. The manganese dioxide acts as a catalyst to speed up the reaction, and is left unchanged at the end. For this reason, the manganese dioxide does not appear in the chemical equation that shows the reaction.

A contact catalyst is one that has a large porous surface to which other substances adhere by a process called adsorption. Atoms or molecules of different substances collect on the surface of the catalyst.

While on the surface they react together and are released in a different form. An example of such a reaction is the formation of ammonia from nitrogen and hydrogen. The catalyst used in this example is the iron ore magnetite. Catalytic converters are used to clean-up exhaust gases in automobiles. It reduces the toxicity of emissions from the internal combustion engines. They are fitted in the motor vehicle exhaust system. A catalytic converter provides an environment for a chemical reaction where in toxic combustion by-products are converted to less toxic substances.

The catalyst itself can be a precious metal, like platinum, palladium or rhodium. Platinum-rhodium catalysts are used as "reduction catalysts" and platinum palladium catalysts are used as "oxidizing catalysts". Zeolites are microporous aluminosilicate minerals, commonly used as commercial adsorbents. Synthetic zeolites are widely used as catalysts in the petrochemical industry.

Indian Catalysis Research

Indian catalysis research and development today is geared to serve the needs of the country on the one hand and the more challenging global requirements on the other. The research is focused, on innovative solutions to the problems of environmental pollution, safety in industrial practice and saving the energy requirement in chemical reactions. Catalysis R & D in Indian institutions now looks forward to the development of competitive catalysts and processes that are required to meet these challenges. Indian per capita consumption of petroleum and petrochemical products was increased considerably during 1970s.

Due to this rapid growth most of the laboratories initiated projects on the development of catalysis for petrochemicals and petroleum refinery processes with an emphasis on self-sufficiency and indigenization rather than on innovation. Catalysis research is expanded to cover topics such as synthesis and application of zeolites, molecular sieves, novel materials and photocatalysis.

The globalisation has resulted in many partnerships between Indian and foreign organisations. For example, a large amount of catalysis research carried out in CSIR's (Council of Scientific & Industrial Research) laboratories. Mention must be made of NCL (National Chemical Laboratory, Pune), IIP (Indian Institute of Petroleum, Dehradun) and IICT (Indian institute of Chemical Technology, Hyderabad) in this regard. These laboratories have collaborations with multinational R & D Centres. Based on these activities, many catalysts/ processes developed in India are expected to go global in coming years. IPCL (Indian Petrochemical Ltd), PDIL (Projects & Development India Ltd.), IIT (Indian Institute of Technology, Mumbai/Kanpur), CFRI (Central Fuel Research Institute, Dhanbad) have also contributed substantially t the area of catalysts.

Mitti ka tel

Petroleum or crude oil is one of the world's most important sources of energy. This makes it very valuable to the countries where it is found. In the beginning, this oil was found to be useful for burning lamps for lighting the homes. In many Western countries, the oil was found to be useful for warming homes. Several business men sent a sample of the thick black oil to Yale University of USA. In 1860s, the sample was analysed by Professor Benjamin Silliman. He reported that the "sample" is full of possibly useful chemicals. In other words, nothing will go as waste from this raw material. It is the chief source of hydrocarbons. Crude oil is generally a dark-coloured liquid with a peculiar smell; found underground either in certain shale rocks or held in huge natural reservoirs under pressure between non-porous layers of rock.

Hydrocarbons: This oil appears to be homogenous. In reality it is a made up of thousands of different combinations of the elements – hydrogen and carbon. These combinations are called hydrocarbons. The hydrogen atoms are arranged in various patterns around the carbon atoms. The more carbon atoms a petroleum molecule contains, the thicker and heavier the hydrocarbon. Gasoline, for example, is made up of light hydrocarbons. Kerosene has slightly heavier hydrocarbons. Lubricating oil is heavier still, and asphalt for paving streets is made of very heavy hydrocarbons.

The word petroleum comes from the Latin words *petra,* meaning rock; *oleum*, means oil. This oil was different than the animal or vegetable oil. People started calling this oil as "*rock-oil*" or in India "*Mitti ka tel*". People started using this oil for mainly for lighting their homes. Hundred years back, the oil was found to be an appropriate fuel for running the internal combustion engines. Cars started running on this liquid which was then an inexpensive fuel. Today, petroleum is much more than just a fuel for automobiles. It is used in the manufacture of thousands of products.... From plastics to hand lotions, from asphalt to wax candles.

In Search of Oil (Background)

Oil began to form deep under the earth millions of years ago when plants and animals died and their skeletons settled at the bottom of seas and lakes. The remains were slowly buried under layers of sediment and in the course of time became crude oil. The oil usually lies trapped between domes of solid rock. Often there is a layer of salt water beneath it. In many cases the richest deposits of oil lie beneath a great amount of rock and can be reached only by expensive drilling operations. Geologists know which rock strata (layers) oil is most likely to be trapped in. They start with experimental data of areas likely to be oil-bearing. Then they begin to gather their evidence.

***Fossils*:** Fossils help to date and identify rocks. Radioisotopes are used to obtain more accurate measurements of their age.

***Magnetometre*:** A special instrument called a magnetometre trails behind an airplane as it flies at an even height a long a pattern of flight lines. As different types of rocks are magnetic to different degrees, the results from the magnetometre give a picture of the magnetic structure and thickness of the rocks deep underground.

***Gravimetre*:** The earth's gravity also changes from place to place. A gravimetre carried across land or mounted on a ship measures the different gravities of the rocks below dense rocks give a higher gravity reading.

***Seismic Tests*:** Then there are seismic tests. These are more exact. They measure the time it takes for shock waves from controlled explosions to reflect upward from rocks underground. The time it takes the shock waves to reach an underground layer and return to the surface shows how deep the layer is. It also gives clues about the type of rock from which the shock waves were reflected. When all these tests are complete and if they show a good chance of oil being in a certain area, a test well is driven down into the rock. It is at this point that large amounts of money are needed to set up the drilling operation. It may cost billions of Rupees to develop a major oilfield before any oil is gotten out of it. Therefore, the exploration be as thorough as possible before any drilling starts. These days the computers are used to gather the information from all tests, but the sure presence of oil is still not known. If there is oil, it probably lies 3000 metres below the surface.

Early Drilling

The Chinese began drilling for oil as long ago as 320 BC, but they did not really have the right equipment to drill deep enough. The first successful oil well was sunk by Edwin L. Drake, who drilled a well about 30 metres deep in USA (Pennsylvania) in 1859. This was the start of today's petroleum oil industry. The story of oil exploration in India began in the dense jungles and swamps of Assam in the 19th century. In India, the first regular oil well was discovered in at Borbhil. The oil was also struck at Makum near Margherita at a depth of just 36 metres (118 feet).

In 1890, noticing the oil seepages around Borbhil, one private company took a decision to drill an oil-well near the village. Mr. W. L. Lake was an oil enthusiastic technologist. He immediately started collecting drilling equipments, portable boilers and local labour. Elephants were used to drag the machinery on sleigh along animal tracks. Lot of digging was done. The legend has it that Mr. Lake used to urge his men "Dig boys... Dig"; hence the name "Digboi"! With such great efforts the oil-well "Digboi Well No. 1" started in the month of September, 1889. Digboi Well No. 2 started in February 1891. The oil production was going on well with 11 wells until 1894. By 1920 there were 80 wells with an average production of 350 bpd (Barrels per day). Thus, there was a modest progress of oil industry in India about 120 years back.

Production of Petrochemicals with Catalysts (Oil Refining)

The oil we used to make "Petrol" (gasoline, or motor spirit) and other products, start out as raw black petroleum. It cannot be used as it is, straight from underground. It must be changed into the specific chemicals that people use. This process of changing petroleum into useful products is called refining. The modern oil refinery is a large and complicated factory. Automatic control of the various stages is efficient and safe. Teams of workers are needed to keep the refinery machinery "running" smoothly.

The crude oil consists of many different chemicals with various chemical and physical properties. Distillation or fractionation is the first stage in petroleum refining. This separates out the various constituents) of the crude oil. Different hydrocarbons vaporize at different temperatures. Here, tall distillation columns are necessary. There are many separate trays that are fitted for collecting various chemicals. As the heating starts, partly vaporized oil is fed into the column part of the way up. The lightest fractions are gases. The gases gather at the top of the tower. Heavier fractions that are slower to evaporate gather at each lower stage. Usually, fractions from several stages are drawn off for further processing. The residue, which is left at the bottom of the tower, contains a wide range of petroleum parts. It can be used for fuel oil without any more processing. It can be further distilled to give bitumen or lubricating oils. A process known as *"cracking"* breaks down large molecules into smaller ones. This is done either by thermal cracking, using high levels of heat and pressure, or by using less heat and pressure and a catalyst. Catalytic cracking is done by the fluid technique. In this, a catalyst in the form of a fine powder is poured like a liquid through the petroleum and out to a regenerator.

In the regenerator, carbon that has become attached to the catalyst is removed. The catalyst is then returned to the refining cycle. Catalytic cracking is normally used to make gasoline. The gasoline produced by this process has a higher octane number than gasoline produced by straight distillation. The octane number is a measure of the tendency of fuels to "knock" (make a knocking noise) when used in automobile engines. The higher the number, the less is knock. Another refining method is called hydrocracking. In this method, hydrogen and catalysts are used under high pressure. The process results in greater amounts of gasoline and less waste than catalytic cracking. During the cracking operation some gases are always formed. These gases are collected and then put under high temperatures and pressures in the presence of a catalyst. This process is called polymerization, joins molecules together to make larger molecules. These larger molecules (polymers) are used in the production of gasoline.

Recent Progress in the Catalytic Petrochemical Processes

The successes of basic research in the field of catalysis have a direct effect on solving many fundamental technological, environmental and social problems that face humanity:

1. The efficient utilisation of raw materials (oil, natural gas etc.)
2. Development of new materials and chemicals
3. Development of systems for environmental protection,
4. Development of new sources of energy,
5. Development of new processes and technologies.

Catalytic processes form the fundamentals of modern chemical and petrochemical industries. In the developed industrial countries catalytic processes create about 20% of the GDP.

The US 2022 advanced biofuels goal is 21 billion gallons. That's 17 percent of the petroleum that the U.S. imported in 2008 for transportation fuels. It could be obtained from algae grown on land roughly the size of South Carolina or 82931km2. To produce that much algal biofuel would be needed 350 gallons of water per gallon of oil or a quarter of what the US currently uses for irrigated agriculture. Estimated growing algae uses anywhere between 8.6 and 50.2 gallons of water per mile driven on algal biofuel. The corn ethanol can be made with less water, but showed a larger usage range: between 0.6 and 61.9 gallons of water per mile driven. Previously published data indicated conventional gas uses between about 0.09 and 0.3 gallons of water per mile.

Basic Features and Tendencies in Chemical Industry Development

1. Chemical industry is developing at a steady rate worldwide. In EU Chemical Industry in 2007 has 2.1% growth. Recent world crisis, however has a strong negative impact
2. The assortment of manufactured articles is constantly enlarged and renewed (Now they are about 200 000). 65% of the chemical production is being transferred to other branches for secondary processing.
3. Special attention is being paid to the development of new technologies and research consuming production at mild working conditions with a high degree of processing, low consumption of materials and high prices.
4. Under the conditions of rapidly outdated technological processes the market for selling licenses and 'know how' becomes very important as a profitable sphere for capital investment.
5. Rapid introduction of environmental friendly waste free technologies

Negative Factors Influencing the Development of Chemical Industry

1. There exists a permanent upward tendency in energy and raw materials prices, accompanied by deteriorated raw materials quality. As a result, desired products with specific properties require new complex technologies, entirely novel technological equipment, which is constantly becoming more complicated and expensive.
2. The necessity to invest considerable money in equipment and technologies for environmental protection. Nowadays the expenses for purifying facilities amount to an average of 20% of total capital invested.
3. There is a continual increase in scope and mostly in value of the research and development activities that ensure a high level and quality of new technologies, materials and information.
4. There are limitations in maximum dimensions of industrial units (reactors, columns etc.), owing to the properties of construction materials.

Overcoming the Influence of Negative Factors can be Achieved By

Developing novel technologies, engineering solutions, methods and products, elaboration and innovation of new highly selective and active catalysts in industry and wide application of automation and computerization. Availability of highly qualified scientists, engineers, industrial experts and industrial personnel.

Importance of the Basic Catalysis Research for Chemical Industry

Industrial catalysis is research intensive industry. In US it spends approximately $17.6 billion annually on Research and Development. The success chemical industry in US and other developed countries is largely attributable to breakthroughs in catalysis science and technology.

Important Directions of Basic Research in Catalysis

- Studies of catalytic reaction mechanisms;
- Catalytic reaction kinetics under steady and non-steady conditions;
- Physical chemistry of surfaces and solid states;
- Application of quantum chemical methods in catalytic research;
- Non linear phenomena in catalysis
- Theory of catalyst selection and prediction of catalyst activity
- Novel catalyst supports, adsorbents;
- Development of catalysts on the basis of new materials with preset qualities: new micro and mesoporous materials, semiconducting materials, perovskites, layered structure compounds, colloidal metals, amorphous metals, superconducting polymers, immobilised complexes, membranes, fibrous materials, solid superacids and bases, intercalation compounds, etc.
- Catalyst deactivation;
- Homogeneous catalysis and single site catalysis.

For all types of chemical catalysts, it is estimated a total 2010 market of almost $2.7 billion, expecting to grow at annual rate of about a 2.5% to more than $3 billion in 2015. Petroleum refining catalysts is a smaller market at $1.4 billion in sales in 2010, but also are predicted to grow at about a 2.5% to more than $1.6 billion in 2015.

Among the catalysts produced for the chemical and petrochemical industry, 49% are used for polymerisation, 18% oxidation processes, whereas 15, 10 and 8% go to organic synthesis, Fischer Tropsch synthesis and hydrogenation? dehydrogenation reactions. Growth is spurred by increasing demand for reformulated and other lesspolluting gasolines, plus new regulations calling for drastic reductions in sulfur content in gasoline and diesel fuel.

Characteristics of the Industry for Catalysts Production

1. Production of catalysts has a strategic importance for the country economy.
2. Catalysts are so called performance chemicals, which enhance the processing of other chemicals.
3. The catalyst production industry has a very complex multistep character.
4. The companies, which produce catalysts, have to support big research teams and extensive multimillion R&D programs.
5. The catalyst production is a very dynamic branch of the chemical industry.
6. In order to be able to produce various types of catalysts, the units for catalyst production should be very versatile.
7. The absolute amount of produced catalysts is not large and the catalyst production industry cannot be very profitable.
8. For selection of suitable catalyst decisive role play catalyst quality and its price has only secondary importance.
9. About 30 % of the newly developed catalysts and catalytic processes are not offered on the market in order to preserve the competitiveness of the owner.
10. Catalyst production industry is not an environmentally friendly industry.
11. A very high qualification of personnel is required. As mentioned above, the catalyst producing companies are located in highly developed industrial countries, since they need highly qualified staff and well-organised R&D facilities.
12. In contrast to other so-called 'high tech' industries, new catalyst producers are a rare phenomenon on the market, due to significant barriers. These involve a risk during introduction into practice associated with very high expenses for development, testing and production of a sample batch, probable losses in case of low quality production, conservative consumers

and necessity of industrial references to catalyst quality, and a lot of expenses for R&D activities.

Basic Tendencies in the Industrial Catalyst Production

Tendencies in the development of industrial catalyst production are determined by the modern state of chemical, petrochemical and oilprocessing industry; it is affected by economical, technological, scientific, and other factors.

1. The endeavour of all big companies to create their own catalyst production is due to the fact that those who dominate in the production of catalysts determine the level of technical progress in industrial branches related to catalytic processes. Today there is no company that controls more than 10% of the catalyst market. The world catalyst trade is about 10 billion US dollars, which is only about 0.15% of the total volume of chemical trade.
2. Greater and greater parts of the catalysts will be produced on special orders of certain consumers. Most often consumers develop catalyst production technologies themselves.
3. The extension of produced catalyst nomenclature includes especially catalysts for new chemicals production and novel processes based on new types of raw materials.
4. In 1984, the producers offered about 700 catalysts to the oil processing industry. About 150 new catalysts were innovated in the next year, as a result of which the total number of catalysts became near 900. Among the newly introduced catalysts, there are 90 types that were designed for the process of fluid catalytic cracking. Other 111 new types of catalysts were introduced in 1986. The largest share falls again to the catalytic cracking catalysts, their number being 71. Now over 1200 types of catalysts are offered.
5. Among the catalysts produced for the chemical and petrochemical industry, 49% are used for polymerisation, 18% oxidation processes, whereas 15, 10 and 8% go to organic synthesis, reactions involving CO and H2, and hydrogenation? dehydrogenation reactions.
6. Expansion of assortment and volume of produced catalysts to meet the demands of environmental protection technologies.
7. Promoting the variety of catalysts for fine chemicals synthesis.
8. Development and production of catalysts for new fields of application: catalytic combustion, gas sensors, fuel components, photocatalysis, etc.

9. Development of catalysts based on the new materials with pre-set qualities: new micro and mesoporous materials, semiconducting materials, perovskites, layered-structure compounds, colloidal metals, amorphous metals, superconducting polymers, immobilised complexes, membranes, fibrous materials, solid superacids and bases, intercalation compounds, etc.
11. The complexity of catalyst manufacture poses very tough requirements to the quality of the raw materials, technological parameter observation, and final product quality.
12. A very high qualification of personnel is required. As mentioned above, the catalyst producing companies are located in highly developed industrial countries, since they need highly qualified staff and well organised R&D facilities.
13. In contrast to other so-called 'high tech' industries, new catalyst producers are a rare phenomenon on the market, due to significant barriers. These involve a risk during introduction into practice associated with very high expenses for development, testing and production of a sample batch, probable losses in case of low quality production, conservative consumers and necessity of industrial references to catalyst quality, and a lot of expenses for R&D activities.
14. All leading chemical companies, which do not produce catalysts as commercial products, have their own research programs in the field of catalysis and production of catalysts, designed for their own processes, which enable them to achieve a break-through on the market.
15. Because catalyst quality is the critical point for installations to operate, some 20% of the used catalysts are produced only to cover producers' own needs. The remaining 80% are offered on the market, half of them being customer ordered or produced in accordance with a recipe, submitted by the user. Preservation of technological secrets is the critical issue in business dealings between user and producer.
16. There are 36 catalyst-supplying companies in the USA. Eight companies from these produce catalysts for chemical and petrochemical industry and only three companies also produce catalysts for environmental protection. A similar situation exists in the European Union.

    The catalyst production in Japan is concentrated in the hands of some ten companies.

17. The great variety of catalysts requires a flexibility of the production facilities. It means that a whole series of catalyst articles should be produced on the same technological line. This circumstance poses very high requirements toward engineering design of technological lines, apparatus performance, and other technological equipment in the production unit.

Major Characteristics of Fine Chemicals Industry

1. Scale of production is below 5000 t/y
2. Expensive raw materials
3. Stringent product specifications
4. Multi step complex technology
5. Big amount of by products > 20 100 kg/kg final product
6. Problems of effluent disposal, safety, health hazard corrosion
7. Product price > US $ 10 per kg product
8. Users buy fine chemicals for what they are than for any function that they perform.
9. Major reasons for high spendings are R&D of technology, raw material production, and development of process chemistry
10. Major sources of profit are production of advanced intermediates and bulk compound production.

Industrial catalysts are "performance chemicals" which should be offered on the market together with information about:

1. Reaction kinetics and kinetic model;
2. Catalytic activity and selectivity;
3. Catalyst pre treatment regimes;
4. Catalyst deactivation kinetics;
5. Catalyst regeneration regimes;
6. Catalyst lifetime: stability, duration of operation, thermal stability;
7. Physical and mechanical properties: strength, abrasion ability, hardness, surface area;
8. Hydrodynamic characteristics of the catalyst grain and of the catalyst bed in the reactor;
9. Safety transition regimes in cases of industrial accidents;
10. Economy of the process.

Sustainable chemistry (green chemistry) characteristics:

1. New products;
2. New feedstock, reagents, products;
3. New catalysts;
4. New engineering approaches and process control.

Principles to follow for sustainable (green) chemistry activity:

1. Do not make waste;
2. Increase carbon efficiency of the desired product;
3. Avoid production of hazardous materials;
4. Products should preserve function at reduced toxicity;
5. Reaction products should be environmentally acceptable;
6. Avoid use of auxiliary substances or use innocuous one;
7. Energy consumption should be minimised;
8. If possible try to use renewable raw materials;
9. Avoid unnecessary modification of the process;
10. Use highly active and selective catalysts;
11. Use most advanced and fast analytical control;
12. Use the less hazardous materials and reaction conditions.

General Scheme of Petrochemical Industry

- Supported noble metals: Pt, Pd, Rh, Ru, Re, Pt-Re
- Supported transition metals: Ni, Co, Fe, Cu, Mo
- Catalyst supports: a-Al2O3, SiO2, TiO2, Activated Carbon, zeolites,
- Raney type metal catalyst: Ni, Cu-Ni
- Oxide catalysts:Cr2O3, Fe2O3, Al2O3-Cr2O3, Fe2O3-K2CO3-Cr2O3, Ca3Ni(PO4)3, Bi2O3MoO3, Bi-Co-Fe-Sb-Mo-K oxide catalysts
- Sulfides catalysts: MoS2/Al2O3, WS2/Al2O3, NiS/Al2O3, CoS/Al2O3
- Micro- and mesoporous materials

Reactors Used

Liquid phase reactions: batch reactors, stirred tank reactors, slurry reactors Gas phase reactions: tubular reactors, multitray fixed bed reactors, moving bed Reactors.

Typical Catalysts Used in Petrochemistry

- Selective oxidation

  The total volume of produced intermediates is 215 million tons per year worth of roughly 100 billion $ US.
- Hydrogenation and dehydrogenation

  These processes are used to produce about 20 % of fine chemicals and pharmaceutical products.
- Organic synthesis
- Ammonia synthesis, Methanol synthesis
- Hydrogen production
- Ammoxidation, Oxychlorination
- Etherification, Esterification

Most Important Processes in Petrochemistry

1. The yields of useful products of many current commercial processes like propylene oxide, formaldehyde, and 1,4 butanediol are low.
2. In some cases the selectivity towards the main product is low and significant amounts of co products are produced (e.g., acetone is produced in phenol production from cumene, and t butanol is produced in propylene oxide production from propylene and isobutane).
3. The economics of the production of many products like acrolein, acrylic acid, methanol, acetic acid, phthalic anhydride, linear alcohols could be considerably improved if they could be produced from parafins instead of olefins.
4. Many important direct catalytic processes can be introduced in the industrial practice if they required low investment and low operational costs and proceed with high yield and selectivity. Methane activation to ethylene, methanol, or formaldehyde; Ethane to ethylene, ethylene glycol, acetic acid, or acetaldehyde; Propane to propylene, acrolein, acrylic acid, or 1,3 propane diol; Butane to butene, 1,4 butane diol, or maleic anhydride; Isobutane to methacrylic acid, Linear long chain alkanes to the alpha olefins or linear alcohols; Benzene to phenol.
5. Due to introduction of new technologies, including new computer developments, time to market for new products is reduced, production schedules and delivery time are shortened and logistics for delivery were improved. Many of these changes

lead to lower production costs and improved environmental performance.

6. The chemical industry, especially the producers of high profit fine chemicals, specialty chemicals and pharmaceuticals are facing with increasing very strong global competition.
7. As a consequence, catalytic scientists are increasingly under great pressure to accelerate the identification of new catalysts and catalyst compositions.

Major goals for improving the process of introduction of catalytic technologies in industry can be formulated as:

- Acceleration of the development new catalysts and new catalytic processes;
- Development of catalysts with selectivity approaching 100 %. Achieving these goals will require following major tools:
- Combinatorial algorithms for high throughput experiments;
- High throughput synthesis of catalysts;
- High throughput catalyst testing;
- New fast analytical methods and instruments;
- New techniques for in situ catalyst characterisation;
- Catalyst design using both empirical and fundamental knowledge;
- Deep understanding of the reaction mechanisms.

Assessment of the applicability of the newly developed catalytic technologies should be based on several criteria:

- Technological characteristics of the new development;
- Engineering viability;
- Impact of technology advances;
- Timeliness of the impact;
- Probability of successful development;
- Cost of investment relative to the potential benefits;
- Market niche for the new products;
- Environmental impact of the technology.

Propane Dehydrogenation Technology Important Problems in Petrochemical Catalysis

Alkenes have substantial value as chemical feedstock, whereas alkanes primarily have fuel value. The single greatest challenge in using hydrocarbon as raw material in petrochemistry is the selective, economical activation of light hydrocarbons to produce a single, more valuable product. One of the most probable approaches is conversion of a low energy material into a high energy material by oxidation. For example introducing novel pathways for the selective conversion of methane/ethane to higher molecular weight products: methane to methanol; methane to ethylene; ethane to ethylene and methane to higher molecular weight molecules.

Hydrocarbon

In organic chemistry, a hydrocarbon is an organic compound consisting entirely of hydrogen and carbon. Hydrocarbons from which one hydrogen atom has been removed are functional groups, called hydrocarbyls. Aromatic hydrocarbons (arenes), alkanes, alkenes, cycloalkanes and alkyne-based compounds are different types of hydrocarbons.

The majority of hydrocarbons found naturally occur in crude oil, where decomposed organic matter provides an abundance of carbon and hydrogen which, when bonded, can catenate to form seemingly limitless chains.

Types of Hydrocarbons

The classifications for hydrocarbons defined by IUPAC nomenclature of organic chemistry are as follows:

1. Saturated hydrocarbons (alkanes) are the simplest of the hydrocarbon species and are composed entirely of single bonds and are saturated with hydrogen. The general formula for saturated hydrocarbons is C_nH_{2n+2} (assuming non-cyclic structures). Saturated hydrocarbons are the basis of petroleum fuels and are found as either linear or branched species. Hydrocarbons with the same molecular formula but different structural formulae are called structural isomers. As given in the example of 3-methylhexane and its higher homologues, branched hydrocarbons can be chiral. Chiral saturated hydrocarbons constitute the side chains of biomolecules such as chlorophyll and tocopherol.
2. Unsaturated hydrocarbons have one or more double or triple bonds between carbon atoms. Those with double bond are called alkenes. Those with one double bond have the formula C_nH_{2n} (assuming non-cyclic structures). Those containing triple bonds are called alkynes, with general formula C_nH_{2n-2}.

3. Cycloalkanes are hydrocarbons containing one or more carbon rings to which hydrogen atoms are attached. The general formula for a saturated hydrocarbon containing one ring is C_nH_{2n}.
4. Aromatic hydrocarbons, also known as arenes, are hydrocarbons that have at least one aromatic ring.

Hydrocarbons can be gases (e.g. methane and propane), liquids (e.g. hexane and benzene), waxes or low melting solids (e.g. paraffin wax and naphthalene) or polymers (e.g. polyethylene, polypropylene and polystyrene).

General Properties

Because of differences in molecular structure, the empirical formula remains different between hydrocarbons; in linear, or "straight-run" alkanes, alkenes and alkynes, the amount of bonded hydrogen lessens in alkenes and alkynes due to the "self-bonding" or catenation of carbon preventing entire saturation of the hydrocarbon by the formation of double or triple bonds.

This inherent ability of hydrocarbons to bond to themselves is referred to as catenation, and allows hydrocarbon to form more complex molecules, such as cyclohexane and in rarer cases, arenes such as benzene. This ability comes from the fact that bond character between carbon atoms is entirely non-polar in that the distribution of electrons between the two elements is somewhat even due to the same electronegativity values of the elements (~0.30), and does not result in the formation of an electrophile.

Generally, with catenation comes the loss of the total amount of bonded hydrocarbons and an increase in the amount of energy required for bond cleavage due to strain exerted upon the molecule; in molecules such as cyclohexane, this is referred to as ring strain, and occurs due to the "destabilised" spatial electron configuration of the atom.

In simple chemistry, as per valence bond theory, the carbon atom must follow the "*4-hydrogen rule*", which states that the maximum number of atoms available to bond with carbon is equal to the number of electrons that are attracted into the outer shell of carbon. In terms of shells, carbon consists of an incomplete outer shell, which comprises 4 electrons, and thus has 4 electrons available for covalent or dative bonding.

Hydrocarbons are hydrophobic and are lipids. Some hydrocarbons also are abundant in the solar system. Lakes of liquid methane and ethane have been found on Titan, Saturn's largest moon, confirmed by the Cassini-Huygens Mission. Hydrocarbons are also abundant in nebulae forming polycyclic aromatic hydrocarbons - PAH compounds.

Simple Hydrocarbons and their Variations

Number of carbon atoms	Alkane	Alkene	Alkyne	Cycloalkane	Alkadiene
1	Methane	–	–	–	–
2	Ethane	Ethene (ethylene)	Ethyne (acetylene)	–	–
3	Propane	Propene (propylene)	Propyne (methylacetylene)	Cyclopropane	Propadiene (allene)
4	Butane	Butene (butylene)	Butyne	Cyclobutane	Butadiene
5	Pentane	Pentene	Pentyne	Cyclopentane	Pentadiene (piperylene)
6	Hexane	Hexene	Hexyne	Cyclohexane	Hexadiene
7	Heptane	Heptene	Heptyne	Cycloheptane	Heptadiene
8	Octane	Octene	Octyne	Cyclooctane	Octadiene
9	Nonane	Nonene	Nonyne	Cyclononane	Nonadiene
10	Decane	Decene	Decyne	Cyclodecane	Decadiene

Alkane

Alkanes (also known as paraffins or saturated hydrocarbons) are chemical compounds that consist only of hydrogen and carbon atoms and are bonded exclusively by single bonds (i.e., they are saturated compounds) without any cycles (or loops; i.e., cyclic structure). Alkanes belong to a homologous series of organic compounds in which the members differ by a constant relative molecular mass of 14.

Each carbon atom must have 4 bonds (either C-H or C-C bonds), and each hydrogen atom must be joined to a carbon atom (H-C bonds). A series of linked carbon atoms is known as the carbon skeleton or carbon backbone. In general, the number of carbon atoms is often used to define the size of the alkane (e.g., C_2-alkane).

An alkyl group, generally abbreviated with the symbol R, is a functional group or side-chain that, like an alkane, consists solely of single-bonded carbon and hydrogen atoms, for example a methyl or ethyl group.

The simplest possible alkane (the parent molecule) is methane, CH_4. There is no limit to the number of carbon atoms that can be linked together, the only limitation being that the molecule is acyclic, is saturated, and is a hydrocarbon. Saturated oils and waxes are examples of larger alkanes where the number of carbons in the carbon backbone tends to be greater than 10.

Alkanes are not very reactive and have little biological activity. Alkanes can be viewed as a molecular tree upon which can be hung the interesting biologically active/reactive portions (functional groups) of the molecule.

Structure Classification

Saturated hydrocarbons can be:

- linear (general formula C_nH_{2n+2}) wherein the carbon atoms are joined in a snake-like structure
- branched (general formula C_nH_{2n+2}, $n > 3$) wherein the carbon backbone splits off in one or more directions
- cyclic (general formula C_nH_{2n}, $n > 2$) wherein the carbon backbone is linked so as to form a loop.

According to the definition by IUPAC, the former two are alkanes, whereas the third group is called cycloalkanes. Saturated hydrocarbons can also combine any of the linear, cyclic (e.g., polycyclic) and branching structures, and they are still alkanes (no general formula) as long as they are acyclic (i.e., having no loops).

Isomerism

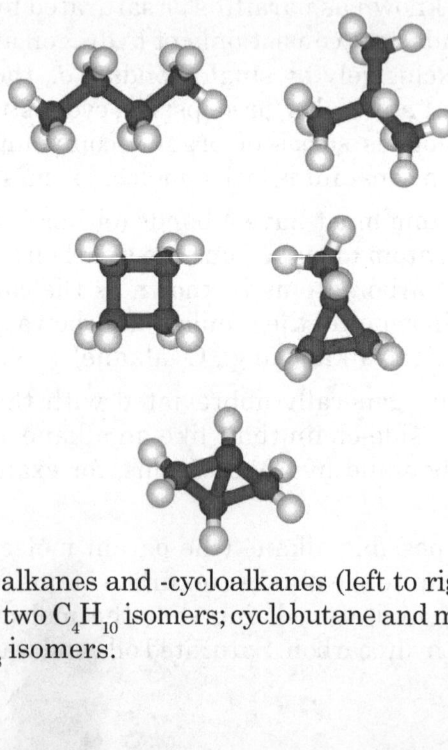

Different C_4-alkanes and -cycloalkanes (left to right): *n*-butane and isobutane are the two C_4H_{10} isomers; cyclobutane and methylcyclopropane are the two C_4H_8 isomers.

Bicyclo[1.1.0]butane is the only C_4H_6 compound and has no isomer; tetrahedrane is the only C_4H_4 compound and has also no isomer.

Alkanes with more than three carbon atoms can be arranged in numerous ways, forming different structural isomers. An isomer, in part, similar to a chemical anagram but unlike an anagram, may contain varying number of atoms and components, for which in a chemical compound can be structurally arranged in a multitude of different combinations and permutations. The simplest isomer of an alkane is the one in which the carbon atoms are arranged in a single chain with no branches. This isomer is sometimes called the *n*-isomer (*n* for "normal", although it is not necessarily the most common). However the chain of carbon atoms may also be branched at one or more points. The number of possible isomers increases rapidly with the number of carbon atoms (sequence A000602 in OEIS). For example:

- C_1: no isomers: methane
- C_2: no isomers: ethane
- C_3: no isomers: propane
- C_4: 2 isomers: *n*-butane & isobutane
- C_5: 3 isomers: pentane, isopentane, neopentane
- C_6: 5 isomers: hexane, 2-Methylpentane, 3-Methylpentane, 2,3-Dimethylbutane & 2,2-Dimethylbutane
- C_{12}: 355 isomers
- C_{32}: 27,711,253,769 isomers
- C_{60}: 22,158,734,535,770,411,074,184 isomers, many of which are not stable.

Branched alkanes can be chiral: 3-methylhexane and its higher homologues are chiral due to their stereogenic centre at carbon atom number 3. Chiral alkanes are of certain importance in biochemistry, as they occur as sidechains in chlorophyll and tocopherol (vitamin E). Chiral alkanes can be resolved into their enantiomers by enantioselective chromatography.

In addition to these isomers, the chain of carbon atoms may form one or more loops. Such compounds are called cycloalkanes.

Nomenclature

The IUPAC nomenclature (systematic way of naming compounds) for alkanes is based on identifying hydrocarbon chains. Unbranched, saturated hydrocarbon chains are named systematically with a Greek numerical prefix denoting the number of carbons and the suffix "-ane".

August Wilhelm von Hofmann suggested systematizing nomenclature by using the whole sequence of vowels a, e, i, o and u to create suffixes -ane, -ene, -ine (or -++yne), -one, -une, for the hydrocarbons. The first three name hydrocarbons with single, double and triple bonds; "-one" represents a ketone; "-ol" represents an alcohol or OH group; "-oxy-" means an ether and refers to oxygen between two carbons, so that methoxy-methane is the IUPAC name for dimethyl ether.

It is difficult or impossible to find compounds with more than one IUPAC name. This is because shorter chains attached to longer chains are prefixes and the convention includes brackets. Numbers in the name, referring to which carbon a group is attached to, should be as low as possible, so that 1- is implied and usually omitted from names of organic compounds with only one side-group; "1-" is implied in nitro-octane. Symmetric compounds will have two ways of arriving at the same name.

Linear Alkanes

Straight-chain alkanes are sometimes indicated by the prefix *n*- (for *normal*) where a non-linear isomer exists. Although this is not strictly necessary, the usage is still common in cases where there is an important difference in properties between the straight-chain and branched-chain isomers, e.g., *n*-hexane or 2- or 3-methylpentane.

The members of the series (in terms of number of carbon atoms) are named as follows:

- methane, CH_4 - one carbon and four hydrogen
- ethane, C_2H_6 - two carbon and six hydrogen
- propane, C_3H_8 - three carbon and 8 hydrogen
- butane, C_4H_{10} - four carbon and 10 hydrogen
- pentane, C_5H_{12} - five carbon and 12 hydrogen
- hexane, C_6H_{14} - six carbon and 14 hydrogen.

These names were derived from methanol, ether, propionic acid and butyric acid, respectively. Alkanes with five or more carbon atoms are named by adding the suffix -ane to the appropriate numerical multiplier prefix with elision of any terminal vowel (*-a* or *-o*) from the basic numerical term.

Hence, pentane, C_5H_{12}; hexane, C_6H_{14}; heptane, C_7H_{16}; octane, C_8H_{18}; etc. The prefix is generally Greek, with the exceptions of nonane which has a Latin prefix, and undecane and tridecane which have mixed-language prefixes.

Branched Alkanes

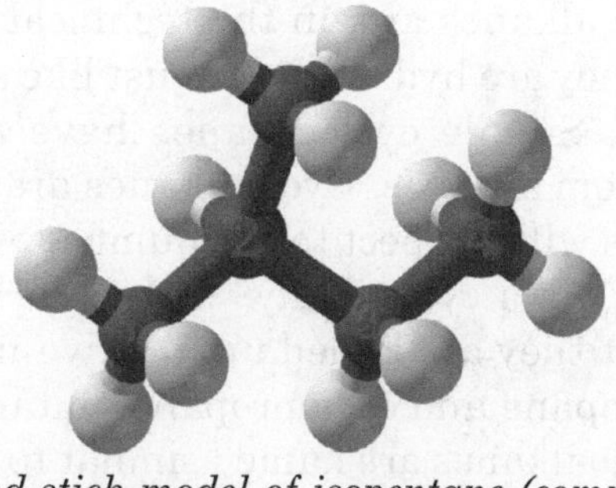

Figure: *Ball-and-stick model of isopentane (common name) or 2-methylbutane (IUPAC systematic name)*

Simple branched alkanes often have a common name using a prefix to distinguish them from linear alkanes, for example *n*-pentane, isopentane, and neopentane.

IUPAC naming conventions can be used to produce a systematic name.

The key steps in the naming of more complicated branched alkanes are as follows:

- Identify the longest continuous chain of carbon atoms
- Name this longest root chain using standard naming rules
- Name each side chain by changing the suffix of the name of the alkane from "-ane" to "-yl"
- Number the root chain so that sum of the numbers assigned to each side group will be as low as possible
- Number and name the side chains before the name of the root chain
- If there are multiple side chains of the same type, use prefixes such as "di-" and "tri-" to indicate it as such, and number each one.
- Add side chain names in alphabetical (disregarding "di-" etc. prefixes) order in front of the name of the root chain

Table: *Comparison of nomenclatures for three isomers of C_5H_{12}*

Common name	***n-pentane***	***isopentane***	***neopentane***
IUPAC name	pentane	2-methylbutane	2,2-dimethylpropane
Structure		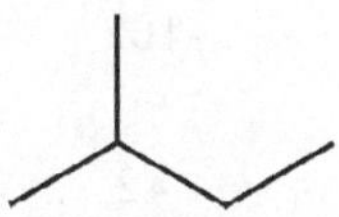	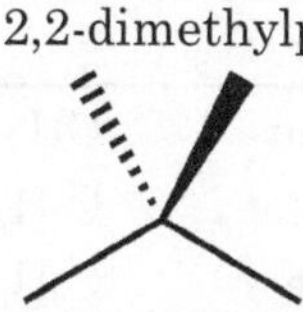

Cyclic Alkanes

So-called cyclic alkanes are, in the technical sense, *not* alkanes, but cycloalkanes. They are hydrocarbons just like alkanes, but contain one or more rings. Simple cycloalkanes have a prefix "cyclo-" to distinguish them from alkanes. Cycloalkanes are named as per their acyclic counterparts with respect to the number of carbon atoms, e.g., cyclopentane (C_5H_{10}) is a cycloalkane with 5 carbon atoms just like pentane (C_5H_{12}), but they are joined up in a five-membered ring. In a similar manner, propane and cyclopropane, butane and cyclobutane, etc. Substituted cycloalkanes are named similar to substituted alkanes — the cycloalkane ring is stated, and the substituents are according to their position on the ring, with the numbering decided by Cahn-Ingold-Prelog rules.

Trivial names

The trivial (non-systematic) name for alkanes is "paraffins." Together, alkanes are known as the *paraffin series*. Trivial names for compounds are usually historical artifacts. They were coined before the development of systematic names, and have been retained due to familiar usage in industry. Cycloalkanes are also called naphthenes. It is almost certain that the term paraffin stems from the petrochemical industry.

Branched-chain alkanes are called *isoparaffins*. The use of the term "paraffin" is a general term and often does not distinguish between a pure compounds and mixtures of isomers with the same chemical formula (i.e., like a chemical anagram), e.g., pentane and isopentane.

Examples

The following trivial names are retained in the IUPAC system:

- isobutane for 2-methylpropane
- isopentane for 2-methylbutane
- neopentane for 2,2-dimethylpropane.

Physical Properties

***Table:** Table of Alkanes*

Alkane	*Formula*	*Boiling point [°C]*	*Melting point [°C]*	*Density [g·cm³] (at 20°C)*
Methane	CH_4	-162	-183	gas
Ethane	C_2H_6	-89	-172	gas
Propane	C_3H_8	-42	-188	gas

Contd...

Alkane	*Formula*	*Boiling point [°C]*	*Melting point [°C]*	*Density [g·cm³] (at 20°C)*
Butane	C_4H_{10}	0	-138	gas
Pentane	C_5H_{12}	36	-130	0.626(liquid)
Hexane	C_6H_{14}	69	-95	0.659(liquid)
Heptane	C_7H_{16}	98	-91	0.684(liquid)
Octane	C_8H_{18}	126	-57	0.703(liquid)
Nonane	C_9H_{20}	151	-54	0.718(liquid)
Decane	$C_{10}H_{22}$	174	-30	0.730(liquid)
Undecane	$C_{11}H_{24}$	196	-26	0.740(liquid)
Dodecane	$C_{12}H_{26}$	216	-10	0.749(liquid)
Icosane	$C_{20}H_{42}$	343	37	solid
Triacontane	$C_{30}H_{62}$	450	66	solid
Tetracontane	$C_{40}H_{82}$	525	82	solid
Pentacontane	$C_{50}H_{102}$	575	91	solid
Hexacontane	$C_{60}H_{122}$	625	100	solid

Methane

Methane is a chemical compound with the chemical formula CH_4. It is the simplest alkane, the principal component of natural gas, and probably the most abundant organic compound on earth. The relative abundance of methane makes it an attractive fuel. Because, however it is a gas at normal conditions, methane is difficult to transport from its source.

Methane is a relatively potent greenhouse gas. The concentration of methane in the Earth's atmosphere in 1998, expressed as a mole fraction, was 1745 nmol/mol (parts per billion, ppb), up from 700 nmol/mol in 1750. By 2008, however, global methane levels, which had stayed mostly flat since 1998, had risen to 1,800 nmol/mol.

Properties and Bonding

Methane is a tetrahedral molecule with four equivalent C-H bonds. Its electronic structure is described by four bonding molecular orbitals (MOs) resulting from the overlap of the valence orbitals on C and H. The lowest energy MO is the result of the overlap of the 2s orbital on carbon with the in-phase combination of the 1s orbitals on the four hydrogen atoms. Above this level in energy is a triply degenerate set of MOs that involve overlap if the 2p orbitals on carbon with various linear combinations of the 1s orbitals on hydrogen. The resulting "three-over-one" bonding scheme is consistent with photoelectron spectroscopic measurements. At room temperature and standard pressure, methane is a colourless, odorless gas. The familiar smell of natural gas as used in homes is a safety measure achieved by the addition of an odorant, often methanethiol or ethanethiol. Methane has a boiling point of "161 °C ("257.8 °F) at a pressure of one atmosphere. As a gas it is flammable only over a narrow range of concentrations (5–15%) in air. Liquid methane does not burn unless subjected to high pressure (normally 4–5 atmospheres).

Chemical Reactions

Main reactions with methane are: combustion, steam reforming to syngas, and halogenation. In general, methane reactions are difficult to control. Partial oxidation to methanol, for example, is challenging because the reaction typically progresses all the way to carbon dioxide and water even with incomplete amounts of oxygen. The enzymes methane monooxygenase can produce methanol from methane, but they cannot be used for industrial scale reactions.

Acid-base Reactions

Like other hydrocarbons, methane is a very weak acid. Its pKa in DMSO is estimated to be 56. It cannot be deprotonated in solution, but the conjugate base methyl lithium is known. Protonation of methane can be achieved with super acids to give CH_5^+, sometimes called the methanium ion. Despite the strength of its C-H bonds, there is intense interest in catalysts that facilitate C–H bond activation in methane (and other low alkanes).

Combustion

In the combustion of methane, several steps are involved. An early intermediate is formaldehyde (HCHO or H_2CO). Oxidation of formaldehyde gives the formyl radical (HCO), which then give carbon monoxide (CO):

$$CH_4 + O_2 \rightarrow CO + H_2 + H_2O$$

The resulting H_2 oxidizes to H_2O, releasing heat. This reaction occurs very quickly, usually in significantly less than a millisecond.

$$2\,H_2 + O_2 \rightarrow 2\,H_2O$$

Finally, the CO oxidizes, forming CO_2 and releasing more heat. This process is generally slower than the other chemical steps, and typically requires a few to several milliseconds to occur.

$$2\,CO + O_2 \rightarrow 2\,CO_2$$

The result of the above is the following total equation:

$CH_4 + 2\,O_2 \rightarrow CO_2 + 2\,H_2O$ (ΔH = “891 kJ/mol (at standard conditions))

Reactions with Halogens

Methane reacts with halogens given appropriate conditions as follows:

$$CH_4 + X_2 \rightarrow CH_3X + HX$$

where X is a halogen: fluorine (F), chlorine (Cl), bromine (Br), or iodine (I). This mechanism for this process is called free radical halogenation, beginning with the attach of Cl · radicals on methane to produce CH_3 ·, which combines with a second Cl · to give methyl chloride (CH_3Cl). Similar reactions will produce dichloromethane (CH_2Cl_2), chloroform($CHCl_3$), and, ulitimately, carbon tetrachloride (CCl_4). The energy required to start this reaction comes from UV radiation or heating.

Uses

Methane is used in industrial chemical processes and may be transported as a refrigerated liquid (liquefied natural gas, or LNG). While leaks from a refrigerated liquid container are initially heavier than air due to the increased density of the cold gas, the gas at ambient temperature is lighter than air. Gas pipelines distribute large amounts of natural gas, of which methane is the principal component.

Fuel

Methane is important for electrical generation by burning it as a fuel in a gas turbine or steam boiler. Compared to other hydrocarbon fuels, burning methane produces less carbon dioxide for each unit of heat released.

At about 891 kJ/mol, methane's heat of combustion is lower than any other hydrocarbon but the ratio of the heat of combustion (891 kJ/mol) to the molecular mass (16.0 g/mol, of which 12.0 g/mol is carbon) shows that methane, being the simplest hydrocarbon, produces more heat per mass unit (55.7 kJ/g) than other complex hydrocarbons. In many cities, methane is piped into homes for domestic heating and cooking purposes.

In this context it is usually known as natural gas, and is considered to have an energy content of 39 megajoules per cubic metre, or 1,000 BTU per standard cubic foot.

Methane in the form of compressed natural gas is used as a vehicle fuel and is claimed to be more environmentally friendly than other fossil fuels such as gasoline/petrol and diesel.

Research into adsorption methods of methane storage for this purpose has been conducted.

Developing Technologies

Research is being conducted by NASA on methane's potential as a rocket fuel.

Methane emitted from coal mines has been converted to electricity.

Chemical Feedstock

Although there would is great interest in converting methane into useful or more easily liquified compounds, the only practical processes are relatively unselective. In the chemical industry, methane is converted to synthesis gas, a mixture of carbon monoxide and hydrogen, by steam reforming. This endergonic process (requiring energy) utilises nickel catalysts and requires high temperatures, around 700–1100 °C:

$$CH_4 + 2\ H_2O \rightarrow CO_2 + 4\ H_2$$

Related chemistries are exploited in the Haber-Bosch Synthesis of ammonia from air, which is reduced with natural gas to a mixture of carbon dioxide, water, and ammonia.

Methane is also subjected to free-radical chlorination in the production of chloromethanes, although methanol is a more typical precursor.

Chapter 5

Methanogenesis

Methanogenesis or biomethanation is the formation of methane by microbes known as methanogens. Organisms capable of producing methane have been identified only from the domain Archaea, a group phylogenetically distinct from both eukaryotes and bacteria, although many live in close association with anaerobic bacteria.

The production of methane is an important and widespread form of microbial metabolism. In most environments, it is the final step in the decomposition of biomass. Recently, some experiments have suggested that leaf tissues of living plants emit methane. Other research has indicated that the plants are not actually generating methane; they are just absorbing methane from the soil and then emitting it through their leaf tissues. There may still be some unknown mechanism by which plants produce methane, but that is by no means certain.

Biochemistry of Methanogenesis

Methanogenesis in microbes is a form of anaerobic respiration. Methanogens do not use oxygen to respire; in fact, oxygen inhibits the growth of methanogens.

The terminal electron acceptor in methanogenesis is not oxygen, but carbon. The carbon can occur in a small number of organic compounds, all with low molecular weights. The two best described pathways involve the use of carbon dioxide and acetic acid as terminal electron acceptors:

$$CO_2 + 4\ H_2 \rightarrow CH_4 + 2H_2O$$

$$CH_3COOH \rightarrow CH_4 + CO_2$$

However, methanogenesis has been shown to use carbon from other small organic compounds, such as formic acid (formate), methanol, methylamines, dimethyl sulfide, and methanethiol.

The biochemistry of methanogenesis is relatively complex, involving the following coenzymes and cofactors: F420, coenzyme B, coenzyme M, methanofuran, and methanopterin.

Importance in Carbon Cycle

Methanogenesis is the final step in the decay of organic matter. During the decay process, electron acceptors (such as oxygen, ferric iron, sulfate, nitrate, and manganese) become depleted, while hydrogen (H_2) and carbon dioxide accumulate. Light organics produced by fermentation also accumulate. During advanced stages of organic decay, all electron acceptors become depleted except carbon dioxide. Carbon dioxide is a product of most catabolic processes, so it is not depleted like other potential electron acceptors.

Only methanogenesis and fermentation can occur in the absence of electron acceptors other than carbon. Fermentation only allows the breakdown of larger organic compounds, and produces small organic compounds. Methanogenesis effectively removes the semi-final products of decay: hydrogen, small organics, and carbon dioxide. Without methanogenesis, a great deal of carbon (in the form of fermentation products) would accumulate in anaerobic environments.

In Ruminants

Methanogenesis occurs in the guts of humans and other animals, especially ruminants. In the rumen, anaerobic organisms, including methanogens, digest cellulose into forms usable by the animal. Without these microorganisms, animals such as cattle would not be able to consume grass. The useful products of methanogenesis are absorbed by the gut, but methane is released from the animal mainly by belching (eructation). The average cow emits around 250 liters of methane per day.

In Humans

Some humans produce flatus that contains methane. In one study of the feces of nine adults, only five of the samples contained archaea capable of producing methane. Similar results are found in samples of gas obtained from within the rectum.

Even among humans whose flatus does contain methane, the amount is only 0% - 10% of the total amount of gas.

Role in Global Warming

Methane in the Earth's atmosphere is an important greenhouse gas with a global warming potential 21 times greater than carbon dioxide (averaged over 100 years), and methanogenesis in livestock and the decay of organic material is thus a considerable contributor to global warming. It may not be a net contributor in the sense that it works on organic material which used up atmospheric carbon dioxide when it was created, but its overall effect is to convert the carbon dioxide into methane which is a much more potent greenhouse gas.

Methanogenesis can also be beneficially exploited, to treat organic waste, to produce useful compounds, and the methane can be collected and used as biogas, a fuel. It is the primary pathway whereby most organic matter disposed of via landfill is broken down.

Methanogenesis and Extra-terrestrial Life

The presence of atmospheric methane has a role in the scientific search for extra-terrestrial life. The argument being that methane in the atmosphere will eventually dissipate, unless something is replenishing it. This can be detected (by using a spectrometre for example) then that means there is, or relatively recently was, life present. This was debated when methane was discovered in the Martian atmosphere by M.J. Mumma of NASA's Goddard Flight Centre, and verified by the Mars Express Orbiter (2004) and in Titan's atmosphere by the Huygens probe (2005). It is also argued that atmospheric methane can come from volcanoes or other fissures in the planet's crust and that without an Isotopic signature it is difficult to say what exactly was the origin.

Biological Routes

Naturally occurring methane is mainly produced by the process of methanogenesis. This multistep process is used by microorganisms as an energy source. The net reaction is:

$$CO_2 + 8\ H^+ + 8\ e^- \rightarrow CH_4 + 2\ H_2O$$

The final step in the process is catalysed by the enzyme methyl-coenzyme M reductase. Methanogenesis is a form of anaerobic respiration used by organisms that occupy landfill, ruminants (e.g., cattle), and the guts of termites. It is uncertain if plants are a source of methane emissions.

Industrial Routes

Natural gas is so abundant that the intentional production of methane would be unusual. Methane can be produced by hydrogenation carbon dioxide through the Sabatier process. It is also a side product of the hydrogenation of carbon monoxide in the Fischer-Tropsch process. This technology is practiced on a large scale to produce longer chain molecules than methane.

Occurrence

Methane was discovered and isolated by Alessandro Volta between 1776 and 1778 when studying marsh gas from Lake Maggiore. It is the major component of natural gas, about 87% by volume. The major source of methane is extraction from geological deposits known as natural gas fields, with coal seam gas extraction becoming a major source. It is associated with other hydrocarbon fuels, and sometimes accompanied by helium and nitrogen. The gas at shallow levels (low pressure) forms by anaerobic decay of organic matter and reworked methane from deep under the Earth's surface. In general, sediments buried deeper and at higher temperatures than those that contain oil generate natural gas.

It is generally transported in bulk by pipeline in its natural gas form, or LNG carriers in its liquefied form; few countries transport it by truck.

Alternative Sources

Apart from gas fields, an alternative method of obtaining methane is via biogas generated by the fermentation of organic matter including manure, wastewater sludge, municipal solid waste (including landfills), or any other biodegradable feedstock, under anaerobic conditions. Rice fields also generate large amounts of methane during plant growth. Methane hydrates/clathrates (ice-like combinations of methane and water on the sea floor, found in vast quantities) are a potential future source of methane.

Cattle belch methane accounts for 16% of the world's annual methane emissions to the atmosphere. The livestock sector in general (primarily cattle, chickens, and pigs) produces 37% of all human-induced methane. Early research has found a number of medical treatments and dietary adjustments that help slightly limit the production of methane in ruminants.

Atmospheric Methane

Methane is created near the Earth's surface, primarily by microorganisms by the process of methanogenesis. It is carried into the stratosphere by rising air in the tropics. Uncontrolled build-up of methane in the atmosphere is naturally checked — although human influence can upset this natural regulation — by methane's reaction with hydroxyl radicals formed from singlet oxygen atoms and with water vapor. It has a net lifetime of about 10 years, and is primarily removed by conversion to carbon dioxide and water

Methane also affects the degradation of the ozone layer.

In addition, there is a large (but unknown) amount of methane in methane clathrates in the ocean floors as well as the Earth's crust. Most methane is the result of biological process called methanogenesis.

In 2010, methane levels in the Arctic were measured at 1850 nmol/mol, a level scientists described as being higher than at any time in the previous 400,000 years. Historically, methane concentrations in the world's atmosphere have ranged between 300 and 400 nmol/mol during glacial periods commonly known as ice ages, and between 600 to 700 nmol/mol during the warm interglacial periods. It has a high global warming potential: 72 times that of carbon dioxide over 20 years, and 25 times over 100 years, and the levels are rising.

Methane in the Earth's atmosphere is an important greenhouse gas with a global warming potential of 25 compared to CO_2 over a 100-year period (although accepted figures probably represents an underestimate). This means that a methane emission will have 25 times the effect on temperature of a carbon dioxide emission of the same mass over the following 100 years.

Methane has a large effect for a brief period (a net lifetime of 8.4 years in the atmosphere), whereas carbon dioxide has a small effect for a long period (over 100 years). Because of this difference in effect and time period, the global warming potential of methane over a 20 year time period is 72.

The Earth's atmospheric methane concentration has increased by about 150% since 1750, and it accounts for 20% of the total radiative forcing from all of the long-lived and globally mixed greenhouse gases (these gases don't include water vapour which is by far the largest component of the greenhouse effect).

Usually, excess methane from landfills and other natural producers of methane is burned so CO_2 is released into the atmosphere instead of methane, because methane is a more effective greenhouse gas. Recently, methane emitted from coal mines has been successfully utilised to generate electricity.

Clathrates

Arctic methane release from permafrost and methane clathrates is an expected consequence of global warming.

Safety

Methane is not toxic; however, it is extremely flammable and may form explosive mixtures with air. Methane is violently reactive with oxidizers, halogens, and some halogen-containing compounds. Methane is also an asphyxiant and may displace oxygen in an enclosed space. Asphyxia may result if the oxygen concentration is reduced to below about 16% by displacement, as most people can tolerate a reduction from 21% to 16% without ill effects.

The concentration of methane at which asphyxiation risk becomes significant is much higher than the 5–15% concentration in a flammable or explosive mixture. Methane off-gas can penetrate the interiors of buildings near landfills and expose occupants to significant levels of methane. Some buildings have specially engineered recovery systems below their basements to actively capture this gas and vent it away from the building.

Ethane

Ethane is a chemical compound with chemical formula C_2H_6. It is the only two-carbon alkane that is an aliphatic hydrocarbon. At standard temperature and pressure, ethane is a colourless, odorless gas. Ethane is isolated on an industrial scale from natural gas, and as a byproduct of petroleum refining. Its chief use is as petrochemical feedstock for ethylene production.

History

Ethane was first synthetically created in 1834 by Michael Faraday, applying electrolysis of a potassium acetate solution. He mistook the hydrocarbon product of this reaction for methane, and did not investigate it further. During the period 1847–1849, in an effort to vindicate the radical theory of organic chemistry, Hermann Kolbe and Edward Frankland produced ethane by the reductions of propionitrile (ethyl cyanide) and ethyl iodide with potassium metal, and, as did Faraday, by the electrolysis of aqueous acetates.

They, however, mistook the product of these reactions for methyl radical, rather than the dimer of methyl, ethane. This error was corrected in 1864 by Carl Schorlemmer, who showed that the product of all these reactions was in fact ethane.

The name *ethane* is derived from the IUPAC nomenclature of organic chemistry. "Eth-" refers to the presence of 2 carbon atoms, and "-ane" refers to the presence of a single bond between them.

Chemistry

In the laboratory, ethane may be conveniently prepared by Kolbe electrolysis. In this technique, an aqueous solution of an acetate salt is electrolysed. At the anode, acetate is oxidized to produce carbon dioxide and methyl radicals, and the highly reactive methyl radicals combine to produce ethane:

$$CH_3COO^- \rightarrow CH_3\cdot + CO_2 + e^-$$

$$CH_3\cdot + \cdot CH_3 \rightarrow C_2H_6$$

Another method, the oxidation of acetic anhydride by peroxides, is conceptually similar.

The chemistry of ethane also involves chiefly free radical reactions. Ethane can react with the halogens, especially chlorine and bromine, by free radical halogenation. This reaction proceeds through the propagation of the ethyl radical:

$$C_2H_5\cdot + Cl_2 \rightarrow C_2H_5Cl + Cl\cdot$$

$$Cl\cdot + C_2H_6 \rightarrow C_2H_5\cdot + HCl$$

Because halogenated ethanes can undergo further free radical halogenation, this process results in a mixture of several halogenated products. In the chemical industry, more selective chemical reactions are used for the production of any particular two-carbon halocarbon.

Combustion

The complete combustion of ethane releases 1559.7 kJ/mol, or 51.9 kJ/g, of heat, and produces carbon dioxide and water according to the chemical equation

$$2\,C_2H_6 + 7\,O_2 \rightarrow 4\,CO_2 + 6\,H_2O + 3170\text{ kJ}$$

Combustion occurs by a complex series of free-radical reactions. Computer simulations of the chemical kinetics of ethane combustion have included hundreds of reactions. An important series of reaction in ethane combustion is the combination of an ethyl radical with oxygen, and the subsequent breakup of the resulting peroxide into ethoxy and hydroxyl radicals.

$$C_2H_5\cdot + O_2 \rightarrow C_2H_5OO\cdot$$

$$C_2H_5OO\cdot + HR \rightarrow C_2H_5OOH + \cdot R$$

$$C_2H_5OOH \rightarrow C_2H_5O\cdot + \cdot OH$$

The principal carbon-containing products of incomplete ethane combustion are single-carbon compounds such as carbon monoxide and formaldehyde. One important route by which the carbon-carbon bond in ethane is broken to yield these single-carbon products is the decomposition of the ethoxy radical into a methyl radical and formaldehyde, which can in turn undergo further oxidation.

$$C_2H_5O\cdot \rightarrow CH_3\cdot + CH_2O$$

Some minor products in the incomplete combustion of ethane include acetaldehyde, methane, methanol, and ethanol. At higher temperatures, especially in the range 600–900 °C, ethylene is a significant product. It arises via reactions like

$$C_2H_5\cdot + O_2 \rightarrow C_2H_4 + \cdot OOH$$

Similar reactions (although with species other than oxygen as the hydrogen abstractor) are involved in the production of ethylene from ethane in steam cracking.

Ethane Barrier

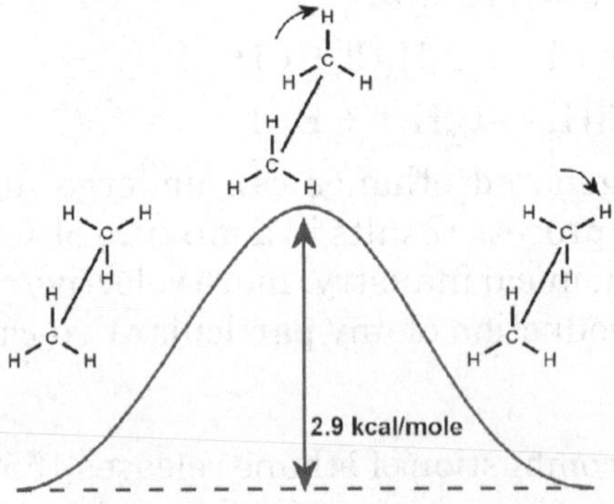

Figure: *Ethane barrier to rotation about the carbon-carbon bond. The curve is potential energy as a function of rotational angle.*

Rotating a molecular substructure about a twistable bond usually requires energy. The minimum energy to produce a 360-degree bond rotation is called the rotational barrier.

Ethane gives a classic, simple example of such a rotational barrier, sometimes called the "ethane barrier." The three hydrogens at each end are free to pinwheel about the central carbon-carbon bond, provided that there is sufficient energy to overcome the barrier of the carbon-hydrogen bonds at each end of the molecule bumping into each other by way of overlap (exchange) repulsion. As far back as 1890-1891 ethane was suggested to prefer a staggered conformation with the two ends of the molecule askew from each other.

Production

After methane, ethane is the second-largest component of natural gas. Natural gas from different gas fields varies in ethane content from less than 1% to over 6% by volume. Prior to the 1960s, ethane and larger molecules were typically not separated from the methane component of natural gas, but simply burnt along with the methane as a fuel. Today, however, ethane is an important petrochemical feedstock, and it is separated from the other components of natural gas in most well-developed gas fields. Ethane can also be separated from petroleum gas, a mixture of gaseous hydrocarbons that arises as a byproduct of petroleum refining. Economics of building and running processing plants can change, however. If the relative value of sending the unprocessed natural gas to a consumer exceeds the value of extracting ethane, then the plant may not be run. This can cause operational issues managing the changing quality of the gas in downstream systems.

Ethane is most efficiently separated from methane by liquefying it at cryogenic temperatures. Various refrigeration strategies exist: the most economical process presently in wide use employs turboexpansion, and can recover over 90% of the ethane in natural gas. In this process, chilled gas expands through a turbine; as it expands, its temperature drops to about -100 °C. At this low temperature, gaseous methane can be separated from the liquefied ethane and heavier hydrocarbons by distillation. Further distillation then separates ethane from the propane and heavier hydrocarbons

Uses

The chief use of ethane is in the chemical industry in the production of ethene (ethylene) by steam cracking. When diluted with steam and briefly heated to very high temperatures (900 °C or more), heavy hydrocarbons break down into lighter hydrocarbons, and saturated hydrocarbons become unsaturated. Ethane is favoured for ethene production because the steam cracking of ethane is fairly selective for ethene, while the steam cracking of heavier hydrocarbons yields a product mixture poorer in ethene, and richer in heavier alkenes (olefins) such as propene (propylene) and butadiene, and in aromatic hydrocarbons.

Experimentally, ethane is under investigation as a feedstock for other commodity chemicals. Oxidative chlorination of ethane has long appeared to be a potentially more economical route to vinyl chloride than ethene chlorination. Many processes for carrying out this reaction have been patented, but poor selectivity for vinyl chloride and corrosive reaction conditions (specifically, a hydrochloric acid-containing reaction mixture at temperatures greater than 500 °C) have discouraged the commercialisation of most of them. Presently, INEOS operates a 1000 t/a ethane-to-vinyl chloride pilot plant at Wilhelmshaven in Germany.

Similarly, the Saudi Arabian firm SABIC has announced construction of a 30,000 t/a plant to produce acetic acid by ethane oxidation at Yanbu. This economic viability of this process may rely on the low cost of ethane near Saudi oil fields, and it may not be competitive with methanol carbonylation elsewhere in the world.

Ethane can be used as a refrigerant in cryogenic refrigeration systems. On a much smaller scale, in scientific research, liquid ethane is used to vitrify water-rich samples for electron microscopy. A thin film of water, quickly immersed in liquid ethane at -150 °C or colder, freezes too quickly for water to crystallize. This rapid freezing does not disrupt the structure of soft objects present in the liquid state, as the formation of ice crystals can do.

Health and Safety

At room temperature, ethane is a flammable gas. When mixed with air at 3.0% – 12.5% by volume, it forms an explosive mixture.

Some additional precautions are necessary where ethane is stored as a cryogenic liquid. Direct contact with liquid ethane can result in severe frostbite. In addition, the vapors evaporating from liquid ethane are, until they warm to room temperature, heavier than air and can creep along the ground or gather in low places, and if they encounter an ignition source, can flash back to the body of ethane from which they evaporated. Containers recently emptied of ethane may contain insufficient oxygen to support life. Beyond this asphyxiation hazard, ethane poses no known acute or chronic toxicological risk. It is not known or suspected to be a carcinogen.

Atmospheric and Extraterrestrial Ethane

Ethane occurs as a trace gas in the Earth's atmosphere, currently having a concentration at sea level of 0.5 ppbv, though its pre-Industrial concentration is likely to have been lower since a significant proportion of the ethane in today's atmosphere may have originated as fossil fuels. Although ethane is a greenhouse gas, it is much less abundant than methane and also less efficient relative to mass. It has also been detected as a trace component in the atmospheres of all four giant planets, and in the atmosphere of Saturn's moon Titan.

Atmospheric ethane results from the Sun's photochemical action on methane gas, also present in these atmospheres: ultraviolet photons of shorter wavelengths than 160 nm can photo-dissociate the methane molecule into a methyl radical and a hydrogen atom. When two methyl radicals recombine, the result is ethane:

$$CH_4 \rightarrow CH_3\bullet + \bullet H$$

$$CH_3\bullet + \bullet CH_3 \rightarrow C_2H_6$$

In the case of Titan, it was once widely hypothesized that ethane produced in this fashion rained back onto the moon's surface, and over time had accumulated into hydrocarbon seas or oceans covering much of the moon's surface. Infrared telescopic observations cast significant doubt on this hypothesis, and the Huygens probe, which landed on Titan in 2005, failed to observe any surface liquids, although it did photograph features that could be presently dry drainage channels. In December 2007 the Cassini probe found at least one lake at Titan's south pole, now called Ontario Lacus because of the lake's similar area to Lake Ontario on Earth (approximately 20,000 km^2). Further analysis of infrared spectroscopic data presented in July 2008 provided stronger evidence for the presence of liquid ethane in Ontario Lacus.

In 1996, ethane was detected in Comet Hyakutake, and it has since been detected in some other comets. The existence of ethane in these distant solar system bodies may implicate ethane as a primordial component of the solar nebula from which the sun and planets are believed to have formed.

In 2006, Dale Cruikshank of NASA/Ames Research Centre (a New Horizons co-investigator) and his colleagues announced the spectroscopic discovery of ethane on Pluto's surface.

Propane

Propane is a three-carbon alkane with the molecular formula C_3H_8, normally a gas, but compressible to a transportable liquid. A by-product of natural gas processing and petroleum refining, it is commonly used as a fuel for engines, oxy-gas torches, barbecues, portable stoves and residential central heating.

A mixture of propane and butane, used mainly as vehicle fuel, is commonly known as liquefied petroleum gas (LPG or LP gas). It may also contain small amounts of propylene and/or butylene. An odorant such as ethanethiol or thiophene is added so that people can easily smell the gas in case of a leak.

History

Propane was first identified as a volatile component in gasoline by Dr. Walter O. Snelling of the U.S. Bureau of Mines in 1910. The volatility of these lighter hydrocarbons caused them to be known as "wild" because of the high vapor pressures of unrefined gasoline. On March 31 the New York Times reported on Dr. Snelling's work with liquefied gas and that "...a steel bottle will carry enough gas to light an ordinary home for three weeks."

It was during this time that Dr. Snelling, in cooperation with Frank P. Peterson, Chester Kerr and Arthur Kerr, created ways to liquefy the LP gases during the refining of natural gasoline. Together they established American Gasol Co., the first commercial marketer of propane. Dr. Snelling had produced relatively pure propane by 1911, and on March 25, 1913 his method of processing and producing LP gases was issued patent #1,056,845. A separate method of producing LP gas through compression was created by Frank Peterson and patented in 1912.

The 1920s saw increased production of LP gas, with the first year of recorded production totaling 223,000 US gallons (840 m^3) in 1922. In 1927, annual marketed LP gas production reached 1 million US gallons (3,800 m^3), and by 1935, the annual sales of LP gas had reached 56 million US gallons (210,000 m^3). Major industry developments in the 1930s included the introduction of railroad tank car transport, gas odorization and the construction of local bottle-filling plants. The year 1945 marked the first year that annual LP gas sales reached a billion gallons. By 1947, 62% of all U.S. homes had been equipped with either natural gas or propane for cooking.

In 1950, 1,000 propane-fueled buses were ordered by the Chicago Transit Authority, and by 1958, sales in the U.S. had reached 7 billion US gallons (26,000,000 m^3) annually. In 2004 it was reported to be a growing $8-billion to $10-billion industry with over 15 billion US gallons (57,000,000 m^3) of propane being used annually in the U.S.

The "prop-" root found in "propane" and names of other compounds with three-carbon chains was derived from "propionic acid".

Sources

Propane is produced as a by-product of two other processes, natural gas processing and petroleum refining. The processing of natural gas involves removal of butane, propane and large amounts of ethane from the raw gas, in order to prevent condensation of these volatiles in natural gas pipelines. Additionally, oil refineries produce some propane as a by-product of cracking petroleum into gasoline or heating oil. The supply of propane cannot easily be adjusted to meet increased demand, because of the by-product nature of propane production. About 90% of U.S. propane is domestically produced. The United States imports about 10% of the propane consumed each year, with about 70% of that coming from Canada via pipeline and rail. The remaining 30% of imported propane comes to the United States from other sources via ocean transport. After it is produced, North American propane is stored in huge salt caverns located in Fort Saskatchewan, Alberta; Mont Belvieu, Texas and Conway, Kansas. These salt caverns were hollowed out in the 1940s, and they can store 80,000,000 barrels (13,000,000 m^3) or more of propane. When the propane is needed, most of it is shipped by pipelines to other areas of the Midwest, the North and the South, for use by customers. Propane is also shipped by barge and railway to selected U.S. areas.

Properties and Reactions

Propane undergoes combustion reactions in a similar fashion to other alkanes. In the presence of excess oxygen, propane burns to form water and carbon dioxide.

$$C_3H_8 + 5\,O_2 \rightarrow 3\,CO_2 + 4\,H_2O + \text{heat}$$

propane + oxygen → carbon dioxide + water

When not enough oxygen is present for complete combustion, incomplete combustion occurs when propane burns and forms water, carbon monoxide, carbon dioxide, and carbon.

$$2\,C_3H_8 + 7\,O_2 \rightarrow 2\,CO_2 + 2\,CO + 2\,C + 8\,H_2O + \text{heat}$$

Propane + Oxygen → Carbon dioxide + Carbon monoxide + Carbon + Water

Unlike natural gas, propane is heavier than air (1.5 times as dense). In its raw state, propane sinks and pools at the floor. Liquid propane will flash to a vapor at atmospheric pressure and appears white due to moisture condensing from the air.

When properly combusted, propane produces about 50 MJ/kg. The gross heat of combustion of one normal cubic metre of propane is around 91 megajoules

Propane is nontoxic; however, when abused as an inhalant it poses a mild asphyxiation risk through oxygen deprivation. Commercial products contain hydrocarbons beyond propane, which may increase risk. Commonly stored under pressure at room temperature, propane and its mixtures expand and cool when released and may cause mild frostbite.

Propane combustion is much cleaner than gasoline combustion, though not as clean as natural gas combustion. The presence of C–C bonds, plus the multiple bonds of propylene and butylene, create organic exhausts besides carbon dioxide and water vapor during typical combustion. These bonds also cause propane to burn with a visible flame.

Greenhouse gas emissions factors for propane are 62.7 kg CO_2/mBTU or 1.55 kg of CO_2 per litre or 73.7 kg/GJ.

Energy Content

The energy density of propane is 46.44 megajoules per kilogram (91,690 BTU per US gallon, 2220 kJ/mol, 50.34 kJ/g).

Weight

The density of compressed liquid propane at 25 °C is 0.493 g/cm^3, which is equivalent to 4.11 pounds per gallon. Thus, compressed liquid propane weighs approximately 4.2 pounds per US liquid gallon, at 60 °F. Propane expands 1.5% per 10 °F.

Uses

The advantage of propane in cars is its liquid state at a moderate pressure. This allows fast refill times, affordable fuel tank construction, and ranges comparable to (though still less than) gasoline. Meanwhile it is noticeably cleaner (both in handling, and in combustion), results in less engine wear (due to carbon deposits) without diluting engine oil (often extending oil-change intervals), and until recently was a relative bargain in North America. Octane rating of propane is relatively high at 110. In the United States the propane fuelling infrastructure is the most developed of all alternative vehicle fuels. Many converted vehicles have provisions for topping off from "barbecue bottles". Purpose-built vehicles are often in commercially owned fleets, and have private fuelling facilities. A further saving for propane fuel vehicle operators, especially in fleets, is that pilferage is much more difficult than with gasoline or diesel fuels. It must be noted that LPG in cars does contain a mixture of butane, where in domestic cylinders it doesn't.

Propane is generally stored and transported in steel cylinders as a liquid with a vapor space above the liquid. The vapor pressure in the cylinder is a function of temperature. When gaseous propane is drawn at a high rate, the latent heat of vaporisation required to create the gas will cause the bottle to cool. (This is why water often condenses on the sides of the bottle and then freezes). In addition, the lightweight, high-octane compounds vaporize before the heavier, low-octane ones. Thus the ignition properties change as the tank empties. For these reasons, the liquid is often withdrawn using a dip tube. Propane is used as fuel in furnaces for heat, in cooking, as a energy source for water heaters, laundry dryers, barbecues, portable stoves, and motor vehicles. Propane remains a popular choice for barbecues and portable stoves because its low boiling point of "42 °C ("44 °F) makes it vaporize as soon as it is released from its pressurized container. Therefore, no carburetor or other vaporizing device is required; a simple metring nozzle suffices. Propane powers some locomotives, buses, forklifts, taxis and ice resurfacing machines and is used for heat and cooking in recreational vehicles and campers.

Commercially available "propane" fuel, or LPG, is not pure. Typically in the USA and Canada, it is primarily propane (at least 90%), with the rest mostly ethane, propylene, butane, and odorants including ethyl mercaptan. This is the HD-5 standard, (Heavy Duty-5% maximum allowable propylene content) written for internal combustion engines. Not all products labelled "propane" conform to this standard. In Mexico, for example, the butane content is much higher.

Domestic and Industrial Fuel

Propane use is growing rapidly in non-industrialised areas of the world. Propane is replacing wood and other traditional fuel sources in such places, where it is now sometimes called "cooking gas". North American barbecue grills powered by propane cannot be used overseas. The "propane" sold overseas is actually a mixture of propane and butane. The warmer the country, the higher the butane content, commonly 50/50 and sometimes reaching 75% butane. Usage is calibrated to the different-sized nozzles found in non-U.S. grills. Americans who take their grills overseas — such as military personnel — can find U.S.-specification propane at AAFES military post exchanges.

North American industries using propane include glass makers, brick kilns, poultry farms and other industries that need portable heat.

In rural areas of North America, as well as northern Australia and some parts of soutern India propane is used to heat livestock facilities, in grain dryers, and other heat-producing appliances. When used for heating or grain drying it is usually stored in a large, permanently placed cylinder which is recharged by a propane-delivery truck. As of 2000, 6.9 million American households use propane as their primary heating fuel.

In North America, local delivery trucks called "bobtails", with an average tank size of 3,000 US gallons (11,000 L), fill up large tanks (sometimes called *pigs*) that are permanently installed on the property, or other service trucks exchange empty cylinders of propane with filled cylinders. Large tractor-trailer trucks called "cargo-liners", with an average tank size of 10,000 US gallons (38,000 L), transport the propane from the pipeline or refinery to the local delivery plant. The bobtail and transport are not unique to the North American market, though the practice is not as common elsewhere, and the vehicles are generally referred to as *tankers*. In many countries, propane is delivered to consumers via small or medium-sized individual tanks.

Refrigeration

Propane is also instrumental in providing off-the-grid refrigeration, usually by means of a gas absorption refrigerator.

Blends of pure, dry "isopropane" (R-290a) (commercial term used to describe isobutane/propane mixtures) and isobutane (R-600a) have negligible ozone depletion potential and very low Global Warming Potential (having a value of 3.3 times the GWP of carbon dioxide) and can serve as a functional replacement for R-12, R-22, R-134a, and other chlorofluorocarbon or hydrofluorocarbon refrigerants in conventional stationary refrigeration and air conditioning systems.

In Motor Vehicles

Such substitution is widely prohibited or discouraged in motor vehicle air conditioning systems, on the grounds that using flammable hydrocarbons in systems originally designed to carry non-flammable refrigerant presents a significant risk of fire or explosion.

Vendors and advocates of hydrocarbon refrigerants argue against such bans on the grounds that there have been very few such incidents relative to the number of vehicle air conditioning systems filled with hydrocarbons. One particular demonstration was conducted by a professor at the University of New South Wales that intentionally tested the result of all the hydrocarbon refrigerant leaking from the air conditioning system into the passenger compartment of a vehicle and then provided with a source of ignition. The professor had stated that no fire or other dangerous event would occur. However, an explosion did in fact result. All four of the car's doors were bent outward, interior materials were burned and melted, and the professor and several observers in and near the car sustained burns to their face, ears, and hands, as well as lacerations from flying shards of broken window glass.

Autogas

Propane is also being used increasingly for vehicle fuels. In the U.S., 190,000 on-road vehicles use propane, and 450,000 forklifts use it for power. It is the third most popular vehicle fuel in America, behind gasoline and diesel. In other parts of the world, propane used in vehicles is known as autogas. About 13 million vehicles worldwide use autogas.

Propane is also used as fuel for small engines, especially those used indoors or in areas with insufficient fresh air and ventillation to carry away the toxic exhaust of an engine running on gasoline or diesel. More recently, there have been lawn care products like string trimmers, lawn mowers and leaf blowers intended for outdoor use but fueled by propane to reduce air pollution.

Other Uses

- Propane is used as a feedstock for the production of base petrochemicals in steam cracking.
- Some propane becomes a feedstock for propyl alcohol, a common solvent.
- Propane is the primary fuel for hot air balloons.
- It is used in semiconductor manufacture to deposit silicon carbide.
- Liquid propane is commonly used in theme parks and in the movie industry as an inexpensive, high-energy fuel for explosions and other special effects.
- Propane is commonly used as a propellant for Paintball and Airsoft guns, relying on the expansion of the gas to fire the projectile.

Propane Risks and Alternate Gas Fuels

Propane is denser than air. If a leak in a propane fuel system occurs, the gas will have a tendency to sink into any enclosed area and thus poses a risk of explosion and fire. The typical scenario is a leaking cylinder stored in a basement; the propane leak drifts across the floor to the pilot light on the furnace or water heater, and results in an explosion or fire. This property makes the use of propane generally unsuitable as a fuel for boats.

Propane is bought and stored in a liquid form (LPG), and thus fuel energy can be stored in a relatively small space. Compressed Natural Gas (CNG), largely methane, is another gas used as fuel, but it cannot be liquefied by compression at normal temperatures, as these are well above its critical temperature. It therefore requires very high pressure to be stored as a liquid, which poses the hazard that, in an accident, just as with any compressed gas tank (such as a CO_2 tank used for a soda concession) a CNG tank may burst with great force, or leak rapidly enough to become a self-propelled missile. Therefore, CNG is much less efficient to store, due to the large tank volume required. Another form of storing natural gas is as a low temperature liquid in insulated containers as Liquefied Natural Gas (LNG). This form of storage is at low pressure and is around 3.5 times as efficient as storing it as CNG. Unlike propane, if a spill occurs LNG will evaporate and dissipate harmlessly because it is lighter than air. Propane is much more commonly used to fuel vehicles than is natural gas because the equipment required costs less. Propane requires just 1,220 kilopascals (177 psi) of pressure to keep it liquid at 37.8 °C (100 °F).

Retail Cost—United States

As of November 2009, the retail cost of propane was approximately US$2.20 per gallon, or roughly $24 per 1 million BTUs.

Propargyl Alcohol

Propargyl alcohol, or 2-propyn-1-ol, is an organic compound which is a simple alcohol containing an alkyne functional group . Propargyl alcohol is a clear colourless viscous liquid that is miscible with water and most polar organic solvents. It is insoluble in most hydrocarbon solvents. It has an odor like *Pelargonium* geraniums.

Propargyl alcohol may be polymerized by either heat or base. It is used as a corrosion inhibitor, a metal complex solution, a solvent stabiliser and an electroplating brightener additive. It is also used as an intermediate in organic synthesis. Propargyl alcohol is irritating to skin and mucous membranes.

Isopropyl Alcohol

Isopropyl alcohol (also Isopropanol, propan-2-ol, 2-propanol or the abbreviation IPA) is a common name for a chemical compound with the molecular formula C_3H_8O. It is a colourless, flammable chemical compound with a strong odor.

Properties

Isopropyl alcohol has a maximum absorbance at 204 nm in an ultraviolet-visible spectrum. Unlike ethanol or methanol, isopropyl alcohol can be separated from aqueous solutions by adding a salt such as sodium chloride, sodium sulfate, or any of several other inorganic salts, since the alcohol is much less soluble in saline solutions than in salt-free water The process is colloquially called salting out, and causes concentrated isopropyl alcohol to separate into a distinct layer.

Being a secondary alcohol, isopropyl alcohol can be oxidized to acetone, which is the corresponding ketone. This can be achieved using oxidizing agents such as chromic acid, or by dehydrogenation of isopropyl alcohol over a heated copper catalyst:

$$(CH_3)_2CH\text{-}OH \rightarrow (CH_3)_2CO + H_2$$

Isopropyl alcohol is often used as both solvent and hydride source in the Meerwein-Ponndorf-Verley reduction and other transfer hydrogenation reactions; it is oxidized to acetone.

Isopropyl alcohol may be converted to 2-bromopropane using phosphorus tribromide, or dehydrated to propene by heating with sulfuric acid.

Like most alcohols, isopropyl alcohol reacts with active metals such as potassium to form alkoxides which can be called *isopropoxides*. The reaction with aluminium (initiated by a trace of mercury) is used to prepare the catalyst aluminium isopropoxide.

Production

In 1994, 1.5 million tonnes of isopropyl alcohol was produced in the United States, Europe, and Japan. This compound is primarily produced by combining water and propene in a hydration reaction. Of minor significance is the hydrogenation of acetone.

There are two routes for the hydration process: indirect hydration via the sulfuric acid process, and direct hydration. The former process, which can use low-quality propene, predominates in the USA while the latter process, which requires high-purity propene, is more commonly used in Europe. These processes give predominantly isopropyl alcohol rather than 1-propanol because the addition of water or sulfuric acid to propene follows Markovnikov's rule.

Indirect Hydration

The indirect process reacts propene with sulfuric acid to form a mixture of sulfate esters. Subsequent hydrolysis of these esters by steam produces isopropyl alcohol, which is distilled. Diisopropyl ether is a significant by-product of this process; it is recycled back to the process and hydrolyzed to give the desired product.

Direct Hydration

Direct hydration reacts propene and water, either in gas or liquid phases, at high pressures in the presence of solid or supported acidic catalysts. Higher purity propylene (> 90 %) tends to be required for this type of process.

Both processes require that the isopropyl alcohol be separated from water and other by-products by distillation. Isopropyl alcohol and water form an azeotrope and simple distillation gives a material which is 87.9% by weight isopropyl alcohol and 12.1% by weight water. Pure (anhydrous) isopropyl alcohol is made by azeotropic distillation of the wet isopropyl alcohol using either diisopropyl ether or cyclohexane as azeotroping agents.

Hydrogenation of Acetone

Crude acetone is hydrogenated in the liquid phase over Raney nickel or a mixture of copper and chromium oxide to give isopropyl alcohol. This process is useful when coupled with excess acetone production, such as the cumene process.

Uses

In 1990, 45 thousand tonnes of isopropyl alcohol was consumed in the United States. The vast majority of isopropyl alcohol was used as a solvent for coatings or for industrial processes. Isopropyl alcohol in particular is popular for pharmaceutical applications, presumably due to the low toxicity of any residues. Some isopropyl alcohol is used as a chemical intermediate. Isopropyl alcohol may be converted to acetone, but the cumene process is more significant. In that year, a significant fraction (5.4 tonnes) was consumed for household use and in personal care products. It is also used as a gasoline additive.

Solvent

Isopropyl alcohol dissolves a wide range of non-polar compounds. It also evaporates quickly and is relatively non-toxic, compared to alternative solvents. Thus it is used widely as a solvent and as a cleaning fluid, especially for dissolving oils.

Examples of this application include cleaning electronic devices such as contact pins (like those on ROM cartridges), magnetic tape and disk heads (such as those in audio and video tape recorders and floppy disk drives), the lenses of lasers in optical disc drives (e.g. CD, DVD) and removing thermal paste from heatsinks and IC packages (such as CPUs.) Isopropyl alcohol is used in keyboard, LCD and laptop cleaning, is sold commercially as a whiteboard cleaner, and is a strong but safer alternative to common household cleaning products.

It is used to clean LCD and glass computer monitor screens (at some risk to the anti-reflection coating on some screens), and used to give second-hand or worn non-vinyl phonograph records newer-looking sheen. Isopropyl alcohol should not be used to clean vinyl records as it may leach plasticizer from the vinyl making it more rigid). It is effective at removing residual glue from some sticky labels although some other adhesives used on tapes and paper labels are resistant to it. It can also be used to remove stains from most fabrics, wood, cotton, etc. In addition it can also be used to clean paintballs or other oil based products so that they may be reused, commonly known as “repainting”.

Intermediate

Isopropyl alcohol is esterified to give isopropyl acetate, another solvent. It reacts with carbon disulfide to give sodium isopropylxanthate, a weed killer. Isopropyl alcohol reacts with titanium tetrachloride and aluminium metal to give titanium and aluminium isopropoxides respectively, the former a catalyst, and the latter a chemical reagent. This compound may serve as a chemical reagent in itself, by acting as a dihydrogen donor in transfer hydrogenation.

Medical

Disinfecting pads typically contain a 60–70% solution of isopropyl alcohol in water. A 75 % v/v solution in water may be used as a hand sanitizer. Isopropyl alcohol is used as a water-drying aid for the prevention of otitis externa, better known as swimmer’s ear.

Automotive

Isopropyl alcohol is a major ingredient in “gas dryer” fuel additives. In significant quantities, water is a problem in fuel tanks, as it separates from the gasoline, and can freeze in the supply lines at cold temperatures. It does not remove water from gasoline; rather, the alcohol solubilizes water in gasoline.

Once soluble, water does not pose the same risk as insoluble water as it will no longer accumulate in the supply lines and freeze. Isopropyl alcohol is often sold in aerosol cans as a windscreen de-icer. Isopropyl alcohol is also used to remove brake fluid traces from hydraulic braking systems, so that the brake fluid (usually DOT 3, DOT 4 or mineral oil) does not contaminate the brake pads, which would result in poor braking.

Laboratory

As a biological specimen preservative, isopropyl alcohol provides a comparatively non-toxic alternative to formaldehyde and other synthetic preservatives. Isopropyl alcohol solutions of 90-99% are optimal for preserving specimens, although concentrations as low as 70% can be used in emergencies.

Isopropyl alcohol is often used in DNA extraction. It is added to a DNA solution in order to precipitate the DNA into a ‘pellet’ after centrifuging the DNA. This is possible because DNA is insoluble in isopropyl alcohol.

Safety

Isopropyl alcohol vapor is denser than air and is highly flammable with a very wide combustible range. It should be kept away from heat and open flame. When mixed with air or other oxidizers it can explode through deflagration. Isopropyl alcohol has also been reported to form explosive peroxides.

Like many organic solvents, long term application to the skin can cause defatting.

Toxicology

Isopropyl alcohol is oxidized by the liver into acetone by alcohol dehydrogenase. Symptoms of isopropyl alcohol poisoning include flushing, headache, dizziness, CNS depression, nausea, vomiting, anesthesia, and coma. Poisoning can occur from ingestion, inhalation, or absorption; therefore, well-ventilated areas and protective gloves are recommended.

Isopropyl alcohol and its metabolite, acetone, act as central nervous system (CNS) depressants. Around 15 g of isopropyl alcohol can have a toxic effect on a 70 kg human if left untreated.

However, it is not nearly as toxic as methanol or ethylene glycol. Isopropyl alcohol does not cause an anion gap acidosis (in which as lowered blood serum pH causes depletion of bicarbonate anion) unlike ethanol and methanol.

Isopropyl alcohol does however, produce an osmolal gap between the calculated and measured osmolalities of serum, as do the other alcohols. Overdoses may cause a fruity odor on the breath as a result of its metabolism to acetone, which is further metabolized to produce the nutrients acetate and glucose.

Butane

Butane is a gas with the formula C_4H_{10} that is an alkane with four carbon atoms.

The term may refer to any of two structural isomers, or to a mixture of them: in the IUPAC nomenclature, however, butane refers only to the unbranched *n*-butane isomer; the other one being called "methylpropane" or isobutane. Butanes are highly flammable, colourless, easily liquefied gases.

The name butane comes from the roots but- (from butyric acid) and -ane.

Isomers

Common name	*normal butane unbranched butane n-butane*	*isobutanei-butane*
IUPAC name	butane	methylpropane
Moleculardiagram	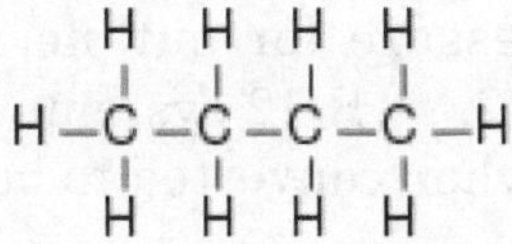	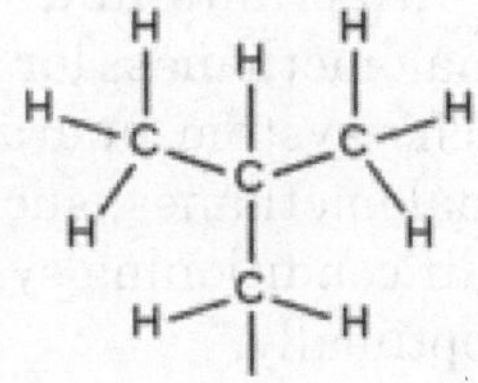
Skeletaldiagram		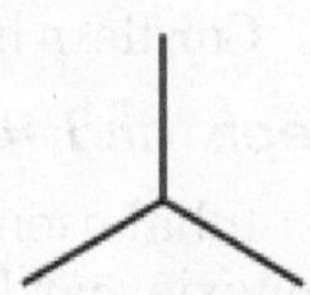

Rotation about the central C-C bond produces two different conformations (*trans* and *gauche*) for *n*-butane.

Reactions

When oxygen is plentiful, butane burns to form carbon dioxide and water vapor; when oxygen is limited, carbon (soot) or carbon monoxide may also be formed.

$$2\ C_4H_{10} + 13\ O_2 \rightarrow 8\ CO_2 + 10\ H_2O$$

The maximum adiabatic flame temperature of butane with air is 2,243 K (1,970 °C; 3,578 °F).

n-Butane is the feedstock for DuPont's catalytic process for the preparation of maleic anhydride:

$$2\ CH_3CH_2CH_2CH_3 + 7\ O_2 \rightarrow 2\ C_2H_2(CO)_2O + 8\ H_2O$$

n-Butane, like all hydrocarbons, undergoes free radical chlorination providing both 1-chloro- and 2-chlorobutanes, as well as more highly chlorinated derivatives. The relative rates of the chlorination is partially explained by the differing bond dissociation energies, 425 and 411 kJ/mol for the two types of C-H bonds. The two central carbon atoms have the slightly weaker C-H bonds.

Uses: The most common use of butane is as lighter fuel for a common lighter or butane torch.

Butane gas is sold bottled as a fuel for cooking and camping. When blended with propane and other hydrocarbons, it is referred to commercially as LPG, for liquified petroleum gas. It is also used as a petrol component, as a feedstock for the production of base petrochemicals in steam cracking, as fuel for cigarette lighters and as a propellant in aerosol sprays such as deodorants.

Very pure forms of butane, especially isobutane, can be used as refrigerants and have largely replaced the ozone layer-depleting halomethanes, for instance in household refrigerators and freezers. The system operating pressure for butane is lower than for the halomethanes, such as R-12, so R-12 systems such as in automotive air conditioning systems, when converted to butane will not function optimally.

Cordless hair irons are usually powered by butane cartridges.

Effects and Health Issues

Inhalation of butane can cause euphoria, drowsiness, narcosis, asphyxia, cardiac arrhythmia, temporary memory loss and frostbite, which can result in death from asphyxiation and ventricular fibrillation. Butane is the most commonly misused volatile substance in the UK, and was the cause of 52% of "solvent related" deaths in 2000. By spraying butane directly into the throat, the jet of fluid can cool rapidly to "20 °C by expansion, causing prolonged laryngospasm. "Sudden sniffer's death" syndrome, first described by Bass in 1970, is the most common single cause of *"solvent related"* death, resulting in 55% of known fatal cases.

The paper "Emission of nitrogen dioxide from butane gas heaters and stoves indoors", from the American Journal of Applied Sciences, indicates that nitrogen dioxide, a toxic gas, results from burning butane gas, and represents a human health hazard from home heaters and stoves.

Isobutane

Isobutane, also known as methylpropane, is an isomer of butane. It is the simplest alkane with a tertiary carbon. Concerns with depletion of the ozone layer by freon gases have led to increased use of isobutane as a gas for refrigeration systems, especially in domestic refrigerators and freezers, and as a propellant in aerosol sprays. When used as a refrigerant or a propellant, isobutane is also known as R-600a. Some portable camp stoves use a mixture of isobutane with propane, usually 80:20. Isobutane is used as a feedstock in the petrochemical industry, for example in the synthesis of isooctane. Its UN number (for hazardous substances see shipping) is UN 1969. Isobutane is the R group for the amino acid leucine.

Nomenclature—Structures of the Two Isomers of Butane

Isobutane is the trivial name retained by the International Union of Pure and Applied Chemistry (IUPAC) in its *1993 Recommendations for the Nomenclature of Organic Chemistry*. Since the longest continuous chain in isobutane is only three carbon atoms, the systematic name is methylpropane. The position number (2-) is unnecessary because it is the only possibility in methylpropane.

Uses

- As a refrigerant.

The use in refrigerators started in 1993 when Greenpeace presented the Greenfreeze project with the German company Foron.

Blends of pure, dry "isobutane" (R-600a) (commercial term used to describe isobutane mixtures) have negligible ozone depletion potential and very low Global Warming Potential (having a value of 3.3 times the GWP of carbon dioxide) and can serve as a functional replacement for R-12, R-22, R-134a, and other chlorofluorocarbon or hydrofluorocarbon refrigerants in conventional stationary refrigeration and air conditioning systems.

- As a propellant for aerosol cans and foam products.

Safety Concerns

Reports surfaced in late 2009 suggesting the use of isobutane as a refrigerant in domestic refrigerators was potentially dangerous. Several explosions resulting from the isobutane leaking into the refrigerator cabinet and a spark from the electrical system have been reported in the United Kingdom. Although unclear how serious this could be, at the time this report came out it was estimated 300 million refrigerators worldwide use isobutane as a refrigerant.

Although these stories are only speculation, the use of a flammable gas as a refrigerant is quite dangerous and encompasses a great deal of risk. The normal risks a CFC or other toxic refrigerant would have when it escapes, are mainly related to depletion of breathable air and frosting at the point of escape. Isobutane has an explosion risk associated also (in addition to depletion of breathable air and frosting). This explosion risk is more dangerous directly to anyone in the vicinity should it accumulate and come into contact with any ignition source. To reduce the risk, the charge of refrigerators was reduced by more than 50% against R134a. The typical charge of a R600a freezer is 45g and of a refrigerator 20g. If 45g are lost at once, 1m^3 of explosive gas can be created. The risk that all gas is lost outside in a short time is very low. For interior refrigeration applications, the R600a appliances should have a hidden condenser, so that the gas can not be lost. 20g is for example approximately 3-4 times the charge of a cigarette lighter. In Europe the limit is 150g for R600a applinces due to safety.

According to an MSDS (Material Safety Data Sheet) for isobutane, this gas should not be exposed to temperatures above 52 Celsius or 151 Fahrenheit while in a closed system such as a storage tank. The range of flammability also factors into safety of this substance as a refrigerant. Flammable ranges for isobutane are between 1.8% - 8.4% and this creates a hazard when a leak forms in the refrigeration system. Many sources of ignition may be nearby so the unit could possibly explode due to the gas leak and cause major damage, not to mention injuries or death to persons nearby. This will create a problem for refrigeration technicians if a leak is reported. The leak must be checked in open air or well ventilated environments and the technician could risk injury to themselves or others. In essence, the refrigeration unit would need full replacement in most cases.

Substitution of this refrigerant for motor vehicle air conditioning systems not originally designed for R600a is widely prohibited or discouraged, on the grounds that using flammable hydrocarbons in systems originally designed to carry non-flammable refrigerant presents a significant risk of fire or explosion.

Vendors and advocates of hydrocarbon refrigerants argue against such bans on the grounds that there have been very few such incidents relative to the number of vehicle air conditioning systems filled with hydrocarbons. One particular test was conducted by a professor at the University of New South Wales that unintentionally tested the scenario of a sudden and complete refrigerant loss into the passenger compartment followed by subsequent ignition. He and several others in the car sustained burns to their face, ears, and hands, and several observers received lacerations from the burst glass of the front passenger window.

Calor

Calor Gas is a brand of bottled butane and propane which is available in Britain and Ireland. It comes in cylinders, which have a special gas regulator.

Calor was formed in 1935. Calor is the UK's leading supplier of Liquefied petroleum gas (LPG). It is currently servicing around 4 million homes and businesses, supplying LPG to power gas appliances from central heating and hot water, to cookers, fires and barbecues. The company predominantly supplies LPG to homes in rural areas where there is no mains gas supply. In the United Kingdom, Calor is part of the SHV Gas Group, a private Dutch company. Calor originally dealt only with cooking and heating appliances, but now covers a whole range of products ranging from barbecues to golf buggies.

Calor provides LPG for three main areas: home, business, and automotive fuels.

History

The company had its origins in the *Modern Gas & Equipment Company Limited* formed in 1935. In 1934, Ritchie Gill, a Cornishman who had been working in the US, returned to England. He spotted an opportunity for a product successfully used in America – *Liquefied Petroleum Gas (LPG).* Gill coined the name Calor Gas and in January 1935 he set up the *Modern Gas & Equipment Co. Ltd*, based in London. In August 1935 the trade was transferred to the Calor (Distributing) Co. Limited.

The business expanded as people in rural areas of the UK realised the benefits of this new fuel - LPG. So by 1939 the company had six offices in the UK: a head office in London and five regional offices. *Calor (Scotland)* had been founded and in Ireland, Calor gas was being distributed by *Messrs. McMullen Ltd.*, which was later purchased by Calor.

In 1947, there was an exceptionally hard winter which led to a major change in the way that Calor was distributed. Up to this point Calor had been relying upon rail transport for the distribution of the LPG. Calor signed a contract with PX (Carriers) Ltd., to deliver cylinders by road.

The first wholly owned filling plant was built at Saxham, near Bury St. Edmunds, Suffolk and a further two filling plants were built in the 1950s so it give Calor an even wider coverage. These two plants were at Millbrook near Southampton (1956) and Port Clarence near Middlesbrough (1959). The Millbrook filling plant was closed in 2001, however the sales and service centre remains selling gas heaters.

The 1950s saw the introduction of the Flavel B500 cooker, which used Calor gas; it became a great success story with caravan owners and smaller households. 1956 saw the birth of Calor's industrial division and installation of the first 'bulk propane' tank at the Meredith & Drew biscuit factory in Newmarket.

In 1963 Calor Ltd. was formed. 1966 saw a new head office being acquired in Slough, which led to over 300 employees moving into this location.

In 1967, Calor was granted a Royal Warrant for the supply of liquefied petroleum gas to Her Majesty the Queen. The company was later granted a second Warrant, this time for supplying Her Majesty Queen Elizabeth The Queen Mother.

In 1969 the Imperial Continental Gas Association acquired all the shares in the Company that it did not already own.

From the early 1970s onwards sales of cabinet heaters, and with them sales of butane in cylinders, escalated enormously, beginning with Super Ser which rapidly replaced the paraffin heater as the nation's number one choice for spot heating. In a decade, the market for cabinet heaters and the gas they burned grew from nothing to £100 million p.a. This led to increasing demand for storage space, which led to Calor investing in the two largest storage facilities. The first was purchased in 1976: a 30,000 tonne refrigerated storage tank at Felixstowe in Suffolk. The second was a joint venture with Conoco to excavate a vast underground cavern on South Humberside, capable of containing some 100,000 tonnes of gas supplied from an adjacent sea terminal. This facility, opened in 1985, gave Calor the ability to overcome short term supply difficulties and to cope with major changes in demand brought on by extremes of weather.

During the 1980s, Calor's domestic central heating and industrial markets continued to expand and in 1982 the company's head office moved to a 40-acre (160,000 m^2) site near Slough.

Despite great efforts to establish a nation-wide network of outlets, Calor Autogas (vehicle propane) never quite reached expectations, due largely to the Government's inability to recognise its considerable environmental advantages by reducing duty. However, the fuel did become very popular for fork lift trucks, helped greatly by the launch, in 1989, of Calor's exhaust purification system, EPS2000, which enabled vehicles to be used both indoors and outside.

In 1986 the Company's parent, Imperial Continental Gas Association plc, rejected a bid from *Gulf Resources & Chemical Corporation*, a company controlled by the Barclay brothers. Instead in 1987 Imperial Continental Gas Association broke itself up into *Calor Group* and *Contibel* (now owned by Tractebel). In 1988 the privately owned Dutch company, SHV became a significant shareholder, acquiring 40% of Calor Group's equity.

In 1991, a joint venture company involving Calor, SHV and Primagaz (in which SHV holds a 50% stake) had been started in both Poland and Slovakia. This was then followed with Hungary in 1992. In 1997 SHV acquired the rest of Calor's shares.

In 2002 Calor opened 12 Customer Operations Centres at Grangemouth, Port Clarence, Stoney Stanton, Elland, Ellesmere Port, Coryton, Cranbrook, Saxham, Fawley, Neath and Newbury.

In December 2005 The Canvey LNG project was announced which commenced with a feasibility study into the development of a strategic LPG import and regasification facility at the existing Calor LNG terminal at Canvey Island, Essex UK. The project is led by Calor Gas and also includes LNG Japan Corporation, (Joint Venture of Sumitomo Corporation and Sojitz Corporation). Following this study Centrica has now been selected as a gas supply partner and would hold equity in the facility, together with capacity rights enabling it to deliver supplies to their British Gas customers from a range of international sources. A full scale planning application and project plan is now being developed.

Products

Calor provide gas both for use in bulk tanks (which may be stored above or below ground) and in smaller gas cylinders.

Bulk Tanks

Calor can provide a variety of sizes in both above- and below-ground formats. Underground tanks are sometimes chosen so as not to detract from areas of natural beauty. Tanks are available with a "Think Tank" monitor, which automatically informs Calor when stocks are low so that the gas can then be refuelled without intervention by the tank's owner.

Gas Cylinders

Smaller 4 x 47 kg gas cylinders are able to supply central heating systems of up to 30 kW as well as cookers and fires. When one pair runs low the valve automatically switches to the second pair so that the supply is continuous.

For leisure use such as caravanning and barbecuing, Calor retail butane, propane and Patio Gas (an alternative brand name for propane) cylinders, all of which come in a number of sizes. A new Calor Lite cylinder, produced from lightweight steel, released in 2008, was targeted at the caravan market.

Autogas

Calor is involved in the supply of the automotive fuel, Autogas (also known as liquefied petroleum gas, or LPG). There are 115,000 vehicles already in the UK that run on LPG, which emits 20% less carbon than petrol, and produces 120 times less harmful particulates than diesel.

Calor Village of the Year

Calor fund and organise the Calor Village of the Year competition in England, Scotland, Wales and Northern Ireland, which over 1,000 villages enter every year. Calor is a member of the Business in the Community 'PerCent Club' investing more than 1% of its pre-tax profit in community projects. The first Calor UK Village of the Year 2008 was announced on 2 December 2008 as the Isle of Barra in the Western Isles of Scotland.

Camping Gaz

Campingaz (pronounced as 'camping gas'), formerly Camping Gaz, is a brand of compressed, mixed butane/propane gas for the small, light weight cans designed for use as a fuel while camping and caravanning. The fuel gas is compressed to a liquid and sold in characteristically blue metal containers. For the larger refillable cylinders the contents are butane. The brand name is also used on appliances manufactured for use with the gas: cookers, lanterns, heaters, grills, refrigerators, etc. as well as more general camping equipment such as sleeping bags.

Company History

The Camping Gaz company was founded in France in 1949. Introducing a small, blue, refillable gas cylinder, directly fitted with special stoves and Welsbach mantle lanterns, the company rapidly expanded to foreign markets. In 1996, *Camping Gaz* became part of Coleman, Inc. (now part of the Jarden Corporation). The brand name was changed to Campingaz in 1998.

Sizes and Availability

The small *Gaz* cartridges are constructed of thin metal and typically used in small cooking rings and gas lamps. Inserting an original *Camping Gaz* cartridge into the device pierces it and allows the fuel gas to flow. The canisters are held in place by a bayonet base, and are discarded when empty. *These must be completely empty* before they are removed from the appliance, otherwise gas will still escape which is a fire hazard.

However, the “Plus” range of self-sealing cartridges, such as the CV 300 Plus, can be disconnected and reconnected even when not empty when they are fitted to compatible devices with the Easy Clic plus connection. The “Plus” range is more expensive than the traditional cartridges but have the advantage of being able to be disconnected.

Larger *Gaz* refillable cylinders come in three sizes. The 901 cylinder contains 0.454 kg of gas with cylinder dimensions of H150mm D110mm, the 904 contains 1.81 kg with cylinder dimensions of H170mm D203mm and the 907 contains 2.72 kg with cylinder dimensions of H250mm D203mm (the largest commonly used size, cylinder type 907, contains 2.72 kg), and are resealable, so they need not be empty when disconnected.

The container is fitted with a M.16x1.5 internal screw thread onto which a valve is connected. This is followed by the appliance (lamp, cooker burners etc.) attached either by a short rigid pipe or a longer flexible hose.

These are typically used for applicances like multi-ring cookers, gas refrigerators, grills or heaters. When empty, the cylinders are reused by returning the empty container to a retail outlet where it is exchanged for a new one for the price of the fuel gas.

Campingaz bottles are sold all over Europe with lower than average availability in Scandinavia and Iceland and certain countries outside Europe (Australia, Canada).

Masses

CV270 mass: 375g. Canister mass: 145g

CV470 mass: 640g. Canister mass: 190g

Travelling with Cartridges or Cylinders

- By air - compressed gas cartridges may not be taken on commercial flights
- By road - Gaz containers can safely be transported by car, but it is advisable to avoid leaving them in direct sunlight. However, the manufacturer states that since they do not explode below 87°C, they would actually be safe in a car parked in sunshine, since cars do not get any hotter than 60°C.
- By tunnel - Eurotunnel permits the transportation of Campingaz through the channel tunnel. Carriage of any form of gas through the Tunnel is subject to a maximum weight limit as detailed on the Eurotunnel website.

Volatile Substance Abuse

Inhalants are a broad range of drugs whose volatile vapours are taken in via the nose and trachea. They are taken by volatilization, and do not include drugs that are inhaled after burning or heating. For example, amyl nitrite and toluene are considered inhalants, but tobacco, marijuana, and crack are not.

While some inhalant drugs are used for medical purposes, as in the case of nitrous oxide (a dental anaesthetic), this article focuses on inhalant abuse as recreational drugs that are used for their intoxicating effect. Inhaling volatile substances because of their intoxicating effect is called huffing.

Inhalant Users

The most serious inhalant abuse occurs among children and teens who "...live on the streets completely without family ties." Inhalant users inhale vapor or aerosol propellant gases using plastic bags held over the mouth or by breathing from a solvent-soaked rag or an open container. The effects of inhalants range from an alcohol-like intoxication and intense euphoria to vivid hallucinations, depending on the substance and the dosage.

Some inhalant users are injured due to the harmful effects of the solvents or gases or due to other chemicals used in the products that they are inhaling. As with any recreational drug, users can be injured due to dangerous behaviour while they are intoxicated, such as driving under the influence. In some cases, users have died from hypoxia (lack of oxygen), pneumonia, cardiac failure or arrest, or aspiration of vomit.

Administration and Effects

Inhalant users inhale vapours or aerosol propellant gases using plastic bags held over the mouth or by breathing from an open container of solvents, such as gasoline or paint thinner. Nitrous oxide gases from whipped cream aerosol cans, aerosol hairspray or non-stick frying spray are sprayed into plastic bags. When inhaling non-stick cooking spray or other aerosol products, some users may filter the aerosolized particles out with a rag. Some gases, such as propane and butane gases, are inhaled directly from the canister.

Once these solvents or gases are inhaled, the extensive capillary surface of the lungs rapidly absorb the solvent or gas, and blood levels peak rapidly. The intoxication effects occur so quickly that the effects of inhalation can resemble the intensity of effects produced by intravenous injection of other psychoactive drugs.

The effects of solvent intoxication can vary widely depending on the dose and what type of solvent or gas is inhaled. A person who has inhaled a small amount of rubber cement or paint thinner vapor may be impaired in a manner resembling alcohol inebriation. A person who has inhaled a larger quantity of solvents or gases, or a stronger chemical, may experience stronger effects such as distortion in perceptions of time and space, hallucinations, and emotional disturbances.

In the short term, many users experience headache, nausea and vomiting, slurred speech, loss of motor coordination, and wheezing. A characteristic "glue sniffer's rash" around the nose and mouth is sometimes seen after prolonged use. An odour of paint or solvents on clothes, skin, and breath is sometimes a sign of inhalant abuse, and paint or solvent residues can sometimes emerge in sweat.

Mechanisms of Action

Inhalants are a large class of drugs and therefore exhibit a variety of mechanisms of action. The mechanisms of action of many non-medical inhalants have not been well elucidated. Anesthetic gases used for surgery, such as nitrous oxide or enflurane, are believed to induce anesthesia primarily by acting as NMDA receptor antagonists, open channel blockers that bind to the inside of the calcium channels on the outer surface of the neuron, and provide high levels of NMDA receptor blockade for a short period of time.

This makes inhaled anesthetic gases different from other NMDA antagonists, such as ketamine, which bind to a regulatory site on the NMDA-sensitive calcium transporter complex and provide slightly lower levels of NMDA blockade, but for a longer and much more predictable duration. This makes a deeper level of anesthesia achievable more easily using anesthetic gases but can also make them more dangerous than other drugs used for this purpose.

Alcohol is known to act as a GABA agonist. The solvent diethyl ether has seen historical episodes of both inhalation and drinking and produces effects suggestive of both NMDA- and GABA-mediated activity.

Dangers and Health Problems

Statistics on deaths caused by inhalant abuse are difficult to determine. It may be severely underreported, because death is often attributed to a discrete event such as a stroke or a heart attack, even if the event happened because of inhalant abuse. Inhalant use or abuse was mentioned on 144 death certificates in Texas during the period

1988–1998 and was reported in 39 deaths in Virginia between 1987 and 1996 from acute voluntary exposure to abused inhalants.

General Risks of Inhalants

Regardless of which inhalant is used, improper administration can lead to death or injury. One major risk is hypoxia, which can occur due to inhaling fumes from a plastic bag, or from using proper equipment but not adding oxygen or room air. When a gas that was stored under high pressure is released, it cools abruptly and can cause frostbite if it is inhaled directly from the container (when nitrous oxide is used as a power adder, its cooling effect is used to make the fuel/air charge denser. Hence more power and torque in part, the rest coming from combustion. In a person, this effect is potentially lethal). Many inhalants are volatile organic chemicals and can catch fire or explode, especially when combined with smoking. As with many other drugs, users may also injure themselves due to loss of coordination or impaired judgment, especially if they attempt to drive.

Solvents have many potential risks in common, including pneumonia, cardiac failure or arrest, and aspiration of vomit. The inhaling of some solvents can cause hearing loss, limb spasms, and damage to the central nervous system and brain. Serious but potentially reversible effects include liver and kidney damage and blood-oxygen depletion. Death from inhalants is generally caused by a very high concentration of fumes. Deliberately inhaling solvents from an attached paper or plastic bag or in a closed area greatly increases the chances of suffocation. Brain damage is typically seen with chronic long-term use as opposed to short-term exposure.

Female inhalant users who are pregnant may have adverse effects on the fetus, and the baby may be smaller when it is born and may need additional health care (similar to those seen with alcohol - Fetal Alcohol Syndrome). There is some evidence of birth defects and disabilities in babies born to women who sniffed solvents such as gasoline.

In the short term, death from solvent abuse occurs most commonly from aspiration of vomit while unconscious or from a combination of respiratory depression and hypoxia, the second cause being especially a risk with heavier-than-air vapours such as butane or gasoline vapor. Deaths typically occur from complications related to excessive sedation and vomiting. Actual overdose from the drug does occur, however, and inhaled solvent abuse is statistically more likely to result in life-threatening respiratory depression than intravenous use of opiates such as heroin. Most deaths from solvent abuse could be prevented if individuals were resuscitated quickly when they stopped breathing and their airway cleared if they vomited. However, most inhalant abuse takes place when people inhale solvents by themselves or in groups of people who are intoxicated. Certain solvents are more hazardous than others, such as gasoline.

In contrast, a few inhalants like amyl nitrate and diethyl ether have medical applications and are less harmful, though they are still dangerous when used recreationally. Nitrous oxide is thought to be particularly non-toxic, though long-term use can lead to a variety of serious health problems linked to destruction of vitamin B12 and folic acid.

Risks Associated with Specific Agents

The hypoxic effect of inhalants can cause damage to many organ systems (particularly the brain, which has a very low tolerance for oxygen deprivation), but there can also be additional toxicity resulting from either the physical properties of the compound itself or additional ingredients present in a product.

Methylene chloride, after being metabolized, can cause carbon monoxide poisoning.

Carbon monoxide poisoning occurs after enough inhalation of carbon monoxide (CO). Carbon monoxide is a toxic gas, but, being colourless, odorless, tasteless, and initially non-irritating, it is very difficult for people to detect. Carbon monoxide is a product of incomplete combustion of organic matter due to insufficient oxygen supply to enable complete oxidation to carbon dioxide (CO_2). It is often produced in domestic or industrial settings by older motor vehicles and other gasoline-powered tools, heaters, and cooking equipment. Exposures at 100 ppm or greater can be dangerous to human health.

Symptoms of mild acute poisoning include lightheadedness, confusion, headaches, vertigo, and flu-like effects; larger exposures can lead to significant toxicity of the central nervous system and heart, and even death. Following acute poisoning, long-term sequelae often occur. Carbon monoxide can also have severe effects on the fetus of a pregnant woman. Chronic exposure to low levels of carbon monoxide can lead to depression, confusion, and memory loss. Carbon monoxide mainly causes adverse effects in humans by combining with hemoglobin to form carboxyhemoglobin (HbCO) in the blood. This prevents oxygen binding to hemoglobin, reducing the oxygen-carrying capacity of the blood, leading to hypoxia. Additionally, myoglobin and mitochondrial cytochrome oxidase are thought to be adversely affected. Carboxyhemoglobin can revert to hemoglobin, but the recovery takes time because the HbCO complex is fairly stable.

Treatment of poisoning largely consists of administering 100% oxygen or providing hyperbaric oxygen therapy, although the optimum treatment remains controversial. Oxygen works as an antidote as it increases the removal of carbon monoxide from hemoglobin, in turn providing the body with normal levels of oxygen. The prevention of poisoning is a significant public health issue. Domestic carbon monoxide poisoning can be prevented by early detection with the use of household carbon monoxide detectors. Carbon monoxide poisoning is the most common type of fatal poisoning in many countries.

Historically, it was also commonly used as a method to commit suicide, usually by deliberately inhaling the exhaust fumes of a running car engine. Modern cars with electronically controlled combustion and catalytic converters produce so little carbon monoxide that this is much less viable. Carbon monoxide poisoning has also been implicated as the cause of apparent haunted houses. Symptoms such as delirium and hallucinations have led people suffering poisoning to think they have seen ghosts or to believe their house is haunted.

Signs and Symptoms

Carbon monoxide is toxic to all aerobic forms of life. It is easily absorbed through the lungs. Carbon monoxide is colourless, odorless, tasteless, and non-irritating, which makes it difficult for humans to detect. Inhaling even relatively small amounts of the gas can lead to hypoxic injury, neurological damage, and even death. Different people and populations may have a different carbon monoxide tolerance level. On average, exposures at 100 ppm or greater is dangerous to human health. In the United States, the OSHA limits long-term workplace exposure levels to less than 50 ppm averaged over an 8-hour period; in addition, employees are to be removed from any confined space if an upper limit ("ceiling") of 100 ppm is reached. Carbon monoxide exposure may lead to a significantly shorter life span due to heart damage. The carbon monoxide tolerance level for any person is altered by several factors, including activity level, rate of ventilation, a pre-existing cerebral or cardiovascular disease, cardiac output, anemia, sickle cell disease and other hematological disorders, barometric pressure, and metabolic rate.

The acute effects produced by carbon monoxide in relation to ambient concentration in parts per million are listed below:

Concentration	*Symptoms*
35 ppm (0.0035%)	Headache and dizziness within six to eight hours of constant exposure
100 ppm (0.01%)	Slight headache in two to three hours
200 ppm (0.02%)	Slight headache within two to three hours; loss of judgment
400 ppm (0.04%)	Frontal headache within one to two hours
800 ppm (0.08%)	Dizziness, nausea, and convulsions within 45 min; insensible within 2 hours
1,600 ppm (0.16%)	Headache, tachycardia, dizziness, and nausea within 20 min; death in less than 2 hours
3,200 ppm (0.32%)	Headache, dizziness and nausea in five to ten minutes. Death within 30 minutes.
6,400 ppm (0.64%)	Headache and dizziness in one to two minutes. Convulsions, respiratory arrest, and death in less than 20 minutes.
12,800 ppm (1.28%)	Unconsciousness after 2-3 breaths. Death in less than three minutes.

Acute Poisoning

The main manifestations of poisoning develop in the organ systems most dependent on oxygen use, the central nervous system and the heart. The initial symptoms of acute carbon monoxide poisoning include headache, nausea, malaise, and fatigue. These symptoms are often mistaken for a virus such as influenza or other illnesses such as food poisoning or gastroenteritis. Headache is the most common symptom of acute carbon monoxide poisoning; it is often described as dull, frontal, and continuous.

Increasing exposure produces cardiac abnormalities including fast heart rate, low blood pressure, and cardiac arrhythmia; central nervous system symptoms include delirium, hallucinations, dizziness, unsteady gait, confusion, seizures, central nervous system depression, unconsciousness, respiratory arrest, and even death. Less common symptoms of acute carbon monoxide poisoning include myocardial ischemia, atrial fibrillation, pneumonia, pulmonary edema, high blood sugar, lactic acidosis, muscle necrosis, acute kidney failure, skin lesions, and visual and auditory problems.

One of the major concerns following acute carbon monoxide poisoning is the severe delayed neurological manifestations that may occur. Problems may include difficulty with higher intellectual functions, short-term memory loss, dementia, amnesia, psychosis, irritability, a strange gait, speech disturbances, Parkinson's disease-like syndromes, cortical blindness, and a depressed mood. Depression may even occur in those who did not have pre-existing depression. These delayed neurological sequelae may occur in up to 50% of poisoned patients after 2 to 40 days. It is difficult to predict who will develop delayed sequelae; however, advancing age, loss of consciousness while poisoned, and initial neurological abnormalities may increase the chance of developing delayed symptoms.

Chronic Poisoning

Chronic exposure to relatively low levels of carbon monoxide may cause persistent headaches, lightheadedness, depression, confusion, memory loss, nausea and vomiting. It is unknown whether low-level chronic exposure may cause permanent neurological damage. Typically, upon removal from exposure to carbon monoxide, symptoms usually resolve themselves, unless there has been an episode of severe acute poisoning.

However, one case noted permanent memory loss and learning problems after a 3-year exposure to relatively low levels of carbon monoxide from a faulty furnace. Chronic exposure may worsen cardiovascular symptoms in some people. Long-term exposures to carbon monoxide present the greatest risk to persons with coronary heart disease and in females who are pregnant.

Table: *Causes*

Concentration	***Source***
0.1 ppm	Natural atmosphere level (MOPITT)
0.5 to 5 ppm	Average level in homes
5 to 15 ppm	Near properly adjusted gas stoves in homes
100 to 200 ppm	Exhaust from automobiles in the Mexico City central area
5,000 ppm	Exhaust from a home wood fire
7,000 ppm	Undiluted warm car exhaust without a catalytic converter

Carbon monoxide is a product of combustion of organic matter under conditions of restricted oxygen supply, which prevents complete oxidation to carbon dioxide (CO_2). Sources of carbon monoxide include cigarette smoke, house fires, faulty furnaces, heaters, wood-burning stoves, internal combustion vehicle exhaust, electrical generators, propane-fueled equipment such as portable stoves, and gasoline-powered tools such as leaf blowers, lawn mowers, high-pressure washers, concrete cutting saws, power trowels, and welders. Exposure typically occurs when equipment is used in buildings or semi-enclosed spaces.

Poisoning may also occur following the use of a self-contained underwater breathing apparatus (SCUBA) due to faulty diving air compressors. Riding in pickup trucks has led to poisoning in children. Idling automobiles with the exhaust pipe blocked by snow has led to the poisoning of car occupants. Generators and propulsion engines on boats, especially houseboats, has resulted in fatal carbon monoxide exposures. Another source of poisoning is exposure to the organic solvent dichloromethane, found in some paint strippers. Dichloromethane is converted into carbon monoxide by the body. In some caves carbon monoxide can build up in enclosed chambers due to decomposing organic matter.

Pathophysiology

The precise mechanisms by which the effects of carbon monoxide are induced upon bodily systems, are complex and not yet fully understood. Known mechanisms include carbon monoxide binding to hemoglobin, myoglobin and mitochondrial cytochrome oxidase, and carbon monoxide causing brain lipid peroxidation.

Hemoglobin

Following absorption, carbon monoxide binds to hemoglobin, which is the principal oxygen-carrying compound in blood; this produces a compound known as carboxyhemoglobin. The traditional belief is that carbon monoxide toxicity arises from the formation of carboxyhemoglobin, which decreases the oxygen-carrying capacity of the blood and inhibits the transport, delivery, and utilisation of oxygen by the body. The affinity between hemoglobin and carbon monoxide is approximately 230 times stronger than the affinity between hemoglobin and oxygen so hemoglobin binds to carbon monoxide in preference to oxygen.

Hemoglobin is a tetramer with four oxygen binding sites. The binding of carbon monoxide at one of these sites increases the oxygen affinity of the remaining three sites, which causes the hemoglobin molecule to retain oxygen that would otherwise be delivered to the tissue. This situation is described as carbon monoxide shifting the oxygen dissociation curve to the left. Because of the increased affinity between hemoglobin and oxygen during carbon monoxide poisoning, the blood oxygen content is increased. But because all the oxygen stays in the hemoglobin, none is delivered to the tissues. This causes hypoxic tissue injury. Hemoglobin acquires a bright red colour when converted into carboxyhemoglobin, so poisoned patients have been described as looking pink-cheeked and healthy. However, this cherry-red appearance is rarely seen in living patients so is not considered a reliable diagnostic sign.

Myoglobin

Carbon monoxide also binds to the hemeprotein myoglobin. It has a high affinity for myoglobin, about 60 times greater than that of oxygen. Carbon monoxide bound to myoglobin may impair its ability to utilise oxygen. This causes reduced cardiac output and hypotension, which may result in brain ischemia.

A delayed return of symptoms have been reported. This results following a recurrence of increased carboxyhemoglobin levels; this effect may be due to a late release of carbon monoxide from myoglobin, which subsequently binds to hemoglobin.

Cytochrome Oxidase

Another mechanism involves effects on the mitochondrial respiratory enzyme chain that is responsible for effective tissue utilisation of oxygen. Carbon monoxide binds to cytochrome oxidase with less affinity than oxygen, so it is possible that it requires significant intracellular hypoxia before binding. This binding interferes with aerobic metabolism and efficient adenosine triphosphate synthesis. Cells respond by switching to anaerobic metabolism, causing anoxia, lactic acidosis, and eventual cell death. The rate of dissociation between carbon monoxide and cytochrome oxidase is slow, causing a relatively prolonged impairment of oxidative metabolism.

Central Nervous System Effects

The mechanism that is thought to have a significant influence on delayed effects involves formed blood cells and chemical mediators, which cause brain lipid peroxidation (degradation of unsaturated fatty acids). Carbon monoxide causes endothelial cell and platelet release of nitric oxide, and the formation of oxygen free radicals including peroxynitrite.

In the brain this causes further mitochondrial dysfunction, capillary leakage, leukocyte sequestration, and apoptosis. The result of these effects is lipid peroxidation, which causes delayed reversible demyelinization of white matter in the central nervous system known as Grinker myelinopathy, which can lead to edema and necrosis within the brain.

This brain damage occurs mainly during the recovery period. This may result in cognitive defects, especially affecting memory and learning, and movement disorders. These disorders are typically related to damage to the cerebral white matter and basal ganglia. Hallmark pathological changes following poisoning are bilateral necrosis of the white matter, globus pallidus, cerebellum, hippocampus and the cerebral cortex.

Pregnancy

Carbon monoxide poisoning in pregnant women may cause severe adverse fetal effects. Poisoning causes fetal tissue hypoxia by decreasing the release of maternal oxygen to the fetus. Carbon monoxide also crosses the placenta and combines with fetal hemoglobin, causing more direct fetal tissue hypoxia. Additionally, fetal hemoglobin has a 10 to 15% higher affinity for carbon monoxide than adult hemoglobin, causing more severe poisoning in the fetus than in the adult. Elimination of carbon monoxide is slower in the fetus, leading to an accumulation of the toxic chemical. The level of fetal morbidity and mortality in acute carbon monoxide poisoning is significant, so despite mild maternal poisoning or following maternal recovery, severe fetal poisoning or death may still occur.

Diagnosis

As many symptoms of carbon monoxide poisoning also occur with many other types of poisonings and infections (such as the flu), the diagnosis is often difficult. A history of potential carbon monoxide exposure, such as being exposed to a residential fire, may suggest poisoning, but the diagnosis is confirmed by measuring the levels of carbon monoxide in the blood. This can be determined by measuring the amount of carboxyhemoglobin compared to the amount of hemoglobin in the blood. Carbon monoxide is produced naturally by the body as a byproduct of converting protoporphyrin into bilirubin. This carbon monoxide also combines with hemoglobin to make carbooxyhemoglobin, but not at toxic levels.

The ratio of carboxyhemoglobin to hemoglobin molecules in an average person may be up to 5%, although cigarette smokers who smoke two packs/day may have levels up to 9%. Serious toxicity occurs with the carboxyhemoglobin to hemoglobin ratio above 25%, and the risk of fatality is high with levels over 70%.

Still, no consistent dose response relationship has been found between carboxyhemoglobin levels and clinical effects. Carboxyhemoglobin levels should be considered as guides to exposure levels rather than guides to effects, as they do not reliably predict clinical course of action or short- and long-term consequences. Patients may continue to experience significant symptoms of CO poisoning long after their blood carboxyhemoglobin concentration has returned to normal. Because of this, for late arriving patients a normal carboxyhemoglobin level does not rule out poisoning.

A CO-oximetre is used to determine carboxyhemoglobin levels. Pulse CO-oximetres estimate carboxyhemoglobin with a non-invasive finger clip similar to a pulse oximetre. These devices function by passing various wavelengths of light through the fingertip and measuring the light absorption of the different types of hemoglobin in the capillaries.

A 2010 study was undertaken to address concerns over the differences in concentration measurements between CO-oximetre results and the results of a standard laboratory blood test. Smaller previous studies had been conducted, however because of specific issues with their study design, the clinical application of CO-oximetry was called into question. In the 2010 study, differences between CO-oximetre results and standard laboratory tests fell outside of the clinically acceptable limits of +/- 5% carboxyhemoglobin in 33.3% of tests. When compared to the laboratory test results, the discrepancies between CO-oximetry tests ranged from -11.6% to +14.4%. Sensitivity in this study was determined to be 48%, while specificity was determined to be 99%. Additionally, because these devices function based on measuring absorbance of light wavelengths, a patient's skin-tone can affect results, which is specifically of concern when these devices are used on black or Hispanic patients. The researchers concluded, based on their results and a lack of peer-reviewed research on the subject, that CO-oximetry cannot replace standard laboratory analysis in clinical care. CO-oximetres were also not recommended for use as an out-of-hospital triage tool.

The use of a regular pulse oximetre is not effective in the diagnosis of carbon monoxide poisoning as patients suffering from carbon monoxide poisoning may have a normal oxygen saturation level on a pulse oximetre. This is due to the carboxyhemoglobin being misrepresented as oxyhemoglobin.

Breath CO monitoring offers a viable alternative to pulse CO-oximetry. Carboxyhemoglobin levels have been shown to have a strong correlation with breath CO concentration . Because of this, monitoring CO in an exhaled sample allows for a non-invasive means to determine carboxyhemoglobin concentrations and to help diagnose patients experiencing symptoms of CO poisoning. Several models of Breath CO monitors are available commercially, which function based on the use of an electrochemical gas sensor to detect CO. These devices are not susceptible to misrepresenting carboxyhemoglobin as oxyhemoglobin. Monitoring CO via exhaled breath has been demonstrated to be useful in emergency medical services when determining the blood CO levels of individuals who are known or suspected to have received exposure.

Differential Diagnosis

There are many conditions to be considered in the differential diagnosis of carbon monoxide poisoning. The earliest symptoms, especially from low level exposures, are often non-specific and readily confused with other illnesses, typically flu-like viral syndromes, depression, chronic fatigue syndrome, chest pain, and migraine or other headaches. Carbon monoxide has been called a "great mimicker" due to the presentation of poisoning being diverse and nonspecific. Other conditions included in the differential diagnosis include acute respiratory distress syndrome, altitude sickness, lactic acidosis, diabetic ketoacidosis, meningitis, methemoglobinemia, or opioid or toxic alcohol poisoning.

Detection in Biological Specimens

Carbon monoxide may be quantitated in blood using spectrophotometric methods or chromatographic techniques in order to confirm a diagnosis of poisoning in hospitalized victims or to assist in the forensic investigation of a case of fatal exposure. Carboxyhemoglobin blood saturations may range up to 8-10% in heavy smokers or persons extensively exposed to automotive exhaust gases. In symptomatic poisoned patients they are often in the 10-30% range, while persons who succumb may have postmortem blood levels of 30-90%.

Carbon Monoxide Detection

Prevention remains a vital public health issue, requiring public education on the safe operation of appliances, heaters, fireplaces, and internal-combustion engines, as well as increased emphasis on the installation of carbon monoxide detectors. In buildings, carbon monoxide detectors are usually installed around heaters and other equipment. If a relatively high level of carbon monoxide is detected, the device sounds an alarm, giving people the chance to evacuate and ventilate the building. Unlike smoke detectors, carbon monoxide detectors do not need to be placed near ceiling level. The United States Consumer Product Safety Commission has stated, "carbon monoxide detectors are as important to home safety as smoke detectors are," and recommends each home have at least one carbon monoxide detector, and preferably one on each level of the building. These devices, which are relatively inexpensive and widely available, are either battery- or AC-powered, with or without battery backup.

Standardization

The use of carbon monoxide detectors has been standardized in many areas. In the USA, NFPA 720-2009, the carbon monoxide detector guidelines published by the National Fire Protection Association, mandates the placement of carbon monoxide detectors/alarms on every level of the residence, including the basement, in addition to outside sleeping areas. In new homes, AC-powered detectors must have battery backup and be interconnected to ensure early warning of occupants at all levels. NFPA 720-2009 is the first national carbon monoxide standard to address devices in non-residential buildings. These guidelines, which now pertain to schools, healthcare centres, nursing homes and other non-residential buildings, includes three main points:

1. A secondary power supply (battery backup) must operate all carbon monoxide notification appliances for at least 12 hours,
2. Detectors must be on the ceiling in the same room as permanently installed fuel-burning appliances, and
3. Detectors must be located on every habitable level and in every HVAC zone of the building.

Treatment

Initial treatment for carbon monoxide poisoning is to immediately remove the person from the exposure without endangering further people. Those who are unconscious may require CPR on site. Administering oxygen via non-rebreather mask shortens the half life of carbon monoxide to 80 minutes from 320 minutes on normal air. Oxygen hastens the dissociation of carbon monoxide from carboxy hemoglobin, thus turning it back into hemoglobin. Due to the possible severe effects in the fetus, pregnant women are treated with oxygen for longer periods of time than non-pregnant patients.

Hyperbaric Oxygen

Hyperbaric oxygen is also used in the treatment of carbon monoxide poisoning, as it may hasten dissociation of CO from carboxy hemoglobin and cytochrome oxidase to a greater extent than normal oxygen. Hyperbaric oxygen at three times atmospheric pressure reduces the half life of carbon monoxide to 23 minutes, compared to 80 minutes for regular oxygen. It may also enhance oxygen transport to the tissues by plasma, partially bypassing the normal transfer through hemoglobin.

However it is controversial whether hyperbaric oxygen actually offers any extra benefits over normal high flow oxygen, in terms of increased survival or improved long-term outcomes. There have been randomized controlled trials in which the two treatment options have been compared; of the six performed, four found hyperbaric oxygen improved outcome and two found no benefit for hyperbaric oxygen. Some of these trials have been criticized for apparent flaws in their implementation. A review of all the literature on carbon monoxide poisoning treatment concluded that the role of hyperbaric oxygen is unclear and the available evidence neither confirms nor denies a medically meaningful benefit. The authors suggested a large, well designed, externally audited, multicentre trial to compare normal oxygen with hyperbaric oxygen.

Other

Further treatment for other complications such as seizure, hypotension, cardiac abnormalities, pulmonary edema, and acidosis may be required. Increased muscle activity and seizures should be treated with dantrolene or diazepam; diazepam should only be given with appropriate respiratory support. Hypotension requires treatment with intravenous fluids; vasopressors may be required to treat myocardial depression. Cardiac dysrhythmias are treated with standard advanced cardiac life support protocols. If severe, metabolic acidosis is treated with sodium bicarbonate. Treatment with sodium bicarbonate is controversial as acidosis may increase tissue oxygen availability. Treatment of acidosis may only need to consist of oxygen therapy. The delayed development of neuropsychiatric impairment is one of the most serious complications of carbon monoxide poisoning. Brain damage is confirmed following MRI or CAT scans. Extensive follow up and supportive treatment is often required for delayed neurological damage. Outcomes are often difficult to predict following poisoning, especially patients who have symptoms of cardiac arrest, coma, metabolic acidosis, or have high carboxy hemoglobin levels. One study reported that approximately 30% of people with severe carbon monoxide poisoning will have a fatal outcome.

Epidemiology

The true number of incidents of carbon monoxide poisoning is unknown, since many non-lethal exposures go undetected. From the available data, carbon monoxide poisoning is the most common cause of injury and death due to poisoning worldwide. Poisoning is typically more common during the winter months. This thought to be due increased domestic use of gas furnaces, gas or kerosene space heaters, and kitchen stoves during the winter months, which if faulty and/or are used without adequate ventilation, may produce excessive carbon monoxide.

It has been estimated that more than 40,000 people per year seek medical attention for carbon monoxide poisoning in the United States. In many industrialised countries carbon monoxide is the cause of more than 50% of fatal poisonings. In the United States, approximately 200 people die each year from carbon monoxide poisoning associated with home fuel-burning heating equipment. Carbon monoxide poisoning contributes to the approximately 5613 smoke inhalation deaths each year in the United States. The CDC reports, "Each year, more than 500 Americans die from unintentional carbon monoxide poisoning, and more than 2,000 commit suicide by intentionally poisoning themselves." For the 10-year period from 1979 to 1988, 56,133 deaths from carbon monoxide poisoning occurred in the United States, with 25,889 of those being suicides, leaving 30,244 unintentional deaths. A report from New Zealand showed that 206 people died from carbon monoxide poisoning in the years of 2001 and 2002. In total carbon monoxide poisoning was responsible for 43.9% of deaths by poisoning in that country. In South Korea, 1,950 people had been poisoned by carbon monoxide with 254 deaths from 2001 through 2003. A report from Jerusalem showed 3.53 per 100,000 people were poisoned annually from 2001 through 2006, in Hubei, China, 218 deaths from poisoning were reported over a 10 year period with 16.5% being from carbon monoxide exposure.

Suicide

Before the 1960s most domestic gas supply in the United Kingdom was coal gas (alternatively known as town gas), which in its unburned form contained high levels of carbon monoxide. Carbon monoxide poisoning by intentionally inhaling coal gas was a common suicide method, accounting for nearly half of all suicides in the United Kingdom in the late 1950s. After the British government phased out coal gas in favour of natural gas in the 1960s, the suicide rate in Britain fell by almost a third and has not risen since.

The use of coal gas as a suicide method has declined as most domestic gas supply worldwide is now natural gas, which lacks carbon monoxide. Until the invention of catalytic converters, suicide has been committed by inhaling the exhaust fumes of a running car engine, particularly in an enclosed space such as a garage. Before 1975, motor car exhaust contained 4-10% carbon monoxide, but newer cars have catalytic converters that eliminate over 99% of the carbon monoxide produced. However even cars with catalytic converters can produce substantial amounts of carbon monoxide if an idling car is left in an enclosed space such as a closed garage.

As carbon monoxide poisoning via car exhaust has become less of a suicide option, there has been an increase in new methods of carbon monoxide poisoning such as burning charcoal or fossil fuels within a confined space. Such incidents have occurred mostly in connection with group suicide pacts in Asian countries such as Japan, Taiwan, and Hong Kong, but are starting to occur in western countries as well, such as the 2007 suicide of Boston lead singer Brad Delp.

Society and Culture—Haunted Houses

Many of the phenomena generally associated with haunted houses, including strange visions and sounds and feelings of dread, can be attributed to carbon monoxide poisoning, as its symptoms include confusion, delirium, emotional disturbances, and hallucinations. In one famous case, carbon monoxide poisoning was clearly identified as the cause of an alleged haunting. Dr. William Wilmer, an ophthalmologist, described the experiences of one of his patients in a 1921 article published in the American Journal of Ophthalmology. "Mr. and Mrs. H." moved into a new home, but soon began to complain of headaches and fatigue. They began to think they were hearing bells and footsteps during the night, accompanied by strange physical sensations and sightings of mysterious figures. When they began to investigate the symptoms, they discovered the previous residents of the house had similar experiences. An examination of their furnace found it to be severely damaged, resulting in incomplete combustion and forcing most of the fumes, including carbon monoxide, into the house rather than up the chimney.

A report published in 2005 described a 23-year old female victim of carbon monoxide poisoning, found delirious and hyperventilating, who thought she saw a ghost while in the shower. A new gas water heater had just been improperly installed in her home, which flooded the house with carbon monoxide when the victim closed all the exterior windows and doors and took a shower.

Gasoline sniffing can cause lead poisoning, though this is less common where leaded gas is banned.

Lead poisoning (also known as plumbism, colica Pictonum, saturnism, Devon colic, or painter's colic) is a medical condition caused by increased levels of the heavy metal lead in the body. Lead interferes with a variety of body processes and is toxic to many organs and tissues including the heart, bones, intestines, kidneys, and reproductive and nervous systems. It interferes with the development of the nervous system and is therefore particularly toxic to children, causing potentially permanent learning and behaviour disorders. Symptoms include abdominal pain, confusion, headache, anemia, irritability, and in severe cases seizures, coma, and death.

Routes of exposure to lead include contaminated air, water, soil, food, and consumer products. Occupational exposure is a common cause of lead poisoning in adults. One of the largest threats to children is lead paint that exists in many homes, especially older ones; thus children in older housing with chipping paint are at greater risk. Prevention of lead exposure can range from individual efforts (e.g. removing lead-containing items such as piping or blinds from the home) to nationwide policies (e.g. laws that ban lead in products or reduce allowable levels in water or soil).

Elevated lead in the body can be detected by the presence of changes in blood cells visible with a microscope and dense lines in the bones of children seen on X-ray. However, the main tool for diagnosis is measurement of the blood lead level; different treatments are used depending on this level. The major treatments are removal of the source of lead and chelation therapy (administration of agents that bind lead so it can be excreted).

Humans have been mining and using this heavy metal for thousands of years, poisoning themselves in the process. Although lead poisoning is one of the oldest known work and environmental hazards, the modern understanding of the small amount of lead necessary to cause harm did not come about until the latter half of the 20th century. No safe threshold for lead exposure has been discovered—that is, there is no known amount of lead that is too small to cause the body harm.

Classification

Classically, "lead poisoning" or "lead intoxication" has been defined as exposure to high levels of lead typically associated with severe health effects. Poisoning is a pattern of symptoms that occur with toxic effects from mid to high levels of exposure; toxicity is a wider spectrum of effects, including subclinical ones (those that do not cause symptoms). However, professionals often use "lead poisoning" and "lead toxicity" interchangeably, and official sources do not always restrict the use of "lead poisoning" to refer only to symptomatic effects of lead.

The amount of lead in the blood and tissues, as well as the time course of exposure, determine toxicity. Lead poisoning may be acute (from intense exposure of short duration) or chronic (from repeat low-level exposure over a prolonged period), but the latter is much more common. Diagnosis and treatment of lead exposure are based on blood lead level (the amount of lead in the blood), measured in micrograms of lead per deciliter of blood (μg/dL). The US Centres for Disease Control and Prevention and the World Health Organisation state that a blood lead level of 10 μg/dL or above is a cause for concern; however, lead may impair development and have harmful health effects even at lower levels, and there is no known safe exposure level. Authorities such as the American Academy of Pediatrics define lead poisoning as blood lead levels higher than 10 μg/dL.

Lead forms a variety of compounds and exists in the environment in various forms. Features of poisoning differ depending on whether the agent is an organic compound (one that contains carbon), or an inorganic one. Organic lead poisoning is now very rare, because countries across the world have phased out the use of organic lead compounds as gasoline additives, but such compounds are still used in industrial settings. Organic lead compounds, which cross the skin and respiratory tract easily, affect the central nervous system predominantly.

Signs and Symptoms

Lead poisoning can cause a variety of symptoms and signs which vary depending on the individual and the duration of lead exposure. Symptoms are nonspecific and may be subtle, and someone with elevated lead levels may have no symptoms. Symptoms usually develop over weeks to months as lead builds up in the body during a chronic exposure, but acute symptoms from brief, intense exposures also occur. Symptoms from exposure to organic lead, which is probably more toxic than inorganic lead due to its lipid solubility, occur rapidly. Poisoning by organic lead compounds has symptoms predominantly in the central nervous system, such as insomnia, delirium, cognitive deficits, tremor, hallucinations, and convulsions.

Symptoms may be different in adults and children; the main symptoms in adults are headache, abdominal pain, memory loss, kidney failure, male reproductive problems, and weakness, pain, or tingling in the extremities. The classic signs and symptoms in children are loss of appetite, abdominal pain, vomiting, weight loss, constipation, anemia, kidney failure, irritability, lethargy, learning disabilities, and behaviour problems. Children may also experience hearing loss, delayed growth, drowsiness, clumsiness, or loss of new abilities, especially speech skills. Symptoms may appear in children at lower blood lead levels than in adults.

Early symptoms of lead poisoning in adults are commonly nonspecific and include depression, loss of appetite, intermittent abdominal pain, nausea, diarrhea, constipation, and muscle pain. Other early signs in adults include malaise, fatigue, decreased libido, and problems with sleep. An unusual taste in the mouth and personality changes are also early signs.

In adults, symptoms can occur at levels above 40 μg/dL, but are more likely to occur only above 50–60 μg/dL. Symptoms begin to appear in children generally at around 60 μg/dL. However, the lead levels at which symptoms appear vary widely depending on unknown characteristics of each individual.

At blood lead levels between 25 and 60 μg/dL, neuropsychiatric effects such as delayed reaction times, irritability, and difficulty concentrating, as well as slowed motor nerve conduction and headache can occur. Anemia may appear at blood lead levels higher than 50 μg/dL. In adults, Abdominal colic, involving paroxysms of pain, may appear at blood lead levels greater than 80 μg/dL.

Signs that occur in adults at blood lead levels exceeding 100 μg/dL include wrist drop and foot drop, and signs of encephalopathy (a condition characterised by brain swelling), such as those that accompany increased pressure within the skull, delirium, coma, seizures, and headache. In children, signs of encephalopathy such as bizarre behaviour, discoordination, and apathy occur at lead levels exceeding 70 μg/dL. For both adults and children, it is rare to be asymptomatic if blood lead levels exceed 100 μg/dL.

Acute Poisoning

In acute poisoning, typical neurological signs are pain, muscle weakness, paraesthesia, and, rarely, symptoms associated with encephalitis. Abdominal pain, nausea, vomiting, diarrhea, and constipation are other acute symptoms. Lead's effects on the mouth include astringency and a metallic taste. Gastrointestinal problems, such as constipation, diarrhea, poor appetite, or weight loss, are common in acute poisoning. Absorption of large amounts of lead over a short time can cause shock (insufficient fluid in the circulatory system) due to loss of water from the gastrointestinal tract.

Hemolysis (the rupture of red blood cells) due to acute poisoning can cause anemia and hemoglobin in the urine. Damage to kidneys can cause changes in urination such as decreased urine output. People who survive acute poisoning often go on to display symptoms of chronic poisoning.

Exposure Routes

Lead is a common environmental pollutant. Causes of environmental contamination include industrial use of lead, such as is found in facilities that process lead-acid batteries or produce lead wire or pipes, and metal recycling and foundries. Children living near facilities that process lead, such as smelters, have been found to have unusually high blood lead levels. In August 2009, parents rioted in China after lead poisoning was found in nearly 2000 children living near zinc and manganese smelters. Lead exposure can occur from contact with lead in air, household dust, soil, water, and commercial products.

Occupational Exposure

In adults, occupational exposure is the main cause of lead poisoning. People can be exposed when working in facilities that produce a variety of lead-containing products; these include radiation shields, ammunition, certain surgical equipment, fetal monitors, plumbing, circuit boards, jet engines, and ceramic glazes.

In addition, lead miners and smelters, plumbers and fitters, auto mechanics, glass manufacturers, construction workers, battery manufacturers and recyclers, firing range instructors, and plastic manufacturers are at risk for lead exposure. Other occupations that present lead exposure risks include welding, manufacture of rubber, printing, zinc and copper smelting, processing of ore, combustion of solid waste, and production of paints and pigments. Parents who are exposed to lead in the workplace can bring lead dust home on clothes or skin and expose their children.

Paint

Some lead compounds are colourful and are used widely in paints, and lead paint is a major route of lead exposure in children. Deteriorating lead paint can produce dangerous lead levels in household dust and soil. Deteriorating lead paint and lead-containing household dust are the main causes of chronic lead poisoning. Many young children display pica, eating things that are not food. Even a small amount of a lead-containing product such as a paint chip or a sip of glaze can contain tens or hundreds of milligrams of lead. Eating chips of lead paint presents a particular hazard to children, generally producing more severe poisoning than occurs from dust. However, removing lead paint from dwellings, e.g. by sanding or torching, can create lead-containing dust and fumes. Therefore, special precautions must be taken when removing lead paint.

Soil

Residual lead in soil contributes to lead exposure in urban areas. Lead content in soil may be caused by broken-down lead paint, residues from lead-containing gasoline or pesticides used in the past, contaminated landfills, or from nearby industries such as foundries or smelters. Although leaded soil is less of a problem in countries that no longer have leaded gasoline, it remains prevalent, raising concerns about the safety of urban agriculture; eating food grown in contaminated soil can present a lead hazard.

Water

Lead from the atmosphere or soil can end up in groundwater and surface water. It is also potentially in drinking water, e.g. from plumbing and fixtures that are either made of lead or have lead solder. Since acidic water breaks down lead in plumbing more readily, chemicals can be added to municipal water to increase the pH and thus reduce the corrosivity of the public water supply. Chloramines, which were adopted as a substitute for chlorine disinfectants due to fewer health concerns, increase corrositivity. In the US, 14–20% of total lead exposure is attributed to drinking water. In 2004, a team of seven reporters from *The Washington Post* discovered high levels of lead in the drinking water in Washington, D.C. and won an award for investigative reporting for a series of articles about this contamination.

In Australia, collecting rainwater from roof runoff used as potable water may contain lead if there are lead contaminates on the roof or in the storage tank. The Australian Drinking Water Guidelines allow a maximum of .01 mg/L lead in water.

Lead-containing Products

Lead can be found in products such as kohl, a South Asian cosmetic, and from some toys. In 2007, millions of toys made in China were recalled from multiple countries owing to safety hazards including lead paint. Vinyl mini-blinds, found especially in older housing, may contain lead. Lead is commonly incorporated into herbal remedies such as Indian Ayurvedic preparations and remedies of Chinese origin. There are also risks of elevated blood lead levels caused by folk remedies like *azarcon* and *greta*, which each contain about 95% lead. Ingestion of metallic lead, such as small lead fishing lures, increases blood lead levels and can be fatal.

Ingestion of lead-contaminated food is also a threat. Ceramic glaze often contains lead, and dishes that have been improperly fired can leach the metal into food, potentially causing severe poisoning. In some places, the solder in cans used for food contains lead. When manufacturing medical instruments and hardware, solder containing lead may be present. People who eat animals hunted with lead bullets may be at risk for lead exposure. Bullets lodged in the body rarely cause significant levels of lead poisoning, but bullets lodged in the joints are the exception, as they deteriorate and release lead into the body over time.

Hunting

Because game animals can be shot using lead bullets, the potential for consumption of game meat to represent an avenue for lead ingestion has seen clinical and epidemiological study. In a recent study conducted by the CDC, a cohort from North Dakota was enrolled and asked to self-report historical consumption of game meat, and participation in other activities that could cause lead exposure. The study found that participants' age, sex, housing age, current hobbies with potential for lead exposure, and game consumption were all associated with blood lead level (PbB).

This study has been cited by popular media as simple evidence that hunting increases exposure to lead poisoning, prompting the University of Illinois Extension to release a statement that there is no such risk. Concerning the CDC report, the authors' conclusion in a related Epi-AID Trip Report notes the small increase associated with game consumption in the study, and urges interpretation with respect to environmental context:

While this study suggests that consumption of wild game meat can adversely affect PbB, no participant had PbB higher than the CDC recommended threshold of 10μg/dl—the level at which CDC recommends case management; and the geometric mean PbB among this study population (1.17μg/dl) was lower than the overall population geometric mean PbB in the United States (1.60 μg/dl). The clinical significance of low PbB in this sample population and the small quantitative increase of 0.30μg/dl in PbB associated with wild game consumption should be interpreted in the context of naturally occurring PbB.

Some hunters may argue that lead-based bullets offer greater accuracy and more humane kills of game animals than might copper-based bullets, which represent the most commonly available, though expensive, alternative to lead. Bullet designs vary greatly, and some lead-based bullets are highly resistant to fragmentation, offering hunters the ability to clean game animals with negligible risk of including lead fragments in prepared meat. Other bullets are prone to fragmentation and exacerbate the risk of lead ingestion from prepared meat. In practice, use of a non-fragmenting bullet, and proper cleaning of the game animal's wound, can eliminate the risk of lead ingestion from eating game; however, isolating such practice to experimentally determine its association with blood lead levels in study is difficult to do.

Pathophysiology

Tetraethyllead, still used as an additive in some fuels, can be absorbed through the skin.

Exposure occurs through inhalation, ingestion or occasionally skin contact. Lead may be taken in through direct contact with mouth, nose, and eyes (mucous membranes), and through breaks in the skin. Tetraethyllead, which was a gasoline additive and is still used in fuels such as aviation fuel, passes through the skin; however inorganic lead found in paint, food, and most lead-containing consumer products is only minimally absorbed through the skin. The main sources of absorption of inorganic lead are from ingestion and inhalation. In adults, about 35–40% of inhaled lead dust is deposited in the lungs, and about 95% of that goes into the bloodstream. Of ingested inorganic lead, about 15% is absorbed, but this percentage is higher in children, pregnant women, and people with deficiencies of calcium, zinc, or iron. Children and infants may absorb about 50% of ingested lead, but little is known about absorption rates in children.

The main body compartments that store lead are the blood, soft tissues, and bone; the half-life of lead in these tissues is measured in weeks for blood, months for soft tissues, and years for bone. Lead in the bones, teeth, hair, and nails is bound tightly and not available to other tissues, and is generally thought not to be harmful. In adults, 94% of absorbed lead is deposited in the bones and teeth, but children only store 70% in this manner, a fact which may partially account for the more serious health effects on children.

The estimated half-life of lead in bone is 20 to 30 years, and bone can introduce lead into the bloodstream long after the initial exposure is gone. The half-life of lead in the blood in men is about 40 days, but it may be longer in children and pregnant women, whose bones are undergoing remodelling, which allows the lead to be continuously re-introduced into the bloodstream. Also, if lead exposure takes place over years, clearance is much slower, partly due to the re-release of lead from bone. Many other tissues store lead, but those with the highest concentrations (other than blood, bone, and teeth) are the brain, spleen, kidneys, liver, and lungs. It is removed from the body very slowly, mainly through urine. Smaller amounts of lead are also eliminated through the feces, and very small amounts in hair, nails, and sweat.

Lead has no known physiologically relevant role in the body, and its harmful effects are myriad. Lead and other heavy metals create reactive radicals which damage cell structures including DNA and cell membranes. Lead also interferes with DNA transcription, enzymes that help in the synthesis of vitamin D, and enzymes that maintain the integrity of the cell membrane. Anemia may result when the cell membranes of red blood cells become more fragile as the result of damage to their membranes.

Lead interferes with metabolism of bones and teeth and alters the permeability of blood vessels and collagen synthesis. Lead may also be harmful to the developing immune system, causing production of excessive inflammatory proteins; this mechanism may mean that lead exposure is a risk factor for asthma in children. Lead exposure has also been associated with a decrease in activity of immune cells such as polymorphonuclear leukocytes. Lead also interferes with the normal metabolism of calcium in cells and causes it to build up within them.

Enzymes

The primary cause of lead's toxicity is its interference with a variety of enzymes because it binds to sulfhydryl groups found on many enzymes. Part of lead's toxicity results from its ability to mimic other metals that take part in biological processes, which act as cofactors in many enzymatic reactions, displacing them at the enzymes on which they act. Lead is able to bind to and interact with many of the same enzymes as these metals but, due to its differing chemistry, does not properly function as a cofactor, thus interfering with the enzyme's ability to catalyse its normal reaction or reactions. Among the essential metals with which lead interacts are calcium, iron, and zinc.

One of the main causes for the pathology of lead is that it interferes with the activity of an essential enzyme called delta-aminolevulinic acid dehydratase, or ALAD, which is important in the biosynthesis of heme, the cofactor found in hemoglobin. Lead also inhibits the enzyme ferrochelatase, another enzyme involved in the formation of heme. Ferrochelatase catalyses the joining of protoporphyrin and Fe to form heme. Lead's interference with heme synthesis results in production of zinc protoporphyrin and the development of anemia. Another effect of lead's interference with heme synthesis is the buildup of heme precursors, such as aminolevulinic acid, which may be directly or indirectly harmful to neurons.

Neurons

Lead interferes with the release of neurotransmitters, chemicals used by neurons to send signals to other cells. It interferes with the release of glutamate, a neurotransmitter important in many functions including learning, by blocking NMDA receptors. The targeting of NMDA receptors is thought to be one of the main causes for lead's toxicity to neurons. A Johns Hopkins report found that in addition to inhibiting the NMDA receptor, lead exposure decreased the amount of the gene for the receptor in part of the brain. In addition, lead has been found in animal studies to cause programmed cell death in brain cells.

Complications

Lead affects every one of the body's organ systems, especially the nervous system, but also the bones and teeth, the kidneys, and the cardiovascular, immune, and reproductive systems. Hearing loss and tooth decay have been linked to lead exposure, as have cataracts. Intrauterine and neonatal lead exposure promote tooth decay. Aside from the developmental effects unique to young children, the health effects experienced by adults are similar to those in children, although the thresholds are generally higher.

Renal System

Kidney damage occurs with exposure to high levels of lead, and evidence suggests that lower levels can damage kidneys as well. The toxic effect of lead causes nephropathy and may cause Fanconi syndrome, in which the proximal tubular function of the kidney is impaired. Long-term exposure at levels lower than those that cause lead nephropathy have also been reported as nephrotoxic in patients from developed countries that had chronic kidney disease or were at risk because of hypertension or diabetes mellitus. Lead poisoning inhibits excretion of the waste product urate and causes a predisposition for gout, in which urate builds up. This condition is known as *saturnine gout*.

Cardiovascular System

Evidence suggests lead exposure is associated with high blood pressure, and studies have also found connections between lead exposure and coronary heart disease, heart rate variability, and death from stroke, but this evidence is more limited. People who have been exposed to higher concentrations of lead may be at a higher risk for cardiac autonomic dysfunction on days when ozone and fine particles are higher.

Reproductive System

Lead affects both the male and female reproductive systems. In men, when blood lead levels exceed 40 μg/dL, sperm count is reduced and changes occur in volume of sperm, their motility, and their morphology. A pregnant woman's elevated blood lead level can lead to miscarriage, prematurity, low birth weight, and problems with development during childhood. Lead is able to pass through the placenta and into breast milk, and blood lead levels in mothers and infants are usually similar. A fetus may be poisoned *in utero* if lead from the mother's bones is subsequently mobilised by the changes in metabolism due to pregnancy; increased calcium intake in pregnancy may help mitigate this phenomenon.

Nervous System

Lead affects the peripheral nervous system (especially motor nerves) and the central nervous system. Peripheral nervous system effects are more prominent in adults and central nervous system effects are more prominent in children. Lead causes the axons of nerve cells to degenerate and lose their myelin coats.

The brain is the organ most sensitive to lead exposure. Lead poisoning interferes with the normal development of a child's brain and nervous system; therefore children are at greater risk of lead neurotoxicity than adults are. In a child's developing brain, lead interferes with synapse formation in the cerebral cortex, neurochemical development (including that of neurotransmitters), and organisation of ion channels. It causes loss of neurons' myelin sheaths, reduces numbers of neurons, interferes with neurotransmission, and decreases neuronal growth.

Lead exposure in young children has been linked to learning disabilities, and children with blood lead concentrations greater than 10 μg/dL are in danger of developmental disabilities. Increased blood lead level in children has been correlated with decreases in intelligence, nonverbal reasoning, short-term memory, attention, reading and arithmetic ability, fine motor skills, emotional regulation, and social engagement. The effect of lead on children's cognitive abilities takes place at very low levels. There is apparently no lower threshold to the dose-response relationship (unlike other heavy metals such as mercury). Reduced academic performance has been associated with lead exposure even at blood lead levels lower than 5 μg/dL. Blood lead levels below 10 μg/dL have been reported to be associated with lower IQ and behaviour problems such as aggression, in proportion with blood lead levels. Between the blood lead levels of 5 and 35 μg/dL, an IQ decrease of 2–4 points for each μg/dL increase is reported in children.

High blood lead levels in adults are also associated with decreases in cognitive performance and with psychiatric symptoms such as depression and anxiety. It was found in a large group of current and former inorganic lead workers in Korea that blood lead levels in the range of 20–50 μg/dL were correlated with neuro-cognitive defects. Increases in blood lead levels from about 50 to about 100 μg/dL in adults have been found to be associated with persistent, and possibly permanent, impairment of central nervous system function.

Lead exposure in children is also correlated with neuropsychiatric disorders such as attention deficit hyperactivity disorder and antisocial behaviour. Elevated lead levels in children are correlated with higher scores on aggression and delinquency measures. A correlation has also been found between prenatal and early childhood lead exposure and violent crime in adulthood. Countries with the highest air lead levels have also been found to have the highest murder rates, after adjusting for confounding factors.

A May 2000 study by economic consultant Rick Nevin theorizes that lead exposure explains 65% to 90% of the variation in violent crime rates in the US. A 2007 paper by the same author claims to show a strong association between preschool blood lead and subsequent crime rate trends over several decades across nine countries. It is believed that the U.S. ban on lead paint in buildings in the late 1970s, as well as the phaseout of leaded gasoline in the 1970s and 1980s, partially helped contribute to the decline of violent crime in the United States since the early 1990s.

Diagnosis

Diagnosis includes determining the clinical signs and the medical history, with inquiry into possible routes of exposure. Clinical toxicologists, medical specialists in the area of poisoning, may be involved in diagnosis and treatment. The main tool in diagnosing and assessing the severity of lead poisoning is laboratory analysis of the blood lead level (BLL). Blood film examination may reveal basophilic stippling of red blood cells (dots in red blood cells visible through a microscope), as well as the changes normally associated with iron-deficiency anemia (microcytosis and hypochromasia). However, basophilic stippling is also seen in unrelated conditions, such as megaloblastic anemia caused by vitamin B12 (colbalamin) and folate deficiencies.

Exposure to lead also can be evaluated by measuring erythrocyte protoporphyrin (EP) in blood samples. EP is a part of red blood cells known to increase when the amount of lead in the blood is high, with a delay of a few weeks. Thus EP levels in conjunction with blood lead levels can suggest the time period of exposure; if blood lead levels are high but EP is still normal, this finding suggests exposure was recent. However, the EP level alone is not sensitive enough to identify elevated blood lead levels below about 35 μg/dL. Due to this higher threshold for detection and the fact that EP levels also increase in iron deficiency, use of this method for detecting lead exposure has decreased.

Blood lead levels are an indicator mainly of recent or current lead exposure, not of total body burden. Lead in bones can be measured noninvasively by X-ray fluorescence; this may be the best measure of cumulative exposure and total body burden. However this method is not widely available and is mainly used for research rather than routine diagnosis. Another radiographic sign of elevated lead levels is the presence of radiodense lines called lead lines at the metaphysis in the long bones of growing children, especially around the knees. These lead lines, caused by increased calcification due to disrupted metabolism in the growing bones, become wider as the duration of lead exposure increases. X-rays may also reveal lead-containing foreign materials such as paint chips in the gastrointestinal tract.

Fecal lead content that is measured over the course of a few days may also be an accurate way to estimate the overall amount of childhood lead intake. This form of measurement may serve as a useful way to see the extent of oral lead exposure from all the diet and environmental sources of lead. Lead poisoning shares symptoms with other conditions and may be easily missed. Conditions that present similarly and must be ruled out in diagnosing lead poisoning include carpal tunnel syndrome, Guillain-Barre syndrome, renal colic, appendicitis, encephalitis in adults, and viral gastroenteritis in children. Other differential diagnoses in children include constipation, abdominal colic, iron deficiency, subdural hematoma, neoplasms of the central nervous system, emotional and behaviour disorders, and mental retardation.

Reference Values

The current reference range for acceptable blood lead concentrations in healthy persons without excessive exposure to environmental sources of lead is less than 10 μg/dL for children and less than 25 μg/dL for adults. The current biological exposure index (a level that should not be exceeded) for lead-exposed workers in the U.S. is 30 μg/dL in a random blood specimen. Blood lead concentrations in poisoning victims have ranged from 30->80 µg/dL in children exposed to lead paint in older houses, 77–104 µg/dL in persons working with pottery glazes, 90–137 µg/dL in individuals consuming contaminated herbal medicines, 109–139 µg/dL in indoor shooting range instructors and as high as 330 µg/dL in those drinking fruit juices from glazed earthenware containers.

Prevention

In most cases, lead poisoning is preventable; the way to prevent it is to prevent exposure to lead. Prevention strategies can be divided into individual (measures taken by a family), preventive medicine (identifying and intervening with high-risk individuals), and public health (reducing risk on a population level).

Recommended steps by individuals to reduce the blood lead levels of children include increasing their frequency of hand washing and their intake of calcium and iron, discouraging them from putting their hands to their mouths, vacuuming frequently, and eliminating the presence of lead-containing objects such as blinds and jewellery in the house. In houses with lead pipes or plumbing solder, these can be replaced. Less permanent but cheaper methods include running water in the morning to flush out the most contaminated water, or adjusting the water's chemistry to prevent corrosion of pipes. Lead testing kits are commercially available for detecting the presence of lead in the household.

Also use only cold water from the tap for drinking, cooking, and for making baby formula. Hot water is more likely than cold water to contain higher amounts of lead. Since most of the lead in household water usually comes from plumbing in the house and not from the local water supply, using cold water can avoid lead exposure. Screening is an important method in preventive medicine strategies. Screening programs exist to test the blood of children at high risk for lead exposure, such as those who live near lead-related industries.

Prevention measures also exist on national and municipal levels. Recommendations by health professionals for lowering childhood exposures include banning the use of lead where it is not essential and strengthening regulations that limit the amount of lead in soil, water, air, household dust, and products.

Regulations exist to limit the amount of lead in paint; for example, a 1978 law in the US restricted the lead in paint for residences, furniture, and toys to 0.06% or less.

In October 2008, the US Environmental Protection Agencyy reduced the allowable lead level by a factor of ten to 0.15 micrograms per cubic metre of air, giving states five years to comply with the standards. The European Union's Restriction of Hazardous Substances Directive limits amounts of lead and other toxic substances in electronics and electrical equipment.

In some places, remediation programs exist to reduce the presence of lead when it is found to be high, for example in drinking water. As a more radical solution, entire towns located near former lead mines have been "closed" by the government, and the population resettled elsewhere, as was the case with Picher, Oklahoma in 2009.

Treatment

***Table:** CDC management guidelines for children with elevated blood levels*

Blood leadlevel (μg/dL)	*Treatment*
10–14	Education, repeat screening
15–19	Repeat screening, case management to abate sources
20–44	Medical evaluation, case management
45–69	Medical evaluation, chelation, case management
>69	Hospitalization, immediate chelation, case management

The mainstays of treatment are removal from the source of lead and, for people who have significantly high blood lead levels or who have symptoms of poisoning, chelation therapy. Treatment of iron, calcium, and zinc deficiencies, which are associated with increased lead absorption, is another part of treatment for lead poisoning. When lead-containing materials are present in the gastrointestinal tract (as evidenced by abdominal X-rays), whole bowel irrigation, cathartics, endoscopy, or even surgical removal may be used to eliminate it from the gut and prevent further exposure. Lead-containing bullets and shrapnel may also present a threat of further exposure and may need to be surgically removed if they are in or near fluid-filled or synovial spaces. If lead encephalopathy is present, anticonvulsants may be given to control seizures, and treatments to control swelling of the brain include corticosteroids and mannitol. Treatment of organic lead poisoning involves removing the lead compound from the skin, preventing further exposure, treating seizures, and possibly chelation therapy for people with high blood lead concentrations.

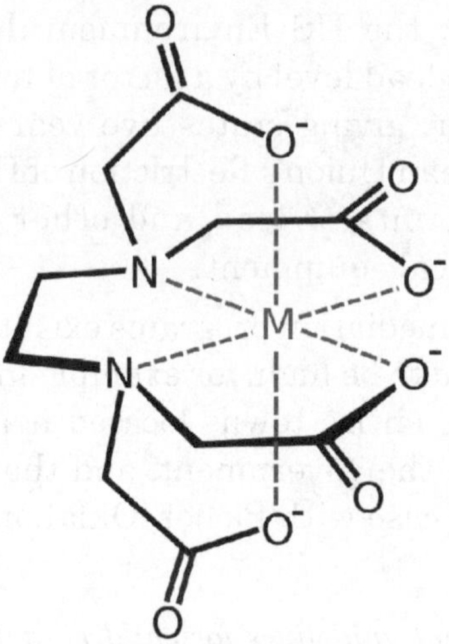

***Figure:** EDTA, a chelating agent, binds a heavy metal, sequestering it.*

A chelating agent is a molecule with at least two negatively charged groups that allow it to form complexes with metal ions with multiple positive charges, such as lead. The chelate that is thus formed is nontoxic and can be excreted in the urine, initially at up to 50 times the normal rate. The chelating agents used for treatment of lead poisoning are edetate disodium calcium ($CaNa_2EDTA$), dimercaprol (BAL), which are injected, and succimer and d-penicillamine, which are administered orally. Chelation therapy is used in cases of acute lead poisoning, severe poisoning, and encephalopathy, and is considered for people with blood lead levels above 25 μg/dL.

While the use of chelation for people with symptoms of lead poisoning is widely supported, use in asymptomatic people with high blood lead levels is more controversial. Chelation therapy is of limited value for cases of chronic exposure to low levels of lead. Chelation therapy is usually stopped when symptoms resolve or when blood lead levels return to premorbid levels. When lead exposure has taken place over a long period, blood lead levels may rise after chelation is stopped because lead is leached into blood from stores in the bone; thus repeated treatments are often necessary.

People receiving dimercaprol need to be assessed for peanut allergies since the commercial formulation contains peanut oil. Calcium EDTA is also effective if administered four hours after the administration of dimercaprol. Administering dimercaprol prior to calcium EDTA is necessary to prevent the redistribution of lead into the central nervous system. An adverse side effect of calcium EDTA is renal toxicity. Succimer is the preferred agent in mild lead poisoning cases. This may be the case in instances where children have a blood lead level >25μg/dL. The most reported adverse side effect for succimer is gastrointestinal disturbances. It is also important to note that chelation therapy only lowers blood lead levels and may not prevent the lead-induced cognitive problems associated with lower lead levels in tissue. This may be because of the inability of these agents to remove sufficient amounts of lead from tissue or inability to reverse preexisting damage. Chelating agents can have adverse effects; for example, chelation therapy can lower the body's levels of necessary nutrients like zinc. Chelating agents taken orally can increase the body's absorption of lead through the intestine.

Chelation challenge, also known as provocation testing, is used to indicate an elevated and mobilisable body burden of heavy metals including lead. This testing involves collecting urine before and after administering a one-off dose of chelating agent to mobilise heavy metals into the urine. Then urine is analysed by a laboratory for levels of heavy metals; from this analysis overall body burden is inferred. Chelation challenge mainly measures the burden of lead in soft tissues, and may not accurately reflect long-term exposure or the amount of lead stored in bone. Although the technique has been used to determine whether chelation therapy is indicated and to diagnose heavy metal exposure, evidence does not support either of these uses as blood levels after chelation are not comparable to the reference range typically used to diagnose heavy metal poisoning. The single chelation dose could also redistribute the heavy metals to more sensitive areas such as central nervous system tissue.

Prognosis

Outcome is related to the extent and duration of lead exposure. Effects of lead on the physiology of the kidneys and blood are generally reversible; its effects on the central nervous system are not. While peripheral effects in adults often go away when lead exposure ceases, evidence suggests that most of lead's effects on a child's central nervous system are irreversible. Children with lead poisoning may thus have adverse health, cognitive, and behavioural effects that follow them into adulthood. Lead encephalopathy is a medical emergency and causes permanent brain damage in 70–80% of children affected by it, even those that receive the best treatment. The mortality rate for people who develop cerebral involvement is about 25%, and of those who survive who had lead encephalopathy symptoms by the time chelation therapy was begun, about 40% have permanent neurological problems such as cerebral palsy.

Exposure to lead may also decrease lifespan and have health effects in the long term. Death rates from a variety of causes have been found to be higher in people with elevated blood lead levels; these include cancer, stroke, and heart disease, and general death rates from all causes. Lead is considered a possible human carcinogen based on evidence from animal studies. Evidence also suggests that age-related mental decline and psychiatric symptoms are correlated with lead exposure. Cumulative exposure over a prolonged period may have a more important effect on some aspects of health than recent exposure. Some health effects, such as high blood pressure, are only significant risks when lead exposure is prolonged (over about one year).

Epidemiology

Since lead has been used widely for centuries, the effects of exposure are worldwide. Environmental lead is ubiquitous, and everyone has some measurable blood lead level. Lead is one of the largest environmental medicine problems in terms of numbers of people exposed and the public health toll it takes. Although regulation reducing lead in products has greatly reduced exposure in the developed world since the 1970s, lead is still allowed in products in many developing countries.

In all countries that have banned leaded gasoline, average blood lead levels have fallen sharply. However, some developing countries still allow leaded gasoline, which is the primary source of lead exposure in most developing countries. Poor children in developing countries are at especially high risk for lead poisoning. Of North American children, 7% have blood lead levels above 10 μg/dL, whereas among Central and South American children, the percentage is 33 to 34%. About one fifth of the world's disease burden from lead poisoning occurs in the Western Pacific, and another fifth is in Southeast Asia.

In developed countries, nonwhite people with low levels of education living in poorer areas are most at risk for elevated lead. In the US, the groups most at risk for lead exposure are the impoverished, city-dwellers, and immigrants. African-American children and those living in old housing have also been found to be at elevated risk for high blood lead levels in the US. Low-income people often live in old housing with lead paint, which may begin to peel, exposing residents to high levels of lead-containing dust.

Risk factors for elevated lead exposure include alcohol consumption and smoking (possibly because of contamination of tobacco leaves with lead-containing pesticides). Adults with certain risk factors might be more susceptible to toxicity; these include calcium and iron deficiencies, old age, disease of organs targeted by lead (e.g. the brain, the kidneys), and possibly genetic susceptibility. Differences in vulnerability to lead-induced neurological damage between males and females have also been found, but some studies have found males to be at greater risk, while others have found females to be.

In adults, blood lead levels steadily increase with increasing age. In adults of all ages, men have higher blood lead levels than women do. Children are more sensitive to elevated blood lead levels than adults are. Children may also have a higher intake of lead than adults; they breathe faster and may be more likely to have contact with and ingest soil. Children ages one to three tend to have the highest blood lead levels, possibly because at that age they begin to walk and explore their environment, and they use their mouths in their exploration. Blood levels usually peak at about 18–24 months old. In many countries including the US, household paint and dust are the major route of exposure in children.

History

Lead poisoning was among the first known and most widely studied work and environmental hazards. One of the first metals to be smelted and used, lead is thought to have been discovered and first mined in Anatolia around 6500 BC. Its density, workability, and corrosion-resistance were among the metal's attractions.

In the 2nd century BC the Greek botanist Nicander described the colic and paralysis seen in lead-poisoned people. Dioscorides, a Greek physician who lived in the 1st century CE, wrote that lead makes the mind "give way".

Lead was used extensively in Roman aqueducts from about 500 BC to 300 AD Julius Caesar's engineer, Vitruvius, reported, "water is much more wholesome from earthenware pipes than from lead pipes. For it seems to be made injurious by lead, because white lead is produced by it, and this is said to be harmful to the human body." Gout, prevalent in affluent Rome, is thought to be the result of lead, or leaded eating and drinking vessels. Sugar of lead (Lead II Acetate) was used to sweeten wine, and the gout that resulted from this was known as "saturnine" gout. It is even hypothesized that lead poisoning may have contributed to the decline of the Roman Empire, a hypothesis thoroughly disputed:

The great disadvantage of lead has always been that it is poisonous. This was fully recognised by the ancients, and Vitruvius specifically warns against its use. Because it was nevertheless used in profusion for carrying drinking water, the conclusion has often been drawn that the Romans must therefore have suffered from lead poisoning; sometimes conclusions are carried even further and it is inferred that this caused infertility and other unwelcome conditions, and that lead plumbing was largely responsible for the decline and fall of Rome. In fact, two things make this otherwise attractive hypothesis impossible. First, the calcium carbonate deposit that formed so thickly inside the aqueduct channels also formed inside the pipes, effectively insulating the water from the lead, so that the two never touched. Second, because the Romans had so few taps and the water was constantly running, it was never actually inside the pipes for more than a few minutes, and certainly not long enough to become contaminated. The thesis that the Romans contracted lead poisoning from the lead pipes in their water systems must therefore be declared completely unfounded.

It has been noted that Romans also consumed quantities of lead through the consumption of defrutum, carenum, and sapa, musts made by boiling down fruit in lead cookware. Defrutum and its relatives were used in Ancient Roman cuisine and cosmetics, including as a food preservative. However, the use of leaden cookware, though popular, was not the general standard and copper cookware was used far more generally. Additionally there is also no indication how often sapa was added or in what quantity.

The consumption of sapa as having a role in the fall of the Roman Empire was used in a theory proposed by geochemist Jerome Nriagu to state that "lead poisoning contributed to the decline of the Roman Empire". John Scarborough, a pharmacologist and classicist, criticized the conclusions drawn by Nriagu's book as "so full of false evidence, miscitations, typographical errors, and a blatant flippancy regarding primary sources that the reader cannot trust the basic arguments."

After antiquity, mention of lead poisoning was absent from medical literature until the end of the Middle Ages. In 1656 the German physician Samuel Stockhausen recognised dust and fumes containing lead compounds as the cause of disease, called since ancient Roman times *morbi metallici*, that were known to afflict miners, smelter workers, potters, and others whose work exposed them to the metal.

The painter Caravaggio might have died of lead poisoning. Bones with high lead levels were recently found in a grave likely to be Caravaggio's grave. Paints used at the time contained high amounts of lead salts. Caravaggio is known to have indulged in violent behaviour, as caused by lead poisoning.

In 17th-century Germany, the physician Eberhard Gockel discovered lead-contaminated wine to be the cause of an epidemic of colic. He had noticed that monks who did not drink wine were healthy, while wine drinkers developed colic, and traced the cause to sugar of lead, made by simmering litharge with vinegar. As a result, Eberhard Ludwig, Duke of Wurttemberg issued an edict in 1696 banning the adulteration of wines with litharge.

In 18th century Boston, lead poisoning was fairly frequent on account of the widespread drinking of rum, which was made in stills with a lead component (the "worm"). Also in the 18th century, "Devonshire colic" was the name given to the symptoms suffered by people of Devon who drank cider made in presses that were lined with lead. Lead was added to cheap wine illegally in the 18th and early 19th centuries as a sweetener. The composer Beethoven, a heavy wine drinker, suffered elevated lead levels (as later detected in his hair) possibly due to this; the cause of his death is controversial, but lead poisoning is a contender as a factor.

With the Industrial Revolution in the 19th century, lead poisoning became common in the work setting. The introduction of lead paint for residential use in the 19th century increased childhood exposure to lead; for millennia before this, most lead exposure had been occupational. An important step in the understanding of childhood lead poisoning occurred when toxicity in children from lead paint was recognised in Australia in 1897. France, Belgium, and Austria banned white lead interior paints in 1909; the League of Nations followed suit in 1922. However, in the United States, laws banning lead house paint were not passed until 1971, and it was phased out and not fully banned until 1978.

The 20th century saw an increase in worldwide lead exposure levels due to the increased widespread use of the metal. Beginning in the 1920s, lead was added to gasoline to improve its combustion; lead from this exhaust persists today in soil and dust in buildings. Blood lead levels worldwide have been declining sharply since the 1980s, when leaded gasoline began to be phased out. In those countries that have banned lead in solder for food and drink cans and have banned leaded gasoline additives, blood lead levels have fallen sharply since the mid-1980s.

The levels found today in most people are orders of magnitude greater than those of pre-industrial society. Due to reductions of lead in products and the workplace, acute lead poisoning is rare in most countries today; however, low level lead exposure is still common. It was not until the second half of the 20th century that subclinical lead exposure became understood to be a problem. During the end of the 20th century, the blood lead levels deemed acceptable steadily declined. Blood lead levels once considered safe are now considered hazardous, with no known safe threshold.

Notable Cases

15,000 people are being relocated from Jiyuan in central Henan province to other locations after 1000 children living around China's largest smelter plant (owned and operated by Yuguang Gold and Lead) were found to have excess lead in their blood. The total cost of this project is estimated to around 1 billion yuan ($150 million). 70% of the cost will be paid by local government and the smelter company, while the rest will be paid by the residents themselves. The government has suspended production at 32 of 35 lead plants. The affected area includes people from 10 different villages.

The Zamfara State lead poisoning epidemic occurred in Nigeria in 2010. As of October 5, 2010 at least 400 children have died from the effects of lead poisoning.

In Other Species

Humans are not alone in suffering from lead's effects; plants and animals are also affected by lead toxicity to varying degrees depending on species. Animals experience many of the same effects of lead exposure as humans do, such as abdominal pain, peripheral neuropathy, and behavioural changes such as increased aggression. Much of what is known about human lead toxicity and its effects is derived from animal studies. Animals are used to test the effects of treatments, such as chelating agents, and to provide information the pathophysiology lead, such as how it is absorbed and distributed in the body.

Farm animals such as cows and horses as well as pet animals are also susceptible to the effects of lead toxicity. Sources of lead exposure in pets can be the same as those that present health threats to humans sharing the environment, such as paint and blinds, and there is sometimes lead in toys made for pets. Lead poisoning in a pet dog may indicate that children in the same household are at increased risk for elevated lead levels.

Wildlife

Lead, one of the leading causes of toxicity in waterfowl, has been known to cause die-offs of wild bird populations. When hunters use lead shot, waterfowl such as ducks can ingest the spent pellets later and be poisoned; predators that eat these birds are also at risk. Lead shot-related waterfowl poisonings were first documented in the US in the 1880s. By 1919, the spent lead pellets from waterfowl hunting was positively identified as the source of waterfowl deaths. Lead shot has been banned for hunting waterfowl in several countries, including the US in 1991 and 1997 in Canada. Other threats to wildlife include lead paint, sediment from lead mines and smelters, and lead weights from fishing lines. Lead in some fishing gear has been banned in several countries.

The critically endangered California Condor has also been affected by lead poisoning. As scavengers, condors eat carcasses of game that have been shot but not retrieved, and with them the fragments from lead bullets; this increases their lead levels. Among condors around the Grand Canyon, lead poisoning due to eating lead shot is the most frequently diagnosed cause of death.

In an effort to protect this species, in areas designated as the California Condor's range the use of projectiles containing lead has been banned to hunt deer, wild pig, elk, pronghorn antelope, coyotes, ground squirrels, and other non-game wildlife. Also, conservation programs exist which routinely capture condors, check their blood lead levels, and treat cases of poisoning.

Methemoglobinemia

Methemoglobinemia (or methaemoglobinaemia) is a disorder characterised by the presence of a higher than normal level of methemoglobin (metHb) in the blood. Methemoglobin is an oxidized form of hemoglobin that has an increased affinity for oxygen, resulting in a reduced ability to release oxygen to tissues. The oxygen–hemoglobin dissociation curve is shifted to the left. When methemoglobin concentration is elevated in red blood cells, tissue hypoxia can occur.

Overview

Normally, methemoglobin levels are <1%, as measured by the co-oximetry test. Elevated levels of methemoglobin in the blood are caused when the mechanisms that defend against oxidative stress within the red blood cell are overwhelmed and the oxygen carrying ferrous ion (Fe2+) of the heme group of the hemoglobin molecule is oxidized to the ferric state (Fe3+). This converts hemoglobin to methemoglobin, resulting in a reduced ability to release oxygen to tissues and thereby hypoxia.

This can give the blood a bluish or chocolate-brown colour. Spontaneous formation of methemoglobin is normally reduced (via electron donation) by protective enzyme systems, e.g. NADH methemoglobin reductase (cytochrome-b5 reductase) (major pathway), NADPH methemoglobin reductase (minor pathway) and to a lesser extent the ascorbic acid and glutathione enzyme systems. Disruptions with these enzyme systems lead to the condition.

Types—Congenital Methemoglobinemia

Due to a deficiency of the enzyme diaphorase I (NADH methemoglobin reductase), methemoglobin levels rise and the blood of met-Hb patients has reduced oxygen-carrying capacity. Instead of being red in colour, the arterial blood of met-Hb patients is brown. This results in the skin of Caucasian patients gaining a bluish hue. Hereditary met-Hb is caused by a recessive gene. If only one parent has this gene, offspring will have normal-hued skin, but if both parents carry the gene there is a chance the offspring will have blue-hued skin.

Another cause of congenital methemoglobinemia is seen in patients with abnormal hemoglobin variants such as hemoglobin M (HbM), or hemoglobin H (HbH), which are not amenable to reduction despite intact enzyme systems.

Methemoglobinemia can also arise in patients with pyruvate kinase deficiency due to impaired production of NADH – the essential cofactor for diaphorase I. Similarly, patients with Glucose-6-phosphate dehydrogenase (G6PD) deficiency may have impaired production of another co-factor, NADPH.

Acquired Methemoglobinemia

Methemoglobinemia can also be acquired. The protective enzyme systems normally present in red blood cells maintain methemoglobin levels at less than one percent of the total hemoglobin in healthy people. Exposure to exogenous oxidizing drugs and their metabolites (such as benzocaine, dapsone and nitrates) may accelerate the rate of formation of methemoglobin up to one-thousandfold, overwhelming the protective enzyme systems and acutely increasing methemoglobin levels.

Other classical drug causes of methemoglobinemia include antibiotics (trimethoprim, sulfonamides and dapsone), local anesthetics (especially articaine and prilocaine), and others such as aniline dyes, metoclopramide, chlorates and bromates. Ingestion of compounds containing nitrates (such as the patina chemical bismuth nitrate) can also cause methemoglobinemia.

Infants under 6 months of age are particularly susceptible to methemoglobinemia caused by nitrates ingested in drinking water (called blue-baby syndrome), dehydration usually caused by gastroenteritis with diarrhea, sepsis, and topical anesthetics containing benzocaine or prilocaine. Nitrates used in agricultural fertilizers may leak into the ground and may contaminate well water. The current EPA standard of 10 ppm nitrate-nitrogen for drinking water is specifically designed to protect infants.

Benzocaine applied to the gums or throat (as commonly used in baby teething gels) can cause methemoglobinemia.

Treatment

Methemoglobinemia can be treated with supplemental oxygen and methylene blue 1% solution (10 mg/ml) 1 to 2 mg/kg administered intravenously slowly over five minutes followed by IV flush with normal saline. Methylene blue restores the iron in hemoglobin to its normal (reduced) oxygen-carrying state. This is achieved by providing an artificial electron acceptor (such as methylene blue, or flavin) for NADPH methemoglobin reductase (RBCs usually don't have one; the presence of methylene blue allows the enzyme to function at 5x normal levels) The NADPH is generated via the hexose monophosphate shunt.

Diaphorase II normally contributes only a small percentage of the red blood cells reducing capacity but is pharmacologically activated by exogenous cofactors, such as methylene blue, to 5 times its normal level of activity. Genetically induced chronic low-level methemoglobinemia may be treated with oral methylene blue daily. Also, vitamin C can occasionally reduce cyanosis associated with chronic methemoglobinemia but has no role in treatment of acute acquired methemoglobinemia.

Symptoms

Signs and symptoms of methemoglobinemia (methemoglobin >1%) include shortness of breath, cyanosis, mental status changes (~50%), headache, fatigue, exercise intolerance, dizziness and loss of consciousness. Arterial blood with elevated methemoglobin levels has a characteristic chocolate-brown colour as compared to normal bright red oxygen containing arterial blood. Severe methemoglobinemia (methemoglobin >50%) patients have dysrhythmias, seizures, coma and death (>70%). Healthy people may not have many symptoms with methemoglobin levels < 15%, however patients with co-morbidities such as anemia, cardiovascular disease, lung disease, sepsis, or presence of other abnormal hemoglobin species (e.g. carboxy hemoglobin, sulfehemoglobin or sickle hemoglobin) may experience moderate to severe symptoms at much lower levels (as low as 5-8%).

Chapter 6

Cycloalkane

Cycloalkanes (also called naphthenes - not to be confused with naphthalene) are types of alkanes which have one or more rings of carbon atoms in the chemical structure of their molecules. Alkanes are types of organic hydrocarbon compounds which have only single chemical bonds in their chemical structure. Cycloalkanes consist of only carbon (C) and hydrogen (H) atoms and are saturated because there are no multiple C-C bonds to hydrogenate (add more hydrogen to). A general chemical formula for cycloalkanes would be $C_nH_{2(n+1-g)}$ where n = number of C atoms and g = number of rings in the molecule. Cycloalkanes with a single ring are named analogously to their normal alkane counterpart of the same carbon count: cyclopropane, cyclobutane, cyclopentane, cyclohexane, etc. The larger cycloalkanes, with greater than 20 carbon atoms are typically called cycloparaffins.

Cycloalkanes are classified into small, common, medium, and large cycloalkanes, where cyclopropane and cyclobutane are the small ones, cyclopentane, cyclohexane, cycloheptane are the common ones, cyclooctane through cyclotridecane are the medium ones, and the rest are the larger ones.

Nomenclature

The naming of polycyclic alkanes such as bicyclic alkanes and spiro alkanes is more complex, with the base name indicating the number of carbons in the ring system, a prefix indicating the number of rings (e.g., "bicyclo"), and a numeric prefix before that indicating the number of carbons in each part of each ring, exclusive of vertices. For instance, a bicyclooctane that consists of a six-member ring and a four-member ring, which share two adjacent carbon atoms that form a shared edge, is [4.2.0]-bicyclooctane. That part of the six-member ring, exclusive of the shared edge has 4 carbons. That part of the four-member ring, exclusive of the shared edge, has 2 carbons. The edge itself, exclusive of the two vertices that define it, has 0 carbons.

There is more than one convention (method or nomenclature) for the naming of compounds, which can be confusing for those who are just learning, and inconvenient for those who are well rehearsed in the older ways. For beginners it is best to learn IUPAC nomenclature from a source that is up to date, because this system is constantly being revised. In the above example [4.2.0]-bicyclooctane would be written bicyclo[4.2.0]octane to fit the conventions for IUPAC naming. It has then got room for an additional numerical prefix if there is the need to include details of other attachments to the molecule such as chlorine or a methyl group. Another convention for the naming of compounds is the common name, which is a shorter name and it gives less information about the compound. An example of a common name is terpineol, the name of which can only really tell us that it is an alcohol (because the suffix 'ol' is in the name) and it should then have a hydroxide (OH) group attached to it.

An example of the IUPAC method is given in the image to the right. In this example the base name is listed first, which indicates the total number of carbons in both rings including the carbons making up the shared edge (e.g., heptane which means 'hept' or 7 carbons and 'ane' which indicates only single bonding between carbons). Then in front of the base name is the numerical prefix, which lists the number of carbons in each ring, excluding the carbons that are shared by each ring, plus the number of carbons on the bridge between the rings. In this case there are two rings with two carbons each and a single bridge with one carbon, excluding the carbons shared by it and the other two rings.

Properties

Cycloalkanes are similar to alkanes in their general physical properties, but they have higher boiling points, melting points, and densities than alkanes. This is due to stronger London forces because the ring shape allows for a larger area of contact. Containing only C-C and C-H bonds, unreactivity of cycloalkanes with little or no ring strain are comparable to non-cyclic alkanes.

Ring Strain

The carbon atoms in cycloalkanes are sp^3 hybridized and are therefore a deviation from the ideal tetrahedral bond angles of 109°28'.

This causes an increase in potential energy and an overall destabilising effect. Eclipsing of hydrogen atoms is an important destabilising effect, as well. The strain energy of a cycloalkane is the theoretical increase in energy caused by the compound's geometry, and is calculated by comparing the experimental standard enthalpy change of combustion of the cycloalkane with the value calculated using average bond energies.

Ring strain is highest for cyclopropane, in which the carbon atoms form a triangle and therefore have 60 degree C-C-C bond angles. There are also three pairs of eclipsed hydrogens. The ring strain is calculated to be around 120 kJ/mol.

Cyclobutane has the carbon atoms in a puckered square with approximately 90-degree bond angles; "puckering" reduces the eclipsing interactions between hydrogen atoms. Its ring strain is therefore slightly less, at around 110 kJ/mol. For a theoretical planar cyclopentane the C-C-C bond angles would be 108 degrees, very close to the measure of the tetrahedral angle. Actual cyclopentane molecules are puckered, but this changes only the bond angles slightly so that angle strain is relatively small. The eclipsing interactions are also reduced, leaving a ring strain of about 25 kJ/mol.

In cyclohexane the ring strain and eclipsing interactions are negligible because the puckering of the ring allows ideal tetrahedral bond angles to be achieved. As well, in the most stable *chair form* of cyclohexane, axial hydrogens on adjacent carbon atoms are pointed in opposite directions, virtually eliminating eclipsing strain.

After cyclohexane, the molecules are unable to take a structure with no ring strain, resulting in an increase in strain energy, which peaks at 9 carbons (around 50 kJ/mol). After that, strain energy slowly decreases until 12 carbon atoms, where it drops significantly; at 14, another significant drop occurs and the strain is on a level comparable with 10 kJ/mol. After 14 carbon atoms, sources disagree on what happens to ring strain, some indicating that it increases steadily, others saying that it disappears entirely. Generally speaking though, bond angle strain and eclipsing strain are only an issue for smaller rings.

Reactions

The simple and the bigger cycloalkanes are very stable, like alkanes, and their reactions, for example, radical chain reactions, are like alkanes.

The small cycloalkanes - particularly cyclopropane - have a lower stability due to Baeyer strain and ring strain. They react similarly to alkenes, though they do not react in electrophilic addition, but in nucleophilic aliphatic substitution. These reactions are ring-opening reactions or ring-cleavage reactions of alkyl cycloalkanes. Cycloalkanes can be formed in a Diels-Alder reaction followed by a catalytic hydrogenation.

Biohydrocarbon

Biomass to Liquid (BtL) or BMtL is a multi-step process which produces liquid biofuels from biomass:

The process uses the whole plant to improve the carbon dioxide balance and increase yield.

- The Fischer–Tropsch process is used to produce synfuels from gasified biomass. Carbonaceous material is gasified and the gas is processed to make purified syngas (a mixture of carbon monoxide and hydrogen). The Fischer-Tropsch polymerizes syngas into diesel-range hydrocarbons. While biodiesel and bio-ethanol production so far only use parts of a plant, i.e. oil, sugar, starch or cellulose, BtL production uses the whole plant which is gasified by gasification. The result is that for BTL, less land area is required per unit of energy produced compared with biodiesel or bioethanol.
- Flash pyrolysis - producing bio-oil (pyrolysis oil), char and gas at temperatures between 350-550°C and residence times < 1 second (also called anhydrous pyrolysis).
- Catalytic depolymerization — using heat and catalysts to separate usable diesel fuel from hydrocarbon wastes.

Energy Storage

Energy storage is accomplished by devices or physical media that store some form of energy to perform some useful operation at a later time. A device that stores energy is sometimes called an accumulator.

All forms of energy are either potential energy (e.g. Chemical, gravitational, electrical energy, etc.) or kinetic energy (e.g. thermal energy). A wind-up clock stores potential energy (in this case mechanical, in the spring tension), a battery stores readily convertible chemical energy to operate a mobile phone, and a hydroelectric dam stores energy in a reservoir as gravitational potential energy. Ice storage tanks store ice (thermal energy) at night to meet peak demand for cooling. Fossil fuels such as coal and gasoline store ancient energy derived from sunlight by organisms that later died, became buried and over time were then converted into these fuels. Even food (which is made by the same process as fossil fuels) is a form of energy stored in chemical form.

Early History

Energy storage as a natural process is as old as the universe itself - the energy present at the initial formation of the universe has been stored in stars such as the Sun, and is now being used by humans directly (e.g. through solar heating), or indirectly (e.g. by growing crops or conversion into electricity in solar cells).

As a purposeful activity, energy storage has existed since pre-history, though it was often not explicitly recognised as such. An example of deliberate mechanical energy storage is the use of logs or boulders as defensive measures in ancient forts—the logs or boulders were collected at the top of a hill or wall, and the energy thus stored used to attack invaders who came within range.

A more recent application is the control of waterways to drive water mills for processing grain or powering machinery. Complex systems of reservoirs and dams were constructed to store and release water (and the potential energy it contained) when required.

Modern Era Developments

Storing energy allows humans to balance the supply and demand of energy. Energy storage systems in commercial use today can be broadly categorized as mechanical, electrical, chemical, biological and thermal. Energy storage became a dominant factor in economic development with the widespread introduction of electricity and refined chemical fuels, such as gasoline, kerosene and natural gas in the late 19th century. Unlike other common energy storage in prior use such as wood or coal, electricity must be used as it is being generated, or converted immediately into another form of energy such as potential, kinetic or chemical. Until recently electrical energy has not been converted and stored on a major scale, however new efforts to that effect began in the 21st century.

In the U.S., the 2009 Stimulus Plan helped finance research into energy storage and its integration with smart electrical grids. Electricity is transmitted in a closed circuit, and for essentially any practical purposes cannot be stored as electrical energy. This means that changes in demand can not be accommodated without either cutting supplies (as by brownouts or blackouts) or by storing the electric energy in another medium.

Even renewable energy must be stored in order to make it reliable. Wind blows intermittently and so some form of storage is required to compensate for calm periods. Solar energy is equally not available on cloudy days and during the nighttime, so stored energy must be available to compensate for the loss of sunlight.

An early solution to the problem of storing energy for electrical purposes was the development of the battery as an electrochemical storage device. Batteries have previously been of limited use in electric power systems due to their relatively small capacity and high cost, however since about the middle of the first decade of the 21st century newer battery technologies have been developed that can now provide significant utility scale load-levelling capabilities. A similar possible solution to deal with the intermittency issue of solar and wind energy is found in the capacitor.

In the 1980s, a number of manufacturers carefully researched thermal energy storage (TES) to meet the growing demand for air conditioning during peak hours. Today, several companies manufacture TES systems. The most popular form of thermal energy storage for cooling is ice storage, since it can store more energy in less space than water storage and it is also less costly than energy recovered via fuel cells or flywheels. Thermal storage has cost-effectively shifted gigawatts of power away from daytime peak usage periods, and in 2009 was used in over 3,300 buildings in over 35 countries. It works by creating ice at night when electricity is usually less costly, and then using the ice to cool the air in buildings during the hotter daytime periods.

Chemical fuels have become the dominant form of energy storage, both in electrical generation and energy transportation. Chemical fuels in common use are processed coal, gasoline, diesel fuel, natural gas, liquefied petroleum gas (LPG), propane, butane, ethanol and biodiesel.

All of these materias are readily converted to mechanical energy and then to electrical energy using heat engines (via turbines or other internal combustion engines, or boilers or other external combustion engines) used for electrical power generation. Heat-engine-powered generators are nearly universal, ranging from small engines producing only a few kilowatts to utility-scale generators with ratings up to 800 megawatts. A key disadvantage to hydrocarbon fuels are their significant emissions of greenhouse gases that contribute to global warming, as well as other significant pollutants emitted by the dirtier fuel sources such as coal and gasoline.

Some areas of the world such as Washington and Oregon in the United States, and Wales in the United Kingdom, have used geographic features to store large quantities of water in elevated reservoirs, using excess electricity at times of low demand to pump water up to the reservoirs, then letting the water pass through turbine generators to retrieve the energy when electrical demands peak.

Liquid hydrocarbon fuels are the most commonly used forms of energy storage for use in transportation, but because the byproducts of the reaction that utilises these liquid fuels' energy (combustion) produce greenhouse gases other energy carriers like hydrogen can be used to avoid production of greenhouse gases.

Advanced Systems

Electrochemical devices called fuel cells were invented about the same time as the battery in the 19th Century. However, for many reasons, fuel cells were not well-developed until the advent of manned spaceflight (such as the Gemini Program in the U.S.) when lightweight, non-thermal (and therefore efficient) sources of electricity were required in spacecraft. Fuel cell development has increased in recent years due to an attempt to increase conversion efficiency of chemical energy stored in hydrocarbon or hydrogen fuels into electricity. Several other technologies have also been investigated, such as flywheels, which can store kinetic energy, and compressed air storage that can be pumped into underground caverns and abandoned mines.

Another method used at the Solar Project and the Solar Tres Power Tower uses molten salt to store solar power and then dispatch that power as needed. The system pumps molten salt through a tower heated by the sun's rays. Insulated containers store the hot salt solution, and when needed water is then used to create steam that is fed to turbines to generate electricity. Research is being conducted on harnessing the quantum effects of nanoscale capacitors to create digital quantum batteries. Although this technology is still in the experimental stage, it theoretically has the potential to provide dramatic increases in energy storage capacity.

Grid Energy Storage

Grid energy storage (or large-scale energy storage) lets energy producers send excess electricity over the electricity transmission grid to temporary electricity storage sites that become energy producers when electricity demand is greater. Grid energy storage is particularly important in matching supply and demand over a 24 hour period of time.

A proposed variant of grid energy storage is called Vehicle-to-Grid energy storage system, where modern electric vehicles that are plugged into the energy grid can release the stored electrical energy in their batteries back into the grid when needed.

Storage Methods

- Chemical
  - — Hydrogen
  - — Biofuels
  - — Liquid nitrogen
  - — Oxyhydrogen
  - — Hydrogen peroxide
- Biological
  - — Starch
  - — Glycogen
- Electrochemical
  - — Batteries
  - — Flow batteries
  - — Fuel cells
- Electrical
  - — Capacitor
  - — Supercapacitor
  - — Superconducting magnetic energy storage (SMES)
- Mechanical
  - — Compressed air energy storage (CAES)
  - — Flywheel energy storage
  - — Hydraulic accumulator
  - — Hydroelectric energy storage
  - — Spring
  - — Gravitational potential energy (device)
- Thermal
  - — Ice Storage
  - — Molten salt
  - — Cryogenic liquid air or nitrogen
  - — Seasonal thermal store
  - — Solar pond
  - — Hot bricks

  - — Steam accumulator
  - — Fireless locomotive
  - — Eutectic system
- Fuel Conservation storage.

Hydrogen

Hydrogen is also being developed as an electrical power storage medium. Hydrogen is not a primary energy source, but a portable energy storage method, because it must first be manufactured by other energy sources in order to be used. However, as a storage medium, it may be a significant factor in using renewable energies.

Underground hydrogen storage is the practice of hydrogen storage in underground caverns, salt domes and depleted oil and gas fields. Large quantities of gaseous hydrogen are stored in underground caverns for many years without any difficulties. The storage of large quantities of hydrogen underground can function as grid energy storage which is essential for the hydrogen economy. By using a turboexpander, the electricity needs for compressed storage at 200 bars amounts to 2.1% of the energy content.

With intermittent renewables such as solar and wind, the output may be fed directly into an electricity grid. At penetrations below 20% of the grid demand, this does not severely change the economics; but beyond about 20% of the total demand, external storage will become important. If these sources are used for electricity to make hydrogen, then they can be utilised fully whenever they are available, opportunistically. Broadly speaking, it does not matter when they cut in or out, the hydrogen is simply stored and used as required. A community based pilot program using wind turbines and hydrogen generators is being undertaken from 2007 for five years in the remote community of Ramea, Newfoundland and Labrador. A similar project has been going on since 2004 on Utsira, a small Norwegian island municipality.

Energy losses are involved in the hydrogen storage cycle of hydrogen production for vehicle applications with electrolysis of water, liquification or compression, and conversion back to electricity and the hydrogen storage cycle of production for the stationary fuel cell applications like microchp at 93 % with biohydrogen or biological hydrogen production, and conversion to electricity.

About 50 kW·h (180 MJ) of solar energy is required to produce a kilogram of hydrogen, so the cost of the electricity clearly is crucial, even for hydrogen uses other than storage for electrical generation. At $0.03/kWh, common off-peak high-voltage line rate in the United States, this means hydrogen costs $1.50 a kilogram for the electricity, equivalent to $1.50 a U.S. gallon for gasoline if used in a fuel cell vehicle. Other costs would include the electrolyzer plant, hydrogen compressors or liquefaction, storage and transportation, which will be significant.

Biofuels

Various biofuels such as biodiesel, straight vegetable oil, alcohol fuels, or biomass can be used to replace hydrocarbon fuels. Various chemical processes can convert the carbon and hydrogen in coal, natural gas, plant and animal biomass, and organic wastes into short hydrocarbons suitable as replacements for existing hydrocarbon fuels. Examples are Fischer-Tropsch diesel, methanol, dimethyl ether, or syngas. This diesel source was used extensively in World War II in Germany, with limited access to crude oil supplies. Today South Africa produces most of the country's diesel from coal for similar reasons. A long term oil price above US$35/bbl may make such synthetic liquid fuels economical on a large scale. Some of the energy in the original source is lost in the conversion process. Historically, coal itself has been used directly for transportation purposes in vehicles and boats using steam engines. And compressed natural gas is being used in special circumstances fuel, for instance in busses for some mass transit agencies.

Synthetic Hydrocarbon Fuel

Carbon dioxide in the atmosphere has been, experimentally, converted into hydrocarbon fuel with the help of energy from another source. To be useful industrially, the energy will probably have to come from sunlight using, perhaps, future artificial photosynthesis technology. Another alternative for the energy is electricity or heat from solar energy or nuclear power.

Compared to hydrogen, many hydrocarbon fuels have the advantage of being immediately usable in existing engine technology and existing fuel distribution infrastructures. Manufacturing synthetic hydrocarbon fuel reduces the amount of carbon dioxide in the atmosphere until the fuel is burned, when the same amount of carbon dioxide returns to the atmosphere.

Methane

Methane is the simplest hydrocarbon with the molecular formula CH_4. Methane could be produced from electricity of renewable energies.

Methane can be stored more easily than hydrogen and the transportation, storage and combustion infrastructure are mature (pipelines, gasometres, power plants).

As hydrogen and oxygen are produced in the electrolysis of water,

$$2H_2O \rightarrow 2H_2 + O_2$$

hydrogen would then be reacted with carbon dioxide in Sabatier process, producing methane and water.

$$CO_2 + 4H_2 \rightarrow CH_4 + 2H_2O$$

Methane would be stored and used to produce electricity later. Produced water would be recycled back to the electrolysis stage, reducing the need for new pure water. In the electrolysis stage oxygen would also be stored for methane combustion in a pure oxygen environment in an adjacent power plant, eliminating e.g. nitrogen oxides. In the combustion of methane, carbon dioxide and water are produced.

$$CH_4 + 2O_2 \rightarrow CO_2 + 2H_2O$$

Produced carbon dioxide would be recycled back to boost the Sabatier process and water would be recycled back to the electrolysis stage. The carbon dioxide produced by methane combustion would be turned back to methane, thus producing no greenhouse gases. Methane production, storage and adjacent combustion would recycle all the reaction products, creating a cycle.

Boron, Silicon, and Zinc

Boron, silicon, lithium, and zinc have been proposed as energy storage solutions.

Mechanical Storage

Energy can be stored in water pumped to a higher elevation using pumped storage methods, in compressed air, or in spinning flywheels.

A mass of 1 kg, elevated to a height of 1000 m stores 9.8 kJ of gravitational energy, which is equivalent to 1 kg mass accelerated to 140 m/s. To store the same mass of water, if increased in temperature by 2.34 Celsius, requires the same amount of energy. Admittedly, this is a bit of an unfair comparison, but it makes it easy to see how it is possible to store more energy in 1 m^3 of cheap rock or sand than 1 m^3 of lead-acid battery, even if the battery is also moved to a higher elevation, not just charged.

Compressed air energy storage technology stores low cost off-peak energy, in the form of compressed air in an underground reservoir.

The air is then released during peak load hours and heated with the exhaust heat of a standard combustion turbine.

This heated air is converted to energy through expansion turbines to produce electricity. A CAES plant has been in existence in McIntosh, Alabama since 1991 and has run successfully.

Other applications are possible. Walker Architects published the first CO_2 gas CAES application, proposing the use of sequestered CO_2 for Energy Storage on October 24, 2008. Several companies have done preliminary design work for vehicles using compressed air power.

Thermal Storage

Thermal storage is the temporary storage or removal of heat for later use. An example of thermal storage is the storage of solar heat energy during the day to be used at a later time for heating at night. In the HVAC/R field, this type of application using thermal storage for heating is less common than using thermal storage for cooling.

An example of the storage of "cold" heat removal for later use is ice made during the cooler night time hours for use during the hot daylight hours. This ice storage is produced when electrical utility rates are lower. This is often referred to as "off-peak" cooling.

When used for the proper application with the appropriate design, off-peak cooling systems can lower energy costs. The U.S. Green Building Council has developed the Leadership in Energy and Environmental Design (LEED) program to encourage the design of high-performance buildings that will help protect our environment. The increased levels of energy performance by utilising off-peak cooling may qualify of credits toward LEED Certification.

The advantages of thermal storage are:

- Commercial electrical rates are lower at night.
- It takes less energy to make ice when the ambient temperature is cool at night. Source energy (energy from the power plant) is saved.
- A smaller, more efficient system can do the job of a much larger unit by running for more hours.

Renewable Energy Storage

Many renewable energy sources (most notably solar and wind) produce intermittent power. Wherever intermittent power sources reach high levels of grid penetration, energy storage becomes one option to provide reliable energy supplies.

Other options include recourse to peaking power plants, methane storage (excess renewable electricity to hydrogen via electrolysis, combine with CO2 (low to neutral CO2 system) to produce methane (synthetic natural gas sabatier process) with stockage in the natural gas network) and smart grids with advanced energy demand management.

The latter involves bringing “prices to devices”, i.e. making electrical equipment and appliances able to adjust their operation to seek the lowest spot price of electricity. On a grid with a high penetration of renewables, low spot prices would correspond to times of high availability of wind and/or sunshine.

Economic Evaluation

For economic evaluation of large scale applications, pumped hydro storage, compressed air, the potential benefits that can be accounted are avoidance of wind curtailment, grid congestion avoidance, price arbitrage and carbon free energy delivery.

Fractional Distillation

Fractional distillation is the separation of a mixture into its component parts, or fractions, such as in separating chemical compounds by their boiling point by heating them to a temperature at which several fractions of the compound will evaporate. It is a special type of distillation. Generally the component parts boil at less than 25 °C from each other under a pressure of one atmosphere (atm). If the difference in boiling points is greater than 25 °C, a simple distillation is used.

Laboratory Setup

Fractional distillation in a laboratory makes use of common laboratory glassware and apparatuses, typically including a Bunsen burner, a round-bottomed flask and a condenser, as well as the single-purpose fractionating column.

Apparatus

Fractional distillation apparatus using a Liebig condenser. A conical flask is used as a receiving flask. Here the distillation head and fractionating column are combined in one piece.

- heat source, such as a hot plate with a bath, and ideally with a magnetic stirrer.
- distilling flask, typically a round-bottom flask
- receiving flask, often also a round-bottom flask

- fractionating column
- distillation head
- thermometre and adapter if needed
- condenser, such as a Liebig condenser, Graham condenser or Allihn condenser
- vacuum adapter (not used in image to the right)
- boiling chips, also known as anti-bumping granules
- Standard laboratory glassware with ground glass joints, e.g. quickfit apparatus.

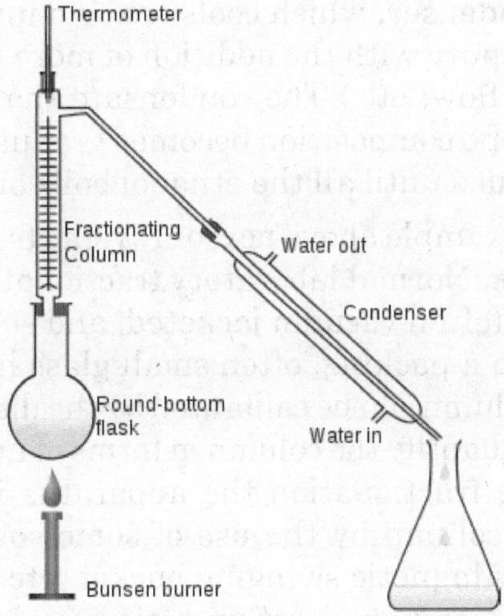

Discussion

As an example, consider the distillation of a mixture of water and ethanol. Ethanol boils at 78.4 °C while water boils at 100 °C. So, by heating the mixture, the most volatile component will concentrate to a greater degree in the vapor leaving the liquid. Some mixtures form azeotropes, where the mixture boils at a lower temperature than either component. In this example, a mixture of 96% ethanol and 4% water boils at 78.2 °C, being more volatile than pure ethanol. For this reason, ethanol cannot be completely purified by direct fractional distillation of ethanol-water mixtures.

The apparatus is assembled as in the diagram. (The diagram represents a batch apparatus, as opposed to a continuous apparatus.) The mixture is put into the round bottomed flask along with a few anti-bumping granules (or a Teflon coated magnetic stirrer bar if using magnetic stirring), and the fractionating column is fitted into the top. As the mixture boils, vapor rises up the column. The vapor condenses on the glass platforms, known as trays, inside the column, and runs back down into the liquid below, refluxing distillate. The column is heated from the bottom. The efficiency in terms of the amount of heating and time required to get fractionation can be improved by insulating the outside of the column in an insulator such as wool, aluminium foil or preferably a vacuum jacket.

The hottest tray is at the bottom and the coolest is at the top. At steady state conditions, the vapor and liquid on each tray are at *equilibrium*. Only the most volatile of the vapours stays in gaseous form all the way to the top. The vapor at the top of the column then passes into the condenser, which cools it down until it liquefies. The separation is more pure with the addition of more trays (to a practical limitation of heat, flow, etc.) The condensate that was initially very close to the azeotrope composition becomes gradually richer in water. The process continues until all the ethanol boils out of the mixture.

Typically the example above now only reflects the theoretical way fractionation works. Normal laboratory fractionation columns will be simple glass tubes (often vacuum jacketed, and sometimes internally silvered) filled with a packing, often small glass helices of 4 to 7 mm diametre. Such a column can be calibrated by the distillation of a known mixture system to quantify the column in terms of number of theoretical plates. To improve fractionation the apparatus is set up to return condensate to the column by the use of some sort of reflux splitter (reflux wire, gago, Magnetic swinging bucket, etc.) - a typical careful fractionation would employ a reflux ratio of around 10:1 (10 parts returned condensate to 1 part condensate take off).

In laboratory distillation, several types of condensers are commonly found. The Liebig condenser is simply a straight tube within a water jacket, and is the simplest (and relatively least expensive) form of condenser. The Graham condenser is a spiral tube within a water jacket, and the Allihn condenser has a series of large and small constrictions on the inside tube, each increasing the surface area upon which the vapor constituents may condense.

Alternate set-ups may utilise a "cow" or "pig" which is connected to three or four receiving flasks. By turning the "cow" or "pig", the distillates can be channelled into the appropriate receiver. A Perkin triangle is versatile piece of apparatus that can also be used to collect distillation fractions which does not require a "cow" or "pig" adapter. A Perkin triangle is most often used where the distillates are air-sensitive or where the fractions distill and are collected under reduced pressure, but can be used for a simple and fractional distillation.

Vacuum distillation systems operate at reduced pressure, thereby lowering the boiling points of the materials. Note that the use of anti-bumping granules will not work at reduced pressures.

Industrial Distillation

Distillation is the most common form of separation technology used in petroleum refineries, petrochemical and chemical plants, natural gas processing and cryogenic air separation plants. In most cases, the distillation is operated at a continuous steady state. New feed is always being added to the distillation column and products are always being removed. Unless the process is disturbed due to changes in feed, heat, ambient temperature, or condensing, the amount of feed being added and the amount of product being removed are normally equal. This is known as continuous, steady-state fractional distillation.

Industrial distillation is typically performed in large, vertical cylindrical columns known as "distillation or fractionation towers" or "distillation columns" with diametres ranging from about 65 centimetres to 6 metres and heights ranging from about 6 metres to 60 metres or more.

The distillation towers have liquid outlets at intervals up the column which allow for the withdrawal of different *fractions* or products having different boiling points or boiling ranges. By increasing the temperature of the product inside the columns, the different hydrocarbons are separated. The "lightest" products (those with the lowest boiling point) exit from the top of the columns and the "heaviest" products (those with the highest boiling point) exit from the bottom of the column.

For example, fractional distillation is used in oil refineries to separate crude oil into useful substances (or fractions) having different hydrocarbons of different boiling points. The crude oil fractions with higher boiling points:

- have more carbon atoms
- have higher molecular weights
- are more branched chain alkanes
- are darker in colour
- are more viscous
- are more difficult to ignite and to burn.

Large-scale industrial towers use reflux to achieve a more complete separation of products. Reflux refers to the portion of the condensed overhead liquid product from a distillation or fractionation tower that is returned to the upper part of the tower as shown in the schematic diagram of a typical, large-scale industrial distillation tower.

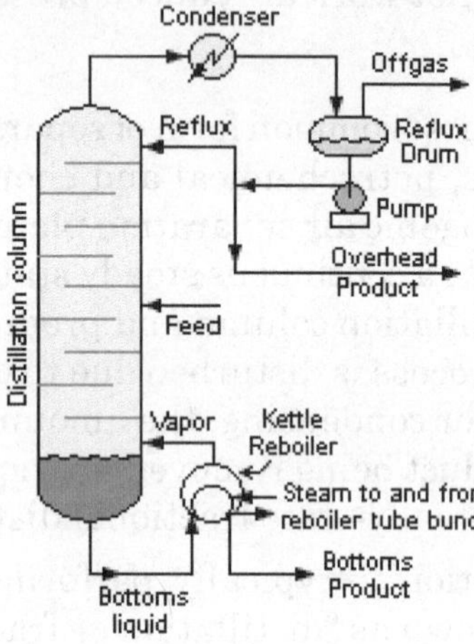

Figure: *Diagram of a typical industrial distillation tower*

Inside the tower, the reflux liquid flowing downwards provides the cooling needed to condense the vapors flowing upwards, thereby increasing the effectiveness of the distillation tower. The more reflux is provided for a given number of theoretical plates, the better the tower's separation of lower boiling materials from higher boiling materials. Alternatively, the more reflux provided for a given desired separation, the fewer theoretical plates are required.

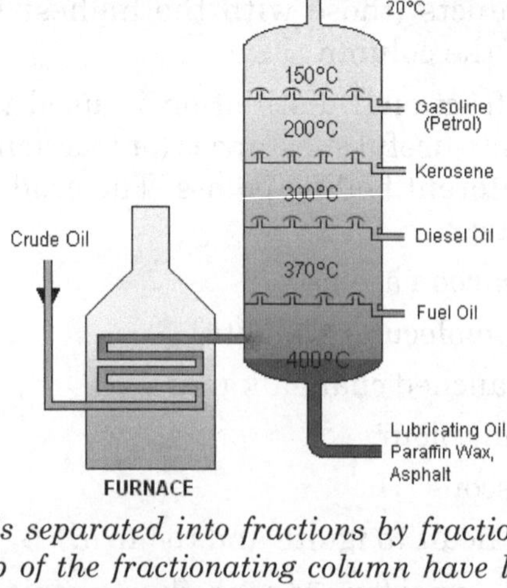

Figure: *Crude oil is separated into fractions by fractional distillation. The fractions at the top of the fractionating column have lower boiling points than the fractions at the bottom. The heavy bottom fractions are often cracked into lighter, more useful products. All of the fractions are processed further in other refining units.*

Fractional distillation is also used in air separation, producing liquid oxygen, liquid nitrogen, and highly concentrated argon. Distillation of chlorosilanes also enable the production of high-purity silicon for use as a semiconductor.

In industrial uses, sometimes a packing material is used in the column instead of trays, especially when low pressure drops across the column are required, as when operating under vacuum. This packing material can either be random dumped packing (1-3" wide) such as Raschig rings or structured sheet metal. Typical manufacturers are Koch, Sulzer and other companies. Liquids tend to wet the surface of the packing and the vapours pass across this wetted surface, where mass transfer takes place. Unlike conventional tray distillation in which every tray represents a separate point of vapor liquid equilibrium the vapor liquid equilibrium curve in a packed column is continuous. However, when modelling packed columns it is useful to compute a number of "theoretical plates" to denote the separation efficiency of the packed column with respect to more traditional trays. Differently shaped packings have different surface areas and void space between packings. Both of these factors affect packing performance.

Design of Industrial Distillation Columns

Design and operation of a distillation column depends on the feed and desired products. Given a simple, binary component feed, analytical methods such as the McCabe-Thiele method or the Fenske equation can be used. For a multi-component feed, simulation models are used both for design and operation.

Moreover, the efficiencies of the vapor-liquid contact devices (referred to as *plates* or *trays*) used in distillation columns, are typically lower than that of a theoretical 100% efficient equilibrium stage. Hence, a distillation column needs more plates than the number of theoretical vapor-liquid equilibrium stages.

An indication of numbers: the separation of two compounds with relative volatility of 1.1 requires a minimum of 130 theoretical plates with a minimum reflux ratio of 20. With a relative volatility of 4, the required number of theoretical plates decreased to 9 with a reflux ratio of 0.66. In another source, a boiling point difference of 30 °C requires 12 theoretical plates and, for a difference of 3 °C, the number of plates increased to 1000.

The reflux ratio is the ratio of the amount of moles returned as refluxed liquid to the fractionating column and the amount of moles of final product, both per unit time.

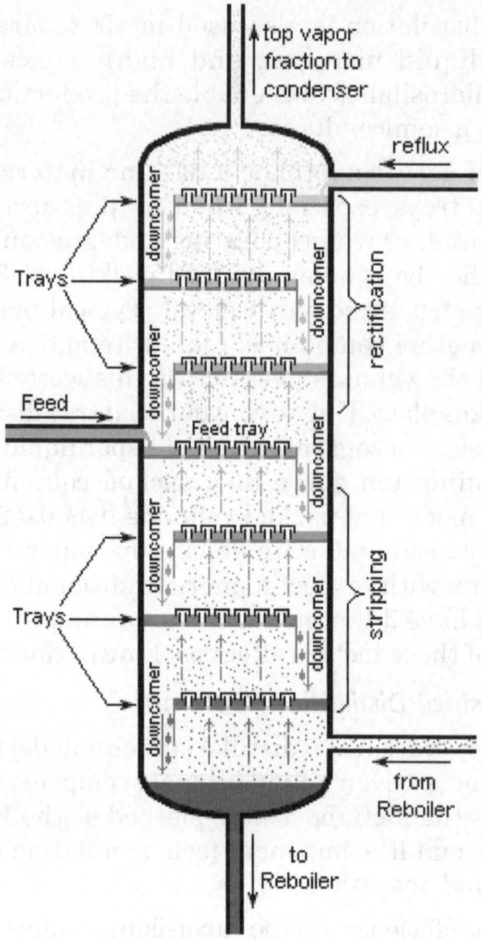

Figure: *Chemical engineering schematic of typical bubble-cap trays in a distillation tower*

Functional Group

In organic chemistry, functional groups are specific groups of atoms within molecules that are responsible for the characteristic chemical reactions of those molecules. THE same kristie same functional group will undergo the same or similar chemical reaction(s) regardless of the size of the molecule it is a part of. However, its relative reactivity can be modified by nearby functional groups.

The word moiety (pronounced) is often used synonymously to "functional group," but, according to the IUPAC definition, a moiety is a part of a molecule that may include functional groups as substructures. For example, an ester (RCOOR') has an ester functional group (COOR) and is composed of an alkoxy moiety (-OR') and an acyl moiety (RCO-), or, equivalently, it may be divided into carboxylate (RCOO-) and alkyl (-R') moieties. Each moiety may contain additional functional groups, for example, methyl para-hydroxybenzoate contains a phenol functional group within the acyl moiety. Combining the names of functional groups with the names of the parent alkanes generates a powerful systematic nomenclature for naming organic compounds.

The atoms of functional groups are linked to each other and to the rest of the molecule by covalent bonds. When the group of covalently bound atoms bears a net charge, the group is referred to more properly as a polyatomic ion or a complex ion. Any subgroup of atoms of a compound also may be called a radical, and if a covalent bond is broken homolytically, the resulting fragment radicals are referred as free radicals. The first carbon atom after the carbon that attaches to the functional group is called the alpha carbon; the second, beta carbon, the third, gamma carbon, etc. If there is another functional group at a carbon, it may be named with the Greek letter, e.g., the gamma-amine in gamma-aminobutanoic acid is on the third carbon of the carbon chain attached to the carboxylic acid group.

Synthetic Chemistry

Organic reactions are facilitated and controlled by the functional groups of the reactants. In general, alkyls are unreactive and difficult to get to react selectively at the desired positions, with few exceptions. In contrast, unsaturated carbon functional groups, and carbon-oxygen and carbon-nitrogen functional groups have a more diverse array of reactions that are also selective.

It may be necessary to create a functional group in the molecule to make it react. For example, to synthesize iso-octane (the 8-carbon ideal gasoline) from the unfunctionalised alkane isobutane (a 4-carbon gas), isobutane is first dehydrogenated into isobutene. This contains the alkene functional group and can now dimerize with another isobutene to give iso-octene, which is then catalytically hydrogenated to iso-octane using pressured hydrogen gas.

Crystallography

The International Union of Crystallography in its Crystallographic Information File dictionary defines "moiety" to represent discrete non-bonded components. Thus, Na_2SO_4 would contain 3 moieties (2 Na^+ and one SO_4^{2-}). The dictionary defines "chemical formula moiety": "Formula with each discrete bonded residue or ion shown as a separate moiety".

Functionalisation

Functionalisation is the addition of functional groups onto the surface of a material by chemical synthesis methods. The functional group added can be subjected to ordinary synthesis methods to attach virtually any kind of organic compound onto the surface.

Functionalisation is employed for surface modification of industrial materials in order to achieve desired surface properties such as water repellent coatings for automobile windshields and non-biofouling, hydrophilic coatings for contact lenses. In addition, functional groups are used to covalently link functional molecules to the surface of chemical and biochemical devices such as microarrays and microelectromechanical systems. Catalysts can be attached to a material that has been functionalised. For example, silica is functionalised with an alkyl silicone, wherein the alkyl contains an amine functional group. A ligand such as an EDTA fragment is synthesized onto the amine, and a metal cation is complexed into the EDTA fragment. The EDTA is *not* adsorbed onto the surface, but connected by a permanent chemical bond. Functional groups are also used to covalently link molecules such as fluorescent dyes, nanoparticles, proteins, DNA, and other compounds of interest for a variety of applications such as sensing and basic chemical research.

Table of Common Functional Groups

The following is a list of common functional groups. In the formulas, the symbols R and R' usually denote an attached hydrogen, or a hydrocarbon side chain of any length, but may sometimes refer to any group of atoms.

Chapter 7

Cheminformatics

Cheminformatics (also known as chemoinformatics, chemioinformatics and chemical informatics) is the use of computer and informational techniques applied to a range of problems in the field of chemistry. These *in silico* techniques are used in, for example, pharmaceutical companies in the process of drug discovery.

These methods can also be used in chemical and allied industries in various other forms.

History

The term chemoinformatics was defined by F.K. Brown in 1998:

Chemoinformatics is the mixing of those information resources to transform data into information and information into knowledge for the intended purpose of making better decisions faster in the area of drug lead identification and optimization.

Since then, both spellings have been used, and some have evolved to be established as Cheminformatics, while European Academia settled in 2006 for Chemoinformatics. The recent establishment of the Journal of Cheminformatics is a strong push towards the shorter variant.

Basics

Cheminformatics combines the scientific working fields of chemistry, computer science and information science for example in the areas of topology, chemical graph theory, information retrieval and data mining in the chemical space.

Cheminformatics can also be applied to data analysis for various industries like paper and pulp, dyes and such allied industries.

Applications

Storage and Retrieval

The primary application of cheminformatics is in the storage, indexing and search of information relating to compounds.

The efficient search of such stored information includes topics that are dealt with in computer science as data mining, information retrieval, information extraction and machine learning. Related research topics include:

- Unstructured data
  - o cInformation retrieval
  - o Information extraction
- Structured Data Mining and mining of Structured data
  - o Database mining
  - o Graph mining
  - o Molecule mining
  - o Sequence mining
  - o Tree mining
- Digital libraries

File Formats

The *in silico* representation of chemical structures uses specialised formats such as the XML-based Chemical Markup Language or SMILES. These representations are often used for storage in large chemical databases. While some formats are suited for visual representations in 2 or 3 dimensions, others are more suited for studying physical interactions, modelling and docking studies.

Virtual Libraries

Chemical data can pertain to real or virtual molecules. Virtual libraries of compounds may be generated in various ways to explore chemical space and hypothesize novel compounds with desired properties.

Virtual libraries of classes of compounds (drugs, natural products, diversity-oriented synthetic products) were recently generated using the FOG (fragment optimized growth) algorithm.

This was done by using cheminformatic tools to train transition probabilities of a Markov chain on authentic classes of compounds, and then using the Markov chain to generate novel compounds that were similar to the training database.

Virtual Screening

Virtual screening (VS) is a computational technique used in drug discovery to search libraries of small molecules in order to identify those structures which are most likely to bind to a drug target, typically a protein receptor or enzyme.

Virtual screening has been defined as the "automatically evaluating very large libraries of compounds" using computer programs. As this definition suggests, VS has largely been a numbers game focusing on how the enormous chemical space of over 10^{60} conceivable compounds to a manageable number that can be synthesized, purchased, and tested. Although searching the entire chemical universe may be a theoretically interesting problem, more practical VS scenarios focus on designing and optimizing targeted combinatorial libraries and enriching libraries of available compounds from in-house compound repositories or vendor offerings. As the accuracy of the method has increased, virtual screening has become an integral part of the drug discovery process.

Methods

There are two broad categories of screening techniques: ligand-based and structure-based.

Ligand-Based

Given a set of structurally diverse ligands that binds to a receptor, a model of the receptor can be built by exploiting the collective information contained in such set of ligands. These are known as pharmacophore models. A candidate ligand can then be compared to the pharmacophore model to determine whether it is compatible with it and therefore likely to bind. A different strategy is to develop logic-based rules describing features of substructures and chemical properties related to activity using support vector inductive logic programming. The logic-based features provide insights into activity which can be understood by medicinal chemists. Support vector machine integrate the features to yield a quantitative QSAR, which is then used to screen a database of molecules.

This approach is well suited to scaffold hopping to identify novel active molecules and is implemented in the package INDDEx. Another approach to ligand-based virtual screening is to use 2D chemical similarity analysis methods to scan a database of molecules against one or more active ligand structure. A popular approach to ligand-based virtual screening is based on searching molecules with shape similar to that of known actives, as such molecules will fit the target's binding site and hence will be likely to bind the target. There are a number of prospective applications of this class of techniques in the literature.

Computing Infrastructure

The computation of pair-wise interactions between atoms, which is a prerequisite for the operation of many virtual screening programs, is of $O(N^2)$ computational complexity, where N is the number of atoms in the system. Because of the quadratic scaling with respect to the number of atoms, the computing infrastructure may vary from a laptop computer for a ligand-based method to a mainframe for a structure-based method.

Ligand-Based

Ligand-based methods typically require a fraction of a second for a single structure comparison operation. A single CPU is enough to perform a large screening within hours. However, several comparisons can be made in parallel in order to expedite the processing of a large database of compounds.

Structure-Based

The size of the task requires a parallel computing infrastructure, such as a cluster of Linux systems, running a batch queue processor to handle the work, such as Sun Grid Engine or Torque PBS.

A means of handling the input from large compound libraries is needed. This requires a form of compound database that can be queried by the parallel cluster, delivering compounds in parallel to the various compute nodes. Commercial database engines may be too ponderous, and a high speed indexing engine, such as Berkeley DB, may be a better choice. Furthermore, it may not be efficient to run one comparison per job, because the ramp up time of the cluster nodes could easily outstrip the amount of useful work. To work around this, it is necessary to process batches of compounds in each cluster job, aggregating the results into some kind of log file. A secondary process, to mine the log files and extract high scoring candidates, can then be run after the whole experiment has been run.

Accuracy

The aim of virtual screening is to identify molecules of novel chemical structure that bind to the macromolecular target of interest. Thus, success of a virtual screen is defined in terms of finding interesting new scaffolds rather than the total number of hits.

Interpretations of virtual screening accuracy should therefore be considered with caution. Low hit rates of interesting scaffolds are clearly preferable over high hit rates of already known scaffolds.

Most tests of virtual screening studies in the literature are retrospective. In these studies, the performance of a VS technique is measured by its ability to retrieve a small set of previously known molecules with affinity to the target of interest (active molecules or just actives) from a library containing a much higher proportion of assumed inactives or decoys. By contrast, in prospective applications of virtual screening, the resulting hits are subjected to experimental confirmation (e.g., IC_{50} measurements). There is consensus that retrospective benchmarks are not good predictors of prospective performance and consequently only prospective studies constitute conclusive proof of the suitability of a technique for a particular target.

Quantitative Structure-Activity Relationship (QSAR)

Quantitative structure–activity relationship models (QSAR models) are regression or classification models used in the chemical and biological sciences and engineering. Like other regression models, QSAR regression models relate a set of "predictor" variables (X) to the potency of the response variable (Y), while classification QSAR models relate the predictor variables to a categorical value of the response variable. In QSAR modelling, the predictors consist of physico-chemical properties or theoretical molecular descriptors of chemicals; the QSAR response-variable could be a biological activity of the chemicals. QSAR models first summarize a supposed relationship between chemical structures and biological activity in a data-set of chemicals. Second QSAR models predict the activities of new chemicals. Related terms include *quantitative structure–property relationships* (*QSPR*) when a chemical property is modelled as the response variable. In a nutshell, QSAR and QSPR tries to discern the relationship between molecular descriptors that describe the unique physico-chemical properties of the set of compounds of interest with their respective biological activity or chemical property.

For example, biological activity can be expressed quantitatively as the concentration of a substance required to give a certain biological response. Additionally, when physicochemical properties or structures are expressed by numbers, one can find a mathematical relationship, or quantitative structure-activity relationship, between the two. The mathematical expression, if carefully validated can then be used to predict the modelled response of other chemical structures, by carefully verifying the Applicability domain (AD).

A QSAR has the form of a mathematical model:

Activity = *f* (physicochemical properties and/or structural properties) + Error

The error includes model error (bias) and observational variability, that is, the variability in observations even on a correct model.

SAR and the SAR Paradox

The basic assumption for all molecule based hypotheses is that similar molecules have similar activities. This principle is also called Structure–Activity Relationship (SAR). The underlying problem is therefore how to define a *small* difference on a molecular level, since each kind of activity, e.g. reaction ability, biotransformation ability, solubility, target activity, and so on, might depend on another difference. Good examples were given in the bioisosterism reviews by Patanie/LaVoie and Brown. In general, one is more interested in finding strong trends. Created hypotheses usually rely on a finite number of chemical data. Thus, the induction principle should be respected to avoid overfitted hypotheses and deriving overfitted and useless interpretations on structural/molecular data.

The SAR paradox refers to the fact that it is not the case that all similar molecules have similar activities.

Types

Fragment Based (Group Contribution)}: Analogously, the "partition coefficient"—a measurement of differential solubility and itself a component of QSAR predictions—can be predicted either by atomic methods (known as "XLogP" or "ALogP") or by chemical fragment methods (known as "CLogP" and other variations). It has been shown that the logP of compound can be determined by the sum of its fragments; fragment-based methods are generally accepted as better predictors than atomic-based methods. Fragmentary values have been determined statistically, based on empirical data for known logP values. This method gives mixed results and is generally not trusted to have accuracy of more than ±0.1 units.

Group or Fragment based QSAR is also known as GQSAR. GQSAR allows flexibility to study various molecular fragments of interest in relation to the variation in biological response. The molecular fragments could be substituents at various substitution sites in congeneric set of molecules or could be on the basis of pre-defined chemical rules in case of non-congeneric set. GQSAR also considers cross-terms fragment descriptors, which could be helpful in identification of key fragment interactions in determining variation of activity.

3D-QSAR

3D-QSAR refers to the application of force field calculations requiring three-dimensional structures, e.g. based on protein crystallography or molecule superimposition. It uses computed potentials, e.g. the Lennard-Jones potential, rather than experimental constants and is concerned with the overall molecule rather than a single substituent. It examines the steric fields (shape of the molecule), the hydrophobic regions (water-soluble surfaces), and the electrostatic fields.

The created data space is then usually reduced by a following feature extraction. The following learning method can be any of the already mentioned machine learning methods, e.g. support vector machines. An alternative approach uses multiple-instance learning by encoding molecules as sets of data instances, each of which represents a possible molecular conformation. A label or response is assigned to each set corresponding to the activity of the molecule, which is assumed to be determined by at least one instance in the set (i.e. some conformation of the molecule).

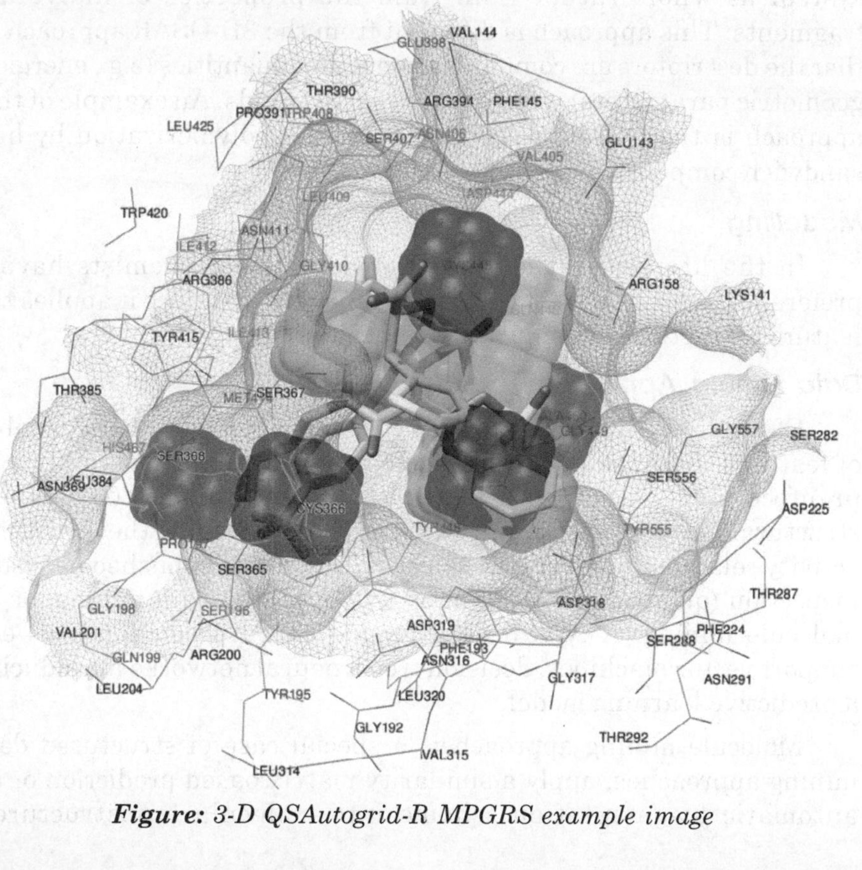

Figure: *3-D QSAutogrid-R MPGRS example image*

On June 18, 2011 the CoMFA patent has dropped any restriction on the use of GRID and PLS technologies and the RCMD team has opened a 3D QSAR web server based on the 3-D QSAutogrid/R engine. 3-D QSAutogrid/R covers all the main features of CoMFA and GRID/GOLPE with implementation by multiprobe/multiregion variable selection (MPGRS) that improves the simplification of interpretation of the 3-D QSAR map. The methodology is based on the integration of the molecular interaction fields as calculated by AutoGrid and the R statistical environment that can be easily coupled with many free graphical molecular interfaces such as UCSF-Chimera, AutoDock Tools, JMol and others.

Chemical Descriptor Based

In this approach, descriptors quantifying various electronic, geometric, or steric properties of a catalyst are computed and used to develop a QSAR. This approach is different from the fragment (or group contribution) approach in that the descriptors are computed for the system as whole rather than from the properties of individual fragments. This approach is different from the 3D-QSAR approach in that the descriptors are computed from scalar quantities (e.g., energies, geometric parameters) rather than from 3D fields. An example of this approach is the QSARs developed for olefin polymerization by half sandwich compounds.

Modelling

In the literature it can be often found that chemists have a preference for partial least squares (PLS) methods, since it applies the feature extraction and induction in one step.

Data Mining Approach

Computer SAR models typically calculate a relatively large number of features. Because those lack structural interpretation ability, the preprocessing steps face a feature selection problem (i.e., which structural features should be interpreted to determine the structure-activity relationship). Feature selection can be accomplished by visual inspection (qualitative selection by a human); by data mining; or by molecule mining. A typical data mining based prediction uses e.g. support vector machines, decision trees, neural networks for inducing a predictive learning model.

Molecule mining approaches, a special case of structured data mining approaches, apply a similarity matrix based prediction or an automatic fragmentation scheme into molecular substructures.

Furthermore there exist also approaches using maximum common subgraph searches or graph kernels.

Evaluation of the Quality of QSAR Models

QSAR modelling produces predictive models derived from application of statistical tools correlating biological activity (including desirable therapeutic effect and undesirable side effects) or physico-chemical properties in QSPR models of chemicals (drugs/toxicants/environmental pollutants) with descriptors representative of molecular structure or properties. QSARs are being applied in many disciplines for example risk assessment, toxicity prediction, and regulatory decisions in addition to drug discovery and lead optimization. Obtaining a good quality QSAR model depends on many factors, such as the quality of input data, the choice of descriptors and statistical methods for modelling and for validation. Any QSAR modelling should ultimately lead to statistically robust and predictive models capable of making accurate and reliable predictions of the modelled response of new compounds.

For validation of QSAR models usually various strategies are adopted:

1. internal validation or cross-validation;
2. external validation by splitting the available data set into training set for model development and prediction set for model predictivity check;
3. blind external validation by application of model on new external data and
4. data randomization or Y-scrambling for verifying the absence of chance correlation between the response and the modelling descriptors.

The success of any QSAR model depends on accuracy of the input data, selection of appropriate descriptors and statistical tools, and most importantly validation of the developed model. Validation is the process by which the reliability and relevance of a procedure are established for a specific purpose; for QSAR models validation must be mainly for robustness, prediction performances and applicability domain of the models. Leave one-out cross-validation generally leads to an overestimation of predictive capacity, and even with external validation, no one can be sure whether the selection of training and test sets was manipulated to maximize the predictive capacity of the model being published. Different aspects of validation of QSAR models that need attention includes methods of selection of training set compounds, setting training set size and impact of variable selection for training set models for determining the quality of prediction. Development of novel validation parameters for judging quality of QSAR models is also important.

Application

Chemical: One of the first historical QSAR applications was to predict boiling points. It is well known for instance that within a particular family of chemical compounds, especially of organic chemistry, that there are strong correlations between structure and observed properties. A simple example is the relationship between the number of carbons in alkanes and their boiling points. There is a clear trend in the increase of boiling point with an increase in the number carbons and this serves as a means for predicting the boiling points of higher alkanes. A still very interesting application is the Hammett equation, Taft equation and pKa prediction methods.

Biological

The biological activity of molecules is usually measured in assays to establish the level of inhibition of particular signal transduction or metabolic pathways. Chemicals can also be biologically active by being toxic. Drug discovery often involves the use of QSAR to identify chemical structures that could have good inhibitory effects on specific targets and have low toxicity (non-specific activity). Of special interest is the prediction of partition coefficient log P, which is an important measure used in identifying "druglikeness" according to Lipinski's Rule of Five. While many quantitative structure activity relationship analyses involve the interactions of a family of molecules with an enzyme or receptor binding site, QSAR can also be used to study the interactions between the structural domains of proteins. Protein-protein interactions can be quantitatively analyzed for structural variations resulted from site-directed mutagenesis. It is part of the machine learning method to reduce the risk for a SAR paradox, especially taking into account that only a finite amount of data is available. In general all QSAR problems can be divided into a coding and learning.

Applications

(Q)SAR models have been used for the risk management of chemicals risk. QSARS are suggested by regulatory authorities; in the European Union, QSARs are suggested by the REACH regulation, where "REACH" abbreviates "Registration, Evaluation, Authorisation and Restriction of Chemicals". The chemical descriptor space whose convex hull is generated by a particular training set of chemicals is called the training set's applicability domain. Prediction of properties of novel chemicals that are located outside the applicability domain uses extrapolation, and so is less reliable (on average) than prediction within the applicability domain. The assessment of the reliability of QSAR predictions remains a research topic.

Computational Fluid Dynamics

Computational fluid dynamics, usually abbreviated as CFD, is a branch of fluid mechanics that uses numerical methods and algorithms to solve and analyze problems that involve fluid flows. Computers are used to perform the calculations required to simulate the interaction of liquids and gases with surfaces defined by boundary conditions.

With high-speed supercomputers, better solutions can be achieved. Ongoing research yields software that improves the accuracy and speed of complex simulation scenarios such as transonic or turbulent flows. Initial experimental validation of such software is performed using a wind tunnel with the final validation coming in full-scale testing, e.g. flight tests.

Background and History

The fundamental basis of almost all CFD problems are the Navier–Stokes equations, which define any single-phase (gas or liquid, but not both) fluid flow. These equations can be simplified by removing terms describing viscous actions to yield the Euler equations. Further simplification, by removing terms describing vorticity yields the full potential equations. Finally, for small perturbations in subsonic and supersonic flows (not transonic or hypersonic) these equations can be linearized to yield the linearized potential equations.

Historically, methods were first developed to solve the linearized potential equations. Two-dimensional (2D) methods, using conformal transformations of the flow about a cylinder to the flow about an airfoil were developed in the 1930s.

One of the earliest type of calculations resembling modern CFD are those by Lewis Fry Richardson, in the sense that these calculations used finite differences and divided the physical space in cells. Although they failed dramatically, these calculations, together with Richardson's book "Weather prediction by numerical process", set the basis for modern CFD and numerical meteorology.

In fact, early CFD calculations during the 1940s using ENIAC used methods close to those in Richardson's 1922 book.

The computer power available paced development of three-dimensional methods. Probably the first work using computers to model fluid flow, as governed by the Navier-Stokes equations, was performed at Los Alamos National Labs, in the T3 group. This group was led by Francis H. Harlow, who is widely considered as one of the pioneers of CFD. From 1957 to late 1960s, this group developed a variety of numerical methods to simulate transient two-dimensional fluid flows, such as Particle-in-cell method (Harlow, 1957), Fluid-in-cell method (Gentry, Martin and Daly, 1966), Vorticity stream function method (Jake Fromm, 1963), and Marker-and-cell method (Harlow and Welch, 1965). Fromm's vorticity-stream-function method for 2D, transient, incompressible flow was the first treatment of strongly contorting incompressible flows in the world.

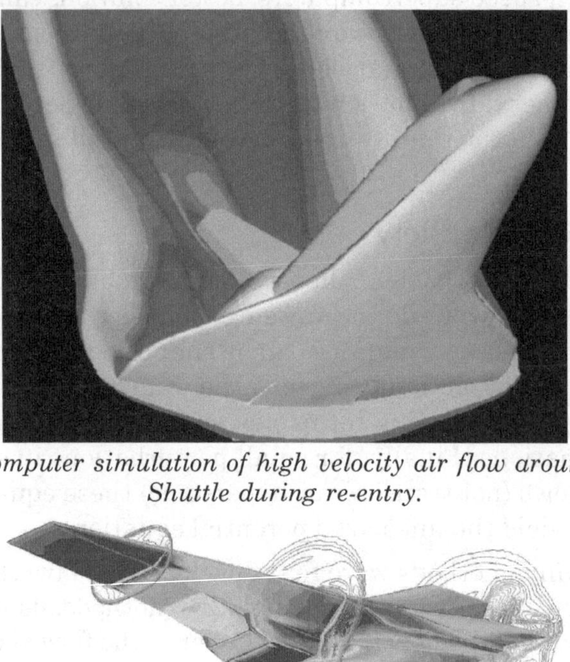

Figure: *A computer simulation of high velocity air flow around the Space Shuttle during re-entry.*

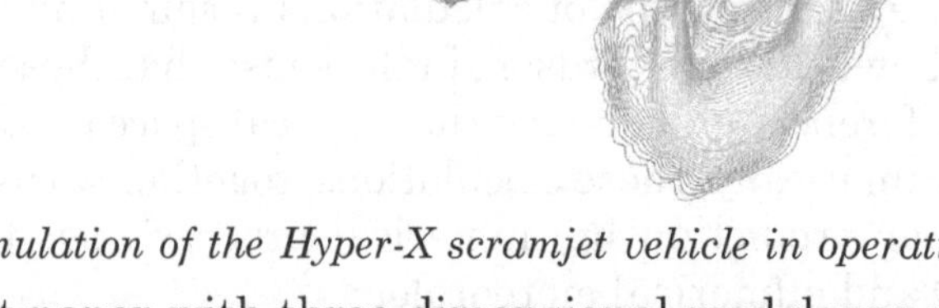

Figure: *A simulation of the Hyper-X scramjet vehicle in operation at Mach-7*

The first paper with three-dimensional model was published by John Hess and A.M.O. Smith of Douglas Aircraft in 1967. This method discretized the surface of the geometry with panels, giving rise to this class of programs being called Panel Methods. Their method itself was simplified, in that it did not include lifting flows and hence was mainly applied to ship hulls and aircraft fuselages. The first lifting Panel Code (A230) was described in a paper written by Paul Rubbert and Gary Saaris of Boeing Aircraft in 1968. In time, more advanced three-dimensional Panel Codes were developed at Boeing (PANAIR, A502), Lockheed (Quadpan), Douglas (HESS), McDonnell Aircraft (MACAERO), NASA (PMARC) and Analytical Methods (WBAERO, USAERO and VSAERO). Some (PANAIR, HESS and MACAERO) were higher order codes, using higher order distributions of surface singularities, while others (Quadpan, PMARC, USAERO and VSAERO) used single singularities on each surface panel. The advantage of the lower order codes was that they ran much faster on the computers of the time. Today, VSAERO has grown to be a multi-order code and is the most widely used program of this class. It has been used in the development of many submarines, surface ships, automobiles, helicopters, aircraft, and more recently wind turbines. Its sister code, USAERO is an unsteady panel method that has also been used for modelling such things as high speed trains and racing yachts. The NASA PMARC code from an early version of VSAERO and a derivative of PMARC, named CMARC, is also commercially available.

In the two-dimensional realm, a number of Panel Codes have been developed for airfoil analysis and design. The codes typically have a boundary layer analysis included, so that viscous effects can be modelled. Professor Richard Eppler of the University of Stuttgart developed the PROFILE code, partly with NASA funding, which became available in the early 1980s. This was soon followed by MIT Professor Mark Drela's XFOIL code. Both PROFILE and XFOIL incorporate two-dimensional panel codes, with coupled boundary layer codes for airfoil analysis work. PROFILE uses a conformal transformation method for inverse airfoil design, while XFOIL has both a conformal transformation and an inverse panel method for airfoil design.

An intermediate step between Panel Codes and Full Potential codes were codes that used the Transonic Small Disturbance equations. In particular, the three-dimensional WIBCO code, developed by Charlie Boppe of Grumman Aircraft in the early 1980s has seen heavy use.

Developers turned to Full Potential codes, as panel methods could not calculate the non-linear flow present at transonic speeds. The first description of a means of using the Full Potential equations was published by Earll Murman and Julian Cole of Boeing in 1970. Frances Bauer, Paul Garabedian and David Korn of the Courant Institute at

New York University (NYU) wrote a series of two-dimensional Full Potential airfoil codes that were widely used, the most important being named Program H. A further growth of Program H was developed by Bob Melnik and his group at Grumman Aerospace as Grumfoil. Antony Jameson, originally at Grumman Aircraft and the Courant Institute of NYU, worked with David Caughey to develop the important three-dimensional Full Potential code FLO22 in 1975. Many Full Potential codes emerged after this, culminating in Boeing's Tranair (A633) code, which still sees heavy use.

The next step was the Euler equations, which promised to provide more accurate solutions of transonic flows. The methodology used by Jameson in his three-dimensional FLO57 code (1981) was used by others to produce such programs as Lockheed's TEAM program and IAI/Analytical Methods' MGAERO program. MGAERO is unique in being a structured cartesian mesh code, while most other such codes use structured body-fitted grids (with the exception of NASA's highly successful CART3D code, Lockheed's SPLITFLOW code and Georgia Tech's NASCART-GT). Antony Jameson also developed the three-dimensional AIRPLANE code which made use of unstructured tetrahedral grids.

In the two-dimensional realm, Mark Drela and Michael Giles, then graduate students at MIT, developed the ISES Euler program (actually a suite of programs) for airfoil design and analysis. This code first became available in 1986 and has been further developed to design, analyze and optimize single or multi-element airfoils, as the MSES program. MSES sees wide use throughout the world. A derivative of MSES, for the design and analysis of airfoils in a cascade, is MISES, developed by Harold "Guppy" Youngren while he was a graduate student at MIT.

The Navier–Stokes equations were the ultimate target of developers. Two-dimensional codes, such as NASA Ames' ARC2D code first emerged. A number of three-dimensional codes were developed (ARC3D, OVERFLOW, CFL3D are three successful NASA contributions), leading to numerous commercial packages.

Detached Eddy Simulation

Detached eddy simulation (DES) is a modification of a RANS model in which the model switches to a subgrid scale formulation in regions fine enough for LES calculations. Regions near solid boundaries and where the turbulent length scale is less than the maximum grid dimension are assigned the RANS mode of solution. As the turbulent length scale exceeds the grid dimension, the regions are solved using the LES mode. Therefore the grid resolution is not as demanding as pure LES, thereby considerably cutting down the cost of the computation.

Though DES was initially formulated for the Spalart-Allmaras model, it can be implemented with other RANS models (Strelets, 2001), by appropriately modifying the length scale which is explicitly or implicitly involved in the RANS model. So while Spalart-Allmaras model based DES acts as LES with a wall model, DES based on other models (like two equation models) behave as a hybrid RANS-LES model. Grid generation is more complicated than for a simple RANS or LES case due to the RANS-LES switch. DES is a non-zonal approach and provides a single smooth velocity field across the RANS and the LES regions of the solution.

Vorticity Confinement Method

Vorticity confinement (VC), a physics-based computational fluid dynamics model analogous to shock capturing methods, was invented by Dr. John Steinhoff, professor at the University of Tennessee Space Institute, in the late 1980s to solve vortex dominated flows. It was first formulated to capture concentrated vortices shed from the wings, and later became popular in a wide range of research areas. During the 1990s and 2000s, it became widely used in the field of engineering.

The Method

VC has a basic familiarity to solitary wave approach which is extensively used in many condensed matter physics applications. The effect of VC is to capture the small scale features over as few as 2 grid cells as they convect through the flow. The basic idea is similar to that of compression discontinuity in Eulerian shock capturing methods. The internal structure is maintained thin and so the details of the internal structure may not be important.

Example

Consider 2D Euler equations, modified using the confinement term, F:

$$\frac{\partial \mathrm{u}}{\partial t} + \mathrm{u}\cdot\nabla\mathrm{u} + \nabla\frac{P}{\rho} = F_D(\mathrm{u}) - F_C(\mathrm{u})$$

The discretized Euler equations with the extra term can be solved on fairly coarse grids, with simple low order accurate numerical methods, but still yield concentrated vortices which convect without spreading. VC has different forms, one of which is VC1. It involves an added dissipation, F_D ,to the partial differential equation, which when balanced with inward convection, F_C, produce stable solutions. Another form is termed as VC2 in which dissipation is balanced with nonlinear anti-diffusion to produce stable solitary wave-like solutions.

F_D : Dissipation

F_C : Inward convection for VC1 and nonlinear anti-diffusion for VC2

The main difference between VC1 and VC2 is that in the latter the centroid of the vortex follows the local velocity moment weighted by vorticity. This should provide greater accuracy than VC1 in cases where the convecting field is weak compared to the self-induced velocity of the vortex. One drawback is that VC2 is not as robust as VC1 because while VC1 involves convection like inward propagation of vorticity balanced by an outward second order diffusion, VC2 involves a second order inward propagation of vorticity balanced by 4th order outward dissipation. This approach has been further extended to solve wave equation and is called Wave confinement (WC).

Immersed Boundary

To enforce no-slip boundary conditions on immersed surfaces, first, the surface is represented implicitly by a smooth "level set" function, "f", defined at each grid point. This is the (signed) distance from each grid point to the nearest point on the surface of an object – positive outside, negative inside. Then, at each time step during the solution, velocities in the interior are set to zero. In a computation using VC, this results in a thin vortical region along the surface, which is smooth in the tangential direction, with no "staircase" effects. The important point is that no special logic is required in the "cut" cells, unlike many conventional schemes: only the same VC equations are applied, as in the rest of the grid, but with a different form for F.

Also, unlike many conventional immersed surface schemes, which are inviscid because of cell size constraints, there is effectively a no-slip boundary condition, which results in a boundary layer with well-defined total vorticity and which, because of VC, remains thin, even after separation. The method is especially effective for complex configurations with separation from sharp corners. Also, even with constant coefficients, it can approximately treat separation from smooth surfaces. General blunt bodies, which typically shed turbulent vorticity that induces a velocity around an upstream body. It is inconsistent to use body fitted grids as the vorticity convects through a non fitted grid.

Linear Eddy Model

The Linear eddy model is a technique used to simulate the convective mixing that takes place in turbulent flow. Specifically, it provides a mathematical way to describe the interactions of a scalar variable within the vector flow field. It is primarily used in one-dimensional representations of turbulent flow, since it can be applied across a wide range of length scales and Reynolds numbers. This model is generally used as a building block for more complicated flow representations, as it provides high resolution predictions that hold across a large range of flow conditions.

Two-Phase Flow

The modelling of two-phase flow is still under development. Different methods have been proposed lately. The Volume of fluid method has received a lot of attention lately, for problems that do not have dispersed particles, but the Level set method and front tracking are also valuable approaches . Most of these methods are either good in maintaining a sharp interface or at conserving mass . This is crucial since the evaluation of the density, viscosity and surface tension is based on the values averaged over the interface. Lagrangian multiphase models, which are used for dispersed media, are based on solving the Lagrangian equation of motion for the dispersed phase.

Solution Algorithms

Discretization in space produces a system of ordinary differential equations for unsteady problems and algebraic equations for steady problems. Implicit or semi-implicit methods are generally used to integrate the ordinary differential equations, producing a system of (usually) nonlinear algebraic equations. Applying a Newton or Picard iteration produces a system of linear equations which is nonsymmetric in the presence of advection and indefinite in the presence of incompressibility. Such systems, particularly in 3D, are frequently too large for direct solvers, so iterative methods are used, either stationary methods such as successive overrelaxation or Krylov subspace methods. Krylov methods such as GMRES, typically used with preconditioning, operate by minimizing the residual over successive subspaces generated by the preconditioned operator.

Multigrid has the advantage of asymptotically optimal performance on many problems. Traditional solvers and preconditioners are effective at reducing high-frequency components of the residual, but low-frequency components typically require many iterations to reduce. By operating on multiple scales, multigrid reduces all components of the residual by similar factors, leading to a mesh-independent number of iterations.

For indefinite systems, preconditioners such as incomplete LU factorization, additive Schwarz, and multigrid perform poorly or fail entirely, so the problem structure must be used for effective preconditioning. Methods commonly used in CFD are the SIMPLE and Uzawa algorithms which exhibit mesh-dependent convergence rates, but recent advances based on block LU factorization combined with multigrid for the resulting definite systems have led to preconditioners that deliver mesh-independent convergence rates.

Unsteady Aerodynamics

CFD made a major break through in late 70s with the introduction of LTRAN2, a 2-D code to model oscillating airfoils based on transonic small perturbation theory by Ballhaus and associates. It uses a Murman-Cole switch algorithm for modelling the moving shock-waves. Later it was extended to 3-D with use of a rotated difference scheme by AFWAL/Boeing that resulted in LTRAN3.

Electrochemistry

Electrochemistry is a branch of chemistry that studies chemical reactions which take place in a solution at the interface of an electron conductor (the electrode: a metal or a semiconductor) and an ionic conductor (the electrolyte). These reactions involve electron transfer between the electrode and the electrolyte or species in solution. Thus electrochemistry deals with interactions between electrical energy and chemical change and vice versa.

If a chemical reaction is driven by an externally applied voltage, as in electrolysis, or if a voltage is created by a chemical reaction as in a battery, it is an *electrochemical* reaction. In contrast, chemical reactions where electrons are transferred between molecules are called oxidation-reduction (redox) reactions. In general, electrochemistry deals with situations where redox reactions are separated in space or time, connected by an external electric circuit.

Oxidation and Reduction

The term "redox" stands for reduction-oxidation. It refers to electrochemical processes involving electron transfer to or from a molecule or ion changing its oxidation state. This reaction can occur through the application of an external voltage or through the release of chemical energy. Oxidation and reduction describe the change of oxidation state that takes place in the atoms, ions or molecules involved in an electrochemical reaction. Formally, oxidation state is the hypothetical charge that an atom would have if all bonds to atoms of different elements were 100% ionic. An atom or ion that gives up an electron to another atom or ion has its oxidation state increase, and the recipient of the negatively charged electron has its oxidation state decrease.

For example, when atomic sodium reacts with atomic chlorine, sodium donates one electron and attains an oxidation state of +1. Chlorine accepts the electron and its oxidation state is reduced to "1. The sign of the oxidation state (positive/negative) actually corresponds to the value of each ion's electronic charge. The attraction of the differently charged sodium and chlorine ions is the reason they then form an ionic bond.

The loss of electrons from an atom or molecule is called oxidation, and the gain of electrons is reduction. This can be easily remembered through the use of mnemonic devices. Two of the most popular are *"OIL RIG"* (Oxidation Is Loss, Reduction Is Gain) and *"LEO"* says *"GER"* (Lose Electrons: Oxidation, Gain Electrons: Reduction). Oxidation and reduction always occur in a paired fashion such that one species is oxidized when another is reduced. For cases where electrons are shared (covalent bonds) between atoms with large differences in electronegativity, the electron is assigned to the atom with the largest electronegativity in determining the oxidation state.

The atom or molecule which loses electrons is known as the *reducing agent*, or *reductant*, and the substance which accepts the electrons is called the *oxidizing agent*, or *oxidant*. Thus, the oxidizing agent is always being reduced in a reaction; the reducing agent is always being oxidized. Oxygen is a common oxidizing agent, but not the only one. Despite the name, an oxidation reaction does not necessarily need to involve oxygen. In fact, a fire can be fed by an oxidant other than oxygen; fluorine fires are often unquenchable, as fluorine is an even stronger oxidant (it has a higher electronegativity and thus accepts electrons even better) than oxygen.

For reactions involving oxygen, the gain of oxygen implies the oxidation of the atom or molecule to which the oxygen is added (and the oxygen is reduced). In organic compounds, such as butane or ethanol, the loss of hydrogen implies oxidation of the molecule from which it is lost (and the hydrogen is reduced). This follows because the hydrogen donates its electron in covalent bonds with non-metals but it takes the electron along when it is lost. Conversely, loss of oxygen or gain of hydrogen implies reduction.

Balancing Redox Reactions

Electrochemical reactions in water are better understood by balancing redox reactions using the ion-electron method where H^+, OH^- ion, H_2O and electrons (to compensate the oxidation changes) are added to cell's half-reactions for oxidation and reduction.

Acidic Medium

In acid medium H^+ ions and water are added to half-reactions to balance the overall reaction. For example, when manganese reacts with sodium bismuthate.

Unbalanced reaction: $Mn^{2+}(aq) + NaBiO_3(s) \rightarrow Bi^{3+}(aq) + MnO_4^-(aq)$

Oxidation: $4\ H_2O(l) + Mn^{2+}(aq) \rightarrow MnO_4^-(aq) + 8\ H^+(aq) + 5\ e^-$

Reduction: $2\ e^- + 6\ H^+(aq) + BiO_3^-(s) \rightarrow Bi^{3+}(aq) + 3\ H_2O(l)$

Finally, the reaction is balanced by multiplying the number of electrons from the reduction half reaction to oxidation half reaction and vice versa and adding both half reactions, thus solving the equation.

$8\ H_2O(l) + 2\ Mn^{2+}(aq) \rightarrow 2\ MnO_4^-(aq) + 16\ H^+(aq) + 10\ e^-$

$10\ e^- + 30\ H^+(aq) + 5\ BiO_3^-(s) \rightarrow 5\ Bi^{3+}(aq) + 15\ H_2O(l)$

Reaction balanced:

$14\ H^+(aq) + 2\ Mn^{2+}(aq) + 5\ NaBiO_3(s) \rightarrow 7\ H_2O(l) + 2\ MnO_4^-(aq) + 5\ Bi^{3+}(aq) + 5\ Na^+(aq)$

Basic Medium

In basic medium OH^- ions and water are added to half reactions to balance the overall reaction. For example, on reaction between potassium permanganate and sodium sulfite.

Unbalanced reaction: $KMnO_4 + Na_2SO_3 + H_2O \rightarrow MnO_2 + Na_2SO_4 + KOH$

Reduction: $3\ e^- + 2\ H_2O + MnO_4^- \rightarrow MnO_2 + 4\ OH^-$

Oxidation: $2\ OH^- + SO_3^{2-} \rightarrow SO_4^{2-} + H_2O + 2\ e^-$

The same procedure as followed on acid medium by multiplying electrons to opposite half reactions solve the equation thus balancing the overall reaction.

$$6\ e^- + 4\ H_2O + 2\ MnO_4^- \rightarrow 2\ MnO_2 + 8\ OH^-$$

$$6\ OH^- + 3\ SO_3^{2-} \rightarrow 3\ SO_4^{2-} + 3\ H_2O + 6e^-$$

Equation balanced:

$2\ KMnO_4 + 3\ Na_2SO_3 + H_2O \rightarrow 2\ MnO_2 + 3\ Na_2SO_4 + 2\ KOH$

Neutral Medium

The same procedure as used on acid medium is applied, for example on balancing using electron ion method to complete combustion of propane.

Unbalanced reaction: $C_3H_8 + O_2 \rightarrow CO_2 + H_2O$

Reduction: $4\ H^+ + O_2 + 4\ e^- \rightarrow 2\ H_2O$

Oxidation: $6\ H_2O + C_3H_8 \rightarrow 3\ CO_2 + 20\ e^- + 20\ H^+$

As in acid and basic medium, electrons which were used to compensate oxidation changes are multiplied to opposite half reactions, thus solving the equation.

$$20\ H^+ + 5\ O_2 + 20\ e^- \rightarrow 10\ H_2O$$

$$6\ H_2O + C_3H_8 \rightarrow 3\ CO_2 + 20\ e^- + 20\ H^+$$

Equation balanced:

$$C_3H_8 + 5\ O_2 \rightarrow 3\ CO_2 + 4\ H_2O$$

Electrochemical Cells

An electrochemical cell is a device that produces an electric current from energy released by a spontaneous redox reaction. This kind of cell includes the Galvanic cell or Voltaic cell, named after Luigi Galvani and Alessandro Volta, both scientists who conducted several experiments on chemical reactions and electric current during the late 18th century. Electrochemical cells have two conductive electrodes (the anode and the cathode). The anode is defined as the electrode where oxidation occurs and the cathode is the electrode where the reduction takes place. Electrodes can be made from any sufficiently conductive materials, such as metals, semiconductors, graphite, and even conductive polymers. In between these electrodes is the electrolyte, which contains ions that can freely move.

The galvanic cell uses two different metal electrodes, each in an electrolyte where the positively charged ions are the oxidized form of the electrode metal. One electrode will undergo oxidation (the anode) and the other will undergo reduction (the cathode). The metal of the anode will oxidize, going from an oxidation state of 0 (in the solid form) to a positive oxidation state and become an ion. At the cathode, the metal ion in solution will accept one or more electrons from the cathode and the ion's oxidation state is reduced to 0. This forms a solid metal that electrodeposits on the cathode. The two electrodes must be electrically connected to each other, allowing for a flow of electrons that leave the metal of the anode and flow through this connection to the ions at the surface of the cathode. This flow of electrons is an electrical current that can be used to do work, such as turn a motor or power a light.

A galvanic cell whose electrodes are zinc and copper submerged in zinc sulfate and copper sulfate, respectively, is known as a Daniell cell.

Half reactions for a Daniell cell are these:

Zinc electrode (anode): $Zn(s) \rightarrow Zn^{2+}(aq) + 2\ e^-$

Copper electrode (cathode): $Cu^{2+}(aq) + 2\ e^- \rightarrow Cu(s)$

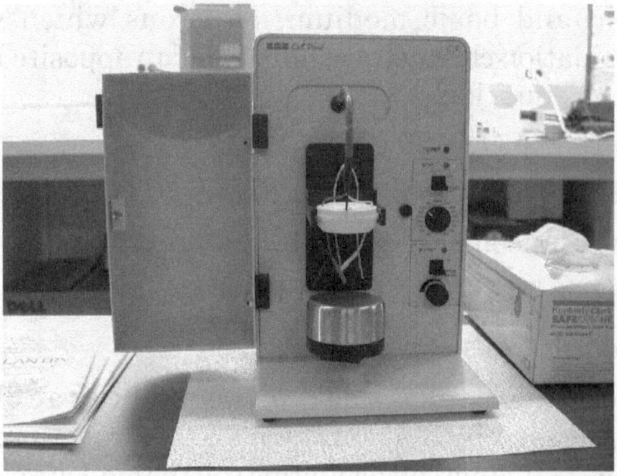

Figure: *A modern cell stand for electrochemical research. The electrodes attach to high-quality metallic wires, and the stand is attached to a potentiostat/galvanostat (not pictured). A shot glass-shaped container is aerated with a noble gas and sealed with the Teflon block.*

In this example, the anode is zinc metal which oxidizes (loses electrons) to form zinc ions in solution, and copper ions accept electrons from the copper metal electrode and the ions deposit at the copper cathode as an electrodeposit. This cell forms a simple battery as it will spontaneously generate a flow of electrical current from the anode to the cathode through the external connection. This reaction can be driven in reverse by applying a voltage, resulting in the deposition of zinc metal at the anode and formation of copper ions at the cathode.

To provide a complete electric circuit, there must also be an ionic conduction path between the anode and cathode electrolytes in addition to the electron conduction path. The simplest ionic conduction path is to provide a liquid junction. To avoid mixing between the two electrolytes, the liquid junction can be provided through a porous plug that allows ion flow while reducing electrolyte mixing. To further minimize mixing of the electrolytes, a salt bridge can be used which consists of an electrolyte saturated gel in an inverted U-tube. As the negatively charged electrons flow in one direction around this circuit, the positively charged metal ions flow in the opposite direction in the electrolyte.

A voltmetre is capable of measuring the change of electrical potential between the anode and the cathode. Electrochemical cell voltage is also referred to as electromotive force or emf.

A cell diagram can be used to trace the path of the electrons in the electrochemical cell. For example, here is a cell diagram of a Daniell cell:

$$Zn(s) \mid Zn^{2+} (1M) \mid\mid Cu^{2+} (1M) \mid Cu(s)$$

First, the reduced form of the metal to be oxidized at the anode (Zn) is written. This is separated from its oxidized form by a vertical line, which represents the limit between the phases (oxidation changes). The double vertical lines represent the saline bridge on the cell. Finally, the oxidized form of the metal to be reduced at the cathode, is written, separated from its reduced form by the vertical line. The electrolyte concentration is given as it is an important variable in determining the cell potential.

Standard Electrode Potential

To allow prediction of the cell potential, tabulations of standard electrode potential are available. Such tabulations are referenced to the standard hydrogen electrode (SHE). The standard hydrogen electrode undergoes the reaction

$$2\ H^+(aq) + 2\ e^- \rightarrow H_2$$

which is shown as reduction but, in fact, the SHE can act as either the anode or the cathode, depending on the relative oxidation/reduction potential of the other electrode/electrolyte combination. The term standard in SHE requires a supply of hydrogen gas bubbled through the electrolyte at a pressure of 1 atm and an acidic electrolyte with H^+ activity equal to 1 (usually assumed to be $[H^+] = 1$ mol/litre).

The SHE electrode can be connected to any other electrode by a salt bridge to form a cell. If the second electrode is also at standard conditions, then the measured cell potential is called the standard electrode potential for the electrode. The standard electrode potential for the SHE is zero, by definition. The polarity of the standard electrode potential provides information about the relative reduction potential of the electrode compared to the SHE. If the electrode has a positive potential with respect to the SHE, then that means it is a strongly reducing electrode which forces the SHE to be the anode (an example is Cu in aqueous $CuSO_4$ with a standard electrode potential of 0.337 V). Conversely, if the measured potential is negative, the electrode is more oxidizing than the SHE (such as Zn in $ZnSO_4$ where the standard electrode potential is "0.76 V).

Standard electrode potentials are usually tabulated as reduction potentials. However, the reactions are reversible and the role of a particular electrode in a cell depends on the relative oxidation/reduction potential of both electrodes. The oxidation potential for a particular electrode is just the negative of the reduction potential. A standard cell potential can be determined by looking up the standard electrode potentials for both electrodes (sometimes called half cell potentials). The one that is smaller will be the anode and will undergo oxidation. The cell potential is then calculated as the sum of the reduction potential for the cathode and the oxidation potential for the anode.

$$E^\circ_{cell} = E^\circ_{red}(\text{cathode}) - E^\circ_{red}(\text{anode}) = E^\circ_{red}(\text{cathode}) + E^\circ_{oxi}(\text{anode})$$

For example, the standard electrode potential for a copper electrode is:

Cell diagram

$$Pt(s) \mid H_2(1\ atm) \mid H^+(1\ M) \mid\mid Cu^{2+}\ (1\ M) \mid Cu(s)$$

$$E^\circ_{cell} = E^\circ_{red}(\text{cathode}) - E^\circ_{red}(\text{anode})$$

At standard temperature, pressure and concentration conditions, the cell's emf (measured by a multimetre) is 0.34 V. By definition, the electrode potential for the SHE is zero. Thus, the Cu is the cathode and the SHE is the anode giving

$$E_{cell} = E^\circ(Cu^{2+}/Cu) - E^\circ(H^+/H_2)$$

Or,

$$E^\circ(Cu^{2+}/Cu) = 0.34\ V$$

Changes in the stoichiometric coefficients of a balanced cell equation will not change E°_{red} value because the standard electrode potential is an intensive property.

Spontaneity of Redox Reaction

During operation of electrochemical cells, chemical energy is transformed into electrical energy and is expressed mathematically as the product of the cell's emf and the electric charge transferred through the external circuit.

$$\text{Electrical energy} = E_{cell} C_{trans}$$

where E_{cell} is the cell potential measured in volts (V) and C_{trans} is the cell current integrated over time and measured in coulombs (C); C_{trans} can also be determined by multiplying the total number of electrons transferred (measured in moles) times Faraday's constant (F).

The emf of the cell at zero current is the maximum possible emf. It is used to calculate the maximum possible electrical energy that could be obtained from a chemical reaction. This energy is referred to as electrical work and is expressed by the following equation:

$$W_{max} = W_{electrical} = -nF \cdot E_{cell},$$

where work is defined as positive into the system.

Since the free energy is the maximum amount of work that can be extracted from a system, one can write:

$$\Delta G = -nF \cdot E_{cell}$$

A positive cell potential gives a negative change in Gibbs free energy. This is consistent with the cell production of an electric current from the cathode to the anode through the external circuit. If the current is driven in the opposite direction by imposing an external potential, then work is done on the cell to drive electrolysis.

A spontaneous electrochemical reaction (change in Gibbs free energy less than zero) can be used to generate an electric current in electrochemical cells. This is the basis of all batteries and fuel cells. For example, gaseous oxygen (O_2) and hydrogen (H_2) can be combined in a fuel cell to form water and energy, typically a combination of heat and electrical energy. Conversely, non-spontaneous electrochemical reactions can be driven forward by the application of a current at sufficient voltage. The electrolysis of water into gaseous oxygen and hydrogen is a typical example.

The relation between the equilibrium constant, *K*, and the Gibbs free energy for an electrochemical cell is expressed as follows:

$$\Delta G° = -RT \ln(K) = -nF \cdot E°_{cell}$$

Rearranging to express the relation between standard potential and equilibrium constant yields

$$E^{o}_{cell} = \frac{RT}{nF} \ln K \; .$$

The previous equation can use Briggsian logarithm as shown below:

$$E^{o}_{cell} = \frac{0.0591V}{n} \log K$$

Cell emf Dependency on Changes in Concentration

Nernst Equation

The standard potential of an electrochemical cell requires standard conditions (ΔG°) for all of the reactants. When reactant concentrations differ from standard conditions, the cell potential will deviate from the standard potential. In the 20th century German chemist Walther Nernst proposed a mathematical model to determine the effect of reactant concentration on electrochemical cell potential.

In the late 19th century, Josiah Willard Gibbs had formulated a theory to predict whether a chemical reaction is spontaneous based on the free energy

$$\Delta G = \Delta G° + RT \cdot \ln(Q)$$

Here ΔG is change in Gibbs free energy, $\Delta G°$ is the cell potential when Q is equal to 1, T is absolute temperature(Kelvin), R is the gas constant and Q is reaction quotient which can be found by dividing products by reactants using only those products and reactants that are aqueous or gaseous.

Gibbs' key contribution was to formalize the understanding of the effect of reactant concentration on spontaneity.

Based on Gibbs' work, Nernst extended the theory to include the contribution from electric potential on charged species. As shown in the previous section, the change in Gibbs free energy for an electrochemical cell can be related to the cell potential. Thus, Gibbs' theory becomes

$$nF\Delta E = nF\Delta E° - RT \ln(Q)$$

Here n is the number of electrons/mole product, F is the Faraday constant (coulombs/mole), and ΔE is cell potential.

Finally, Nernst divided through by the amount of charge transferred to arrive at a new equation which now bears his name:

$$\Delta E = \Delta E° - (RT/nF)\ln(Q)$$

Assuming standard conditions (T = 25 °C) and R = 8.3145 J/(K·mol), the equation above can be expressed on base—10 logarithm as shown below:

$$\Delta E = \Delta E^o - \frac{0.05916 \text{ V}}{\text{n}} \log \text{Q}$$

Concentration Cells

A concentration cell is an electrochemical cell where the two electrodes are the same material, the electrolytes on the two half-cells involve the same ions, but the electrolyte concentration differs between the two half-cells.

An example is an electrochemical cell, where two copper electrodes are submerged in two copper(II) sulfate solutions, whose concentrations are 0.05 M and 2.0 M, connected through a salt bridge. This type of cell will generate a potential that can be predicted by the Nernst equation. Both can undergo the same chemistry (although the reaction proceeds in reverse at the anode)

$$Cu^{2+}(aq) + 2\ e^- \rightarrow Cu(s)$$

Le Chatelier's principle indicates that the reaction is more favourable to reduction as the concentration of Cu^{2+} ions increases. Reduction will take place in the cell's compartment where concentration is higher and oxidation will occur on the more dilute side.

The following cell diagram describes the cell mentioned above:

$$Cu(s)\ |\ Cu^{2+}\ (0.05\ M)\ ||\ Cu^{2+}\ (2.0\ M)\ |\ Cu(s)$$

Where the half cell reactions for oxidation and reduction are:

Oxidation: $Cu(s) \rightarrow Cu^{2+}\ (0.05\ M) + 2\ e^-$

Reduction: $Cu^{2+}\ (2.0\ M) + 2\ e^- \rightarrow Cu(s)$

Overall reaction: $Cu^{2+}\ (2.0\ M) \rightarrow Cu^{2+}\ (0.05\ M)$

The cell's emf is calculated through Nernst equation as follows:

$$E = E^o - \frac{0.05916V}{2}\log\frac{[Cu^{2+}]_{diluted}}{[Cu^{2+}]_{concentrated}}$$

The value of $E°$ in this kind of cell is zero, as electrodes and ions are the same in both half-cells.

After replacing values from the case mentioned, it is possible to calculate cell's potential:

$$E = 0 - \frac{0.05916V}{2}\log\frac{0.05}{2.0} = 0.0474V$$

or by:

$$E = 0 - \frac{0.0257V}{2}\ln\frac{0.05}{2.0} = 0.0474V$$

However, this value is only approximate, as reaction quotient is defined in terms of ion activities which can be approximated with the concentrations as calculated here. The Nernst equation plays an important role in understanding electrical effects in cells and organelles. Such effects include nerve synapses and cardiac beat as well as the resting potential of a somatic cell.

Battery

Many types of battery have been commercialized and represent an important practical application of electrochemistry. Early wet cells powered the first telegraph and telephone systems, and were the source of current for electroplating. The zinc-manganese dioxide dry cell was the first portable, non-spillable battery type that made flashlights and other portable devices practical. The mercury battery using zinc and mercuric oxide provided higher levels of power and capacity than the original dry cell for early electronic devices, but has been phased out of common use due to the danger of mercury pollution from discarded cells.

The lead acid battery was the first practical secondary (rechargeable) battery that could have its capacity replenished from an external source. The electrochemical reaction that produced current was (to a useful degree) reversible, allowing electrical energy and chemical energy to be interchanged as needed. Common lead acid batteries contain a mixture of acid and water, as well as lead plates. The most common mixture used today is 30% acid. One problem however is if left uncharged acid will crystallize within the lead plates of the battery rendering it useless. These batteries last an average of 3 years with daily use however it is not unheard of for a lead acid battery to still be functional after 7-10 years. Lead-acid cells continue to be widely used in automobiles.

All the preceding types have water-based electrolytes, which limits the maximum voltage per cell. The freezing of water limits low temperature performance. The lithium battery, which does not (and cannot) use water in the electrolyte, provides improved performance over other types; a rechargeable lithium ion battery is an essential part of many mobile devices.

The flow battery, an experimental type, offers the option of vastly larger energy capacity because its reactants can be replenished from external reservoirs. The fuel cell can turn the chemical energy bound in hydrocarbon gases or hydrogen directly into electrical energy with much higher efficiency than any combustion process; such devices have powered many spacecraft and are being applied to grid energy storage for the public power system.

Corrosion

Corrosion is the term applied to steel rust caused by an electrochemical process. Most people are likely familiar with the corrosion of iron, in the form of reddish rust. Other examples include the black tarnish on silver, and red or green corrosion that may appear on copper and its alloys, such as brass. The cost of replacing metals lost to corrosion is in the multi-billions of dollars per year.

Iron Corrosion

For iron rust to occur the metal has to be in contact with oxygen and water, although chemical reactions for this process are relatively complex and not all of them are completely understood, it is believed the causes are the following: Electron transferring (reduction-oxidation)

One area on the surface of the metal acts as the anode, which is where the oxidation (corrosion) occurs. At the anode, the metal gives up electrons.

$$Fe(s) \rightarrow Fe^{2+}(aq) + 2\ e^-$$

Electrons are transferred from iron reducing oxygen in the atmosphere into water on the cathode, which is placed in another region of the metal.

$$O_2(g) + 4\ H^+(aq) + 4\ e^- \rightarrow 2\ H_2O(l)$$

Global reaction for the process:

$$2\ Fe(s) + O_2(g) + 4\ H^+(aq) \rightarrow 2\ Fe^{2+}(aq) + 2\ H_2O(l)$$

Standard emf for iron rusting:

$$E° = E°_{cathode} - E°_{anode}$$

$$E° = 1.23V - (\text{“}0.44\ V) = 1.67\ V$$

Iron corrosion takes place on acid medium; H^+ ions come from reaction between carbon dioxide in the atmosphere and water, forming carbonic acid. Fe^{2+} ions oxides, following this equation:

$$4\ Fe^{2+}(aq) + O_2(g) + (4{+}2x)H_2O(l) \rightarrow 2\ Fe_2O_3{\cdot}xH_2O + 8\ H^+(aq)$$

Iron(III) oxide hydrated is known as rust. The concentration of water associated with iron oxide varies, thus chemical representation is presented as $Fe_2O_3{\cdot}xH_2O$. The electric circuit works as passage of electrons and ions occurs, thus if an electrolyte is present it will facilitate oxidation, this explains why rusting is quicker on salt water.

Corrosion of Common Metals

Coinage metals, such as copper and silver, slowly corrode through use. A patina of green-blue copper carbonate forms on the surface of copper with exposure to the water and carbon dioxide in the air. Silver coins or cutlery that are exposed to high sulfur foods such as eggs or the low levels of sulfur species in the air develop a layer of black Silver sulfide.

Gold and platinum are extremely difficult to oxidize under normal circumstances, and require exposure to a powerful chemical oxidizing agent such as aqua regia.

Some common metals oxidize extremely rapidly in air. Titanium and aluminium oxidize instantaneously in contact with the oxygen in the air. These metals form an extremely thin layer of oxidized metal on the surface. This thin layer of oxide protects the underlying layers of the metal from the air preventing the entire metal from oxidizing. These metals are used in applications where corrosion resistance is important. Iron, in contrast, has an oxide that forms in air and water, called rust, that does not stop the further oxidation of the iron. Thus iron left exposed to air and water will continue to rust until all of the iron is oxided.

Prevention of Corrosion

Attempts to save a metal from becoming anodic are of two general types. Anodic regions dissolve and destroy the structural integrity of the metal.

While it is almost impossible to prevent anode/cathode formation, if a non-conducting material covers the metal, contact with the electrolyte is not possible and corrosion will not occur.

Coating

Metals can be coated with paint or other less conductive metals (*passivation*). This prevents the metal surface from being exposed to electrolytes. Scratches exposing the metal substrate will result in corrosion. The region under the coating adjacent to the scratch acts as the anode of the reaction.

Sacrificial Anodes

A method commonly used to protect a structural metal is to attach a metal which is more anodic than the metal to be protected. This forces the structural metal to be cathodic, thus spared corrosion. It is called *"sacrificial"* because the anode dissolves and has to be replaced periodically.

Zinc bars are attached to various locations on steel ship hulls to render the ship hull cathodic. The zinc bars are replaced periodically. Other metals, such as magnesium, would work very well but zinc is the least expensive useful metal.

To protect pipelines, an ingot of buried or exposed magnesium (or zinc) is buried beside the pipeline and is connected electrically to the pipe above ground. The pipeline is forced to be a cathode and is protected from being oxidized and rusting. The magnesium anode is sacrificed. At intervals new ingots are buried to replace those lost.

Electrolysis

The spontaneous redox reactions of a conventional battery produce electricity through the different chemical potentials of the cathode and anode in the electrolyte. However, electrolysis requires an external source of electrical energy to induce a chemical reaction, and this process takes place in a compartment called an electrolytic cell.

Electrolysis of Molten Sodium Chloride

When molten, the salt sodium chloride can be electrolyzed to yield metallic sodium and gaseous chlorine. Industrially this process takes place in a special cell named Down's cell. The cell is connected to an electrical power supply, allowing electrons to migrate from the power supply to the electrolytic cell.

Reactions that take place at Down's cell are the following:

Anode (oxidation): $2\ Cl^- \rightarrow Cl_2(g) + 2\ e^-$

Cathode (reduction): $2\ Na^+(l) + 2\ e^- \rightarrow 2\ Na(l)$

Overall reaction: $2\ Na^+ + 2\ Cl^-(l) \rightarrow 2\ Na(l) + Cl_2(g)$

This process can yield large amounts of metallic sodium and gaseous chlorine, and is widely used on mineral dressing and metallurgy industries.

The emf for this process is approximately "4 V indicating a (very) non-spontaneous process. In order for this reaction to occur the power supply should provide at least a potential of 4 V. However, larger voltages must be used for this reaction to occur at a high rate.

Electrolysis of Water

Water can be converted to its component elemental gasses, H_2 and O_2 through the application of an external voltage. Water doesn't decompose into hydrogen and oxygen spontaneously as the Gibbs free energy for the process at standard conditions is about 474.4 kJ. The decomposition of water into hydrogen and oxygen can be performed in an electrolytic cell. In it, a pair of inert electrodes usually made of platinum immersed in water act as anode and cathode in the electrolytic process. The electrolysis starts with the application of an external voltage between the electrodes. This process will not occur except at extremely high voltages without an electrolyte such as sodium chloride or sulfuric acid (most used 0.1 M).

Bubbles from the gases will be seen near both electrodes. The following half reactions describe the process mentioned above:

Anode (oxidation): $2\ H_2O(l) \rightarrow O_2(g) + 4\ H^+(aq) + 4\ e^-$

Cathode (reduction): $2\ H_2O(g) + 2\ e^- \rightarrow H_2(g) + 2\ OH^-(aq)$

Overall reaction: $2\ H_2O(l) \rightarrow 2\ H_2(g) + O_2(g)$

Although strong acids may be used in the apparatus, the reaction will not net consume the acid. While this reaction will work at any conductive electrode at a sufficiently large potential, platinum catalyzes both hydrogen and oxygen formation, allowing for relatively mild voltages (~2 V depending on the pH).

Electrolysis of Aqueous Solutions

Electrolysis in an aqueous is a similar process as mentioned in electrolysis of water. However, it is considered to be a complex process because the contents in solution have to be analyzed in half reactions, whether reduced or oxidized.

Electrolysis of a Solution of Sodium Chloride

The presence of water in a solution of sodium chloride must be examined in respect to its reduction and oxidation in both electrodes. Usually, water is electrolysed as mentioned in electrolysis of water yielding *gaseous oxygen in the anode* and gaseous hydrogen in the cathode. On the other hand, sodium chloride in water dissociates in Na^+ and Cl^- ions, cation, which is the positive ion, will be attracted to the cathode (-), thus reducing the sodium ion. The anion will then be attracted to the anode (+) oxidizing chloride ion.

The following half reactions describes the process mentioned:

1. Cathode: $Na^+(aq) + e^- \rightarrow Na(s)$ $E^\circ_{red} = -2.71$ V
2. Anode: $2\ Cl^-(aq) \rightarrow Cl_2(g) + 2\ e^-$ $E^\circ_{red} = +1.36$ V
3. Cathode: $2\ H_2O(l) + 2\ e^- \rightarrow H_2(g) + 2\ OH^-(aq)$ $E^\circ_{red} = -0.83$ V
4. Anode: $2\ H_2O(l) \rightarrow O_2(g) + 4\ H^+(aq) + 4\ e^-$ $E^\circ_{red} = +1.23$ V

Reaction 1 is discarded as it has the most negative value on standard reduction potential thus making it less thermodynamically favourable in the process.

When comparing the reduction potentials in reactions 2 and 4, the reduction of chloride ion is favoured. Thus, if the Cl^- ion is favoured for reduction, then the water reaction is favoured for oxidation producing gaseous oxygen, however experiments show gaseous chlorine is produced and not oxygen.

Although the initial analysis is correct, there is another effect that can happen, known as the overvoltage effect. Additional voltage is sometimes required, beyond the voltage predicted by the E°_{cell}. This may be due to kinetic rather than thermodynamic considerations. In fact, it has been proven that the activation energy for the chloride ion is very low, hence favourable in kinetic terms. In other words, although the voltage applied is thermodynamically sufficient to drive electrolysis, the rate is so slow that to make the process proceed in a reasonable time frame, the voltage of the external source has to be increased (hence, overvoltage).

Finally, reaction 3 is favourable because it describes the proliferation of OH^- ions thus letting a probable reduction of H^+ ions less favourable an option.

The overall reaction for the process according to the analysis would be the following:

Anode (oxidation): $2\ Cl^-(aq) \rightarrow Cl_2(g) + 2\ e^-$

Cathode (reduction): $2\ H_2O(l) + 2\ e^- \rightarrow H_2(g) + 2\ OH^-(aq)$

Overall reaction: $2\ H_2O + 2\ Cl^-(aq) \rightarrow H_2(g) + Cl_2(g) + 2\ OH^-(aq)$

As the overall reaction indicates, the concentration of chloride ions is reduced in comparison to OH^- ions (whose concentration increases). The reaction also shows the production of gaseous hydrogen, chlorine and aqueous sodium hydroxide.

Quantitative Electrolysis and Faraday's Laws

Quantitative aspects of electrolysis were originally developed by Michael Faraday in 1834. Faraday is also credited to have coined the terms *electrolyte*, electrolysis, among many others while he studied quantitative analysis of electrochemical reactions. Also he was an advocate of the law of conservation of energy.

First Law

Faraday concluded after several experiments on electrical current in non-spontaneous process, the mass of the products yielded on the electrodes was proportional to the value of current supplied to the cell, the length of time the current existed, and the molar mass of the substance analyzed. In other words, the amount of a substance deposited on each electrode of an electrolytic cell is directly proportional to the quantity of electricity passed through the cell.

Below is a simplified equation of Faraday's first law:

$$m = \frac{1}{96485\,(\mathrm{C{\cdot}mol^{-1}})} \cdot \frac{QM}{n}$$

Where

m is the mass of the substance produced at the electrode (in grams),

Q is the total electric charge that passed through the solution (in coulombs),

n is the valence number of the substance as an ion in solution (electrons per ion),

M is the molar mass of the substance (in grams per mole).

Second Law

Faraday devised the laws of chemical electrodeposition of metals from solutions in 1857. He formulated the second law of electrolysis stating *"the amounts of bodies which are equivalent to each other in their ordinary chemical action have equal quantities of electricity naturally associated with them."* In other words, the quantities of different elements deposited by a given amount of electricity are in the ratio of their chemical equivalent weights.

An important aspect of the second law of electrolysis is electroplating which together with the first law of electrolysis, has a significant number of applications in the industry, as when used to protect metals to avoid corrosion.

Applications

There are various extremely important electrochemical processes in both nature and industry, like the coating of objects with metals or metal oxides through electrodeposition and the detection of alcohol in drunken drivers through the redox reaction of ethanol. The generation of chemical energy through photosynthesis is inherently an electrochemical process, as is production of metals like aluminum and titanium from their ores. Certain diabetes blood sugar metres measure the amount of glucose in the blood through its redox potential.

The action potentials that travel down neurons are based on electric current generated by the movement of sodium and potassium ions into and out of cells. Specialised cells in certain animals like the electric eel can generate electric currents powerful enough to disable much larger animals.

Electrochemiluminescence

Electrochemiluminescence or electrogenerated chemiluminescence (ECL) is a kind of luminescence produced during electrochemical reactions in solutions. In electrogenerated chemiluminescence, electrochemically generated intermediates undergo a highly exergonic reaction to produce an electronically excited state that then emits light upon relaxation to a lower-level state. This wavelength of the emitted photon of light corresponds to the energy gap between these two states. ECL excitation can be caused by energetic electron transfer (redox) reactions of electrogenerated species. Such luminescence excitation is a form of chemiluminescence where one/all reactants are produced electrochemically on the electrodes.

ECL is usually observed during application of potential (several volts) to electrodes of electrochemical cell that contains solution of luminescent species (polycyclic aromatic hydrocarbons, metal complexes, Quantum Dots or Nanoparticles) in aprotic organic solvent (ECL composition). In organic solvents both oxidized and reduced forms of luminescent species can be produced at different electrodes simultaneously or at a single one by sweeping its potential between oxidation and reduction. The excitation energy is obtained from recombination of oxidized and reduced species.

In aqueous medium which is mostly used for analytical applications simultaneous oxidation and reduction of luminescent species is difficult to achieve due to electrochemical splitting of water itself so the ECL reaction with the coreactants is used. In the later case luminescent species are oxidized at the electrode together with the coreactant which gives a strong reducing agent after some chemical transformations (the oxidative reduction mechanism).

ECL proved to be very useful in analytical applications as a highly sensitive and selective method. It combines analytical advantages of chemiluminescent analysis (absence of background optical signal) with ease of reaction control by applying electrode potential. As an analytical technique it presents outstanding advantages over other common analytical methods due to its versatility, simplified optical setup compared with photoluminescence (PL), and good temporal and spatial control compared with chemiluminescence (CL). Enhanced selectivity of ECL analysis is reached by variation of electrode potential thus controlling species that are oxidized/reduced at the electrode and take part in ECL reaction.

It generally uses Ruthenium complexes, especially $[Ru\ (Bpy)3]^{2+}$ (which releases a photon at ~620 nm) regenerating with TPA (Tripropylamine) in liquid phase or liquid–solid interface. It can be used as monolayer immobilized on an electrode surface (made e.g. of nafion, or special thin films made by Langmuir–Blogett technique or self-assembly technique) or as a coreactant or more commonly as a tag and used in HPLC, Ru tagged antibody based immunoassays, Ru Tagged DNA probes for PCR etc., NADH or H_2O_2 generation based biosensors, oxalate and organic amine detection and many other applications and can be detected from picomolar sensitivity to dynamic range of more than six orders of magnitude. Photon detection is done with photomultiplier tubes (PMT) or silicon photodiode or gold coated fibre-optic sensors. The importance of ECL techniques detection for bio-related applications has been well established. ECL is heavily used commercially for many clinical lab applications.

Environmental Engineering

Chemical engineers meet environmental challenges. Their unique expertise enables them to develop advanced technologies, monitoring devices, modelling techniques, and operating strategies that

- Reduce the volume and toxicity of pollutants allowed to enter the air, waterways, and soil;
- Significantly reduce the negative environmental impact of industrial facilities, power plants, and transportation vehicles; and
- Allow greater reuse of post-consumer and post-industrial waste streams.

Chapter 8

Ceramics

A ceramic is an inorganic, nonmetallic solid prepared by the action of heat and subsequent cooling. Ceramic materials may have a crystalline or partly crystalline structure, or may be amorphous (e.g., a glass). Because most common ceramics are crystalline, the definition of ceramic is often restricted to inorganic crystalline materials, as opposed to the noncrystalline glasses, a distinction followed here. The earliest ceramics made by humans were pottery objects, including 27,000 year old figurines, made from clay, either by itself or mixed with other materials, hardened in fire. Later ceramics were glazed and fired to create a coloured, smooth surface. Ceramics now include domestic, industrial and building products and a wide range of ceramic art. In the 20th century, new ceramic materials were developed for use in advanced ceramic engineering; for example, in semiconductors.

The word "ceramic" comes from the Greek word êåñαìéêüò (*keramikos*), "of pottery" or "for pottery", from êÝñαìïò (*keramos*), "potter's clay, tile, pottery". The earliest mention of the root "ceram-" is the Mycenaean Greek *ke-ra-me-we*, "workers of ceramics", written in Linear B syllabic script. "Ceramic" may be used as an adjective describing a material, product or process; or as a singular noun, or, more commonly, as a plural noun, "ceramics".

Types of Ceramic Product

For convenience, ceramic products are usually divided into four sectors; these are shown below with some examples:

- Structural, including bricks, pipes, floor and roof tiles
- Refractories, such as kiln linings, gas fire radiants, steel and glass making crucibles

- White Wares, including tableware, cookware, wall tiles, pottery products and sanitary ware
- Technical, is also known as engineering, advanced, special, and in Japan, fine ceramics. Such items include tiles used in the Space Shuttle program, gas burner nozzles, ballistic protection, nuclear fuel uranium oxide pellets, biomedical implants, coatings of jet engine turbine blades, ceramic disk brake, missile nose cones, bearing (mechanical), etc. Frequently, the raw materials do not include clays.

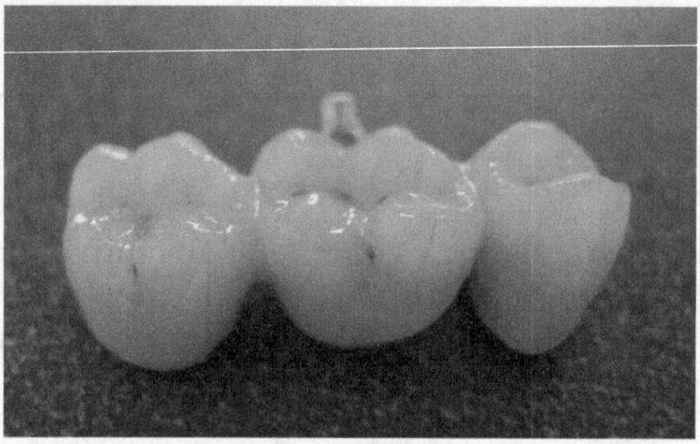

***Figure:** Fixed partial porcelain denture, or "bridge"*

Examples of White Ware Ceramics

- Earthenware, which is often made from clay, quartz and feldspar.
- Stoneware
- Porcelain, which is often made from kaolin
- Bone china

Classification of Technical Ceramics

Technical ceramics can also be classified into three distinct material categories:

- Oxides: alumina, beryllia, ceria, zirconia
- Nonoxides: carbide, boride, nitride, silicide
- Composite materials: particulate reinforced, fibre reinforced, combinations of oxides and nonoxides.

Each one of these classes can develop unique material properties because ceramics tend to be crystalline.

Other Applications of Ceramics

- Knife blades: the blade of a ceramic knife will stay sharp for much longer than that of a steel knife, although it is more brittle and can snap from a fall onto a hard surface.

- Ceramic brake disks for vehicles are resistant to abrasion at high temperatures.
- Advanced composite ceramic and metal matrices have been designed for most modern armoured fighting vehicles because they offer superior penetrating resistance against shaped charges (such as HEAT rounds) and kinetic energy penetrators.
- Ceramics such as alumina and boron carbide have been used in ballistic armored vests to repel large-caliber rifle fire. Such plates are known commonly as small arms protective inserts, or SAPIs. Similar material is used to protect the cockpits of some military airplanes, because of the low weight of the material.
- Ceramics can be used in place of steel for ball bearings. Their higher hardness means they are much less susceptible to wear and typically last for triple the lifetime of a steel part. They also deform less under load, meaning they have less contact with the bearing retainer walls and can roll faster. In very high speed applications, heat from friction during rolling can cause problems for metal bearings, which are reduced by the use of ceramics. Ceramics are also more chemically resistant and can be used in wet environments where steel bearings would rust. In some cases, their electricity-insulating properties may also be valuable in bearings. Two drawbacks to ceramic bearings are a significantly higher cost and susceptibility to damage under shock loads.
- In the early 1980s, Toyota researched production of an adiabatic engine using ceramic components in the hot gas area. The ceramics would have allowed temperatures of over 3000°F (1650°C). The expected advantages would have been lighter materials and a smaller cooling system (or no need for one at all), leading to a major weight reduction. The expected increase of fuel efficiency of the engine (caused by the higher temperature, as shown by Carnot's theorem) could not be verified experimentally; it was found that the heat transfer on the hot ceramic cylinder walls was higher than the transfer to a cooler metal wall as the cooler gas film on the metal surface works as a thermal insulator. Thus, despite all of these desirable properties, such engines have not succeeded in production because of costs for the ceramic components and the limited advantages. (Small imperfections in the ceramic material with its low fracture toughness lead to cracks, which can lead to

potentially dangerous equipment failure.) Such engines are possible in laboratory settings, but mass production is not feasible with current technology.

- Work is being done in developing ceramic parts for gas turbine engines. Currently, even blades made of advanced metal alloys used in the engines' hot section require cooling and careful limiting of operating temperatures. Turbine engines made with ceramics could operate more efficiently, giving aircraft greater range and payload for a set amount of fuel.
- Recent advances have been made in ceramics which include bioceramics, such as dental implants and synthetic bones. Hydroxyapatite, the natural mineral component of bone, has been made synthetically from a number of biological and chemical sources and can be formed into ceramic materials. Orthopedic implants coated with these materials bond readily to bone and other tissues in the body without rejection or inflammatory reactions so are of great interest for gene delivery and tissue engineering scaffolds. Most hydroxyapatite ceramics are very porous and lack mechanical strength, and are used to coat metal orthopedic devices to aid in forming a bond to bone or as bone fillers. They are also used as fillers for orthopedic plastic screws to aid in reducing the inflammation and increase absorption of these plastic materials. Work is being done to make strong, fully dense nanocrystalline hydroxyapatite ceramic materials for orthopedic weight bearing devices, replacing foreign metal and plastic orthopedic materials with a synthetic, but naturally occurring, bone mineral. Ultimately, these ceramic materials may be used as bone replacements or with the incorporation of protein collagens, synthetic bones.
- High-tech ceramic is used in watchmaking for producing watch cases. The material is valued by watchmakers for its light weight, scratch resistance, durability and smooth touch. IWC is one of the brands that initiated the use of ceramic in watchmaking. The case of the IWC 2007 Top Gun edition of the Pilot's Watch double chronograph is crafted in black ceramic.

Types of Ceramic Material

A ceramic material is an inorganic, non-metallic, often crystalline oxide, nitride or carbide material. Some elements, such as carbon or silicon, may be considered ceramics. Ceramic materials are brittle, hard, strong in compression, weak in shearing and tension. They withstand chemical erosion that occurs in other materials subjected to acidic or caustic environments.

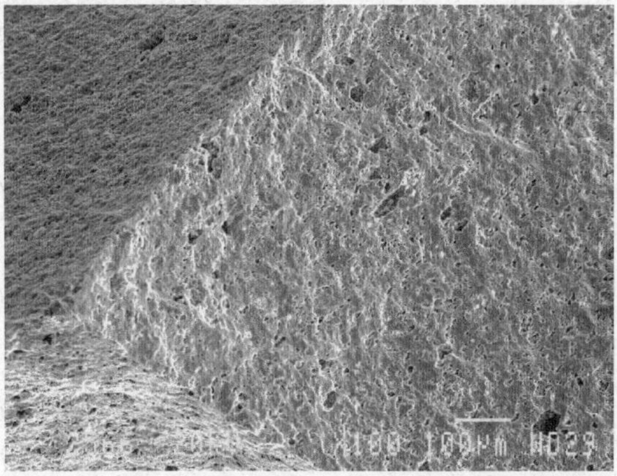

Figure: *A low magnification SEM micrograph of an advanced ceramic material. The properties of ceramics make fracturing an important inspection method.*

Ceramics generally can withstand very high temperatures, such as temperatures that range from 1,000 °C to 1,600 °C (1,800 °F to 3,000 °F). A glass is often not understood as a ceramic because of its amorphous (noncrystalline) character. However, glassmaking involves several steps of the ceramic process and its mechanical properties are similar to ceramic materials. Traditional ceramic raw materials include clay minerals such as kaolinite, whereas more recent materials include aluminium oxide, more commonly known as alumina. The modern ceramic materials, which are classified as advanced ceramics, include silicon carbide and tungsten carbide. Both are valued for their abrasion resistance, and hence find use in applications such as the wear plates of crushing equipment in mining operations. Advanced ceramics are also used in the medicine, electrical and electronics industries.

Crystalline Ceramics

Crystalline ceramic materials are not amenable to a great range of processing. Methods for dealing with them tend to fall into one of two categories – either make the ceramic in the desired shape, by reaction *in situ*, or by "forming" powders into the desired shape, and then sintering to form a solid body. Ceramic forming techniques include shaping by hand (sometimes including a rotation process called "throwing"), slip casting, tape casting (used for making very thin ceramic capacitors, e.g.), injection molding, dry pressing, and other variations. Details of these processes are described in the two books listed below. A few methods use a hybrid between the two approaches.

Noncrystalline Ceramics

Noncrystalline ceramics, being glass, tend to be formed from melts. The glass is shaped when either fully molten, by casting, or when in a state of toffee-like viscosity, by methods such as blowing into a mold. If later heat treatments cause this glass to become partly crystalline, the resulting material is known as a glass-ceramic, widely used as cook-top and also as a glass composite material for nuclear waste disposal.

Ceramics in Archaeology

Ceramic artifacts have an important role in archaeology for understanding the culture, technology and behaviour of peoples of the past. They are among the most common artifacts to be found at an archaeological site, generally in the form of small fragments of broken pottery called sherds. Processing of collected sherds can be consistent with two main types of analysis: technical and traditional. Traditional analysis involves sorting ceramic artifacts, sherds and larger fragments into specific types based on style, composition, manufacturing and morphology.

By creating these typologies it is possible to distinguish between different cultural styles, the purpose of the ceramic and technological state of the people among other conclusions. In addition, by looking at stylistic changes of ceramics over time is it possible to separate (seriate) the ceramics into distinct diagnostic groups (assemblages). A comparison of ceramic artifacts with known dated assemblages allows for a chronological assignment of these pieces.

The technical approach to ceramic analysis involves a finer examination of the composition of ceramic artifacts and sherds to determine the source of the material and through this the possible manufacturing site. Key criteria are the composition of the clay and the temper used in the manufacture of the article under study: temper is a material added to the clay during the initial production stage, and it is used to aid the subsequent drying process.

Types of temper include shell pieces, granite fragments and ground sherd pieces called 'grog'. Temper is usually identified by microscopic examination of the temper material. Clay identification is determined by a process of refiring the ceramic, and assigning a colour to it using Munsell Soil Colour notation. By estimating both the clay and temper compositions, and locating a region where both are known to occur, an assignment of the material source can be made. From the source assignment of the artifact further investigations can be made into the site of manufacture.

Ceramic Chemistry

Ceramic chemistry (or, more specifically, glaze chemistry) studies the relationship between the physical properties of fired ceramic glazes and their chemistry. Although ceramic technicians have long understood many of these relationships, the advent of computer software to automate the conversion from batch to formula and analysis has brought this science within the reach of many more people. Physical properties of glazes in fired products (like thermal expansion, hardness, index of refraction, colour and melting temperature or range) are directly (but not solely) related to the chemistry. Properties of glass melts like viscosity and surface tension are also principally products of chemistry.

Technicians in the ceramic tile, tableware, artware, pottery, sanitaryware, glass, fibreglass, bottle glass, optical and related industries all use this science. In ceramic chemistry, fired glazes are viewed as composed of oxides (examples are SiO_2, Al_2O_3, B_2O_3, Na_2O, K_2O, CaO, Li_2O, MgO, ZnO, MnO, Fe_2O_3, CoO).

Each oxide is known to contribute specific properties to the fired glass. Materials suppliers publish chemical analyses of their products that cite percentages of these oxides as well as volatiles (oxides that burn away during firing like H_2O, CO_2, SO_3).

For example, in traditional ceramics here are some examples of what the application of ceramic chemistry can accomplish.

- Fix glaze defects like crazing, blistering, pin-holing, settling, clouding, leaching, crawling, marking, scratching, powdering.
- Substitute frits or incorporate better, cheaper materials, replace no-longer-available ones.
- Adjust glaze melting temperature, gloss, surface character and colour (in ceramic colour is a matter of chemistry).
- Identify weaknesses (e.g. proximity to volatile firing zones, use of unreliable materials) in glazes to avoid problems.
- Creating and optimizing base glazes to work with difficult colours or stains and for special effects dependent on opacification, crystallization or variegation.
- Create glazes from scratch and use native materials in the highest possible percentage.

In ceramic bodies the physical properties of the final fired product are more related to the firing curve, the physical properties (e.g. particle size and shape, decomposition history) of the ingredient materials and the mineralogy and interaction between the different particle types

Ceramic Engineering

Ceramic engineering is the science and technology of creating objects from inorganic, non-metallic materials. This is done either by the action of heat, or at lower temperatures using precipitation reactions from high-purity chemical solutions. The term includes the purification of raw materials, the study and production of the chemical compounds concerned, their formation into components and the study of their structure, composition and properties.

Figure: *Ceramic bread knife*

Ceramic materials may have a crystalline or partly crystalline structure, with long-range order on atomic scale. Glass ceramics may have an amorphous or glassy structure, with limited or short-range atomic order. They are either formed from a molten mass that solidifies on cooling, formed and matured by the action of heat, or chemically synthesized at low temperatures using, for example, hydrothermal or sol-gel synthesis.

The special character of ceramic materials gives rise to many applications in materials engineering, electrical engineering, chemical engineering and mechanical engineering. As ceramics are heat resistant, they can be used for many tasks for which materials like metal and polymers are unsuitable. Ceramic materials are used in a wide range of industries, including mining, aerospace, medicine, refinery, food and chemical industries, packaging science, electronics, industrial and transmission electricity, and guided lightwave transmission.

Military

The military requirements of World War II encouraged developments, which created a need for high-performance materials and helped speed the development of ceramic science and engineering. Throughout the

1960s and 1970s, new types of ceramics were developed in response to advances in atomic energy, electronics, communications, and space travel. The discovery of ceramic superconductors in 1986 has spurred intense research to develop superconducting ceramic parts for electronic devices, electric motors, and transportation equipment.

Figure: *Soldiers pictured during the 2003 Iraq War seen through IR transparent Night Vision Goggles*

There is an increasing need in the military sector for high-strength, robust materials which have the capability to transmit light around the visible (0.4–0.7 micrometres) and mid-infrared (1–5 micrometres) regions of the spectrum. These materials are needed for applications requiring transparent armour. Transparent armor is a material or system of materials designed to be optically transparent, yet protect from fragmentation or ballistic impacts. The primary requirement for a transparent armour system is to not only defeat the designated threat but also provide a multi-hit capability with minimized distortion of surrounding areas. Transparent armour windows must also be compatible with night vision equipment. New materials that are thinner, lightweight, and offer better ballistic performance are being sought. Such solid-state components have found widespread use for various applications in the electro-optical field including: optical fibres for guided lightwave transmission, optical switches, laser amplifiers and lenses, hosts for solid-state lasers and optical window materials for gas lasers, and infrared (IR) heat seeking devices for missile guidance systems and IR night vision.

Modern Industry

Now a multi-billion dollar a year industry, ceramic engineering and research has established itself as an important field of science. Applications continue to expand as researchers develop new kinds of ceramics to serve different purposes.

- Zirconium dioxide ceramics are used in the manufacture of knives. The blade of the ceramic knife will stay sharp for much longer than that of a steel knife, although it is more brittle and can be snapped by dropping it on a hard surface.
- Ceramics such as alumina, boron carbide and silicon carbide have been used in bulletproof vests to repel large-calibre rifle fire. Such plates are known commonly as small-arms protective inserts (SAPI). Similar material is used to protect cockpits of some military aircraft, because of the low weight of the material.
- Silicon nitride parts are used in ceramic ball bearings. Their higher hardness means that they are much less susceptible to wear and can offer more than triple lifetimes. They also deform less under load meaning they have less contact with the bearing retainer walls and can roll faster. In very high speed applications, heat from friction during rolling can cause problems for metal bearings; problems which are reduced by the use of ceramics. Ceramics are also more chemically resistant and can be used in wet environments where steel bearings would rust. The major drawback to using ceramics is a significantly higher cost. In many cases their electrically insulating properties may also be valuable in bearings.
- In the early 1980s, Toyota researched production of an adiabatic ceramic engine which can run at a temperature of over 6000 °F (3300 °C). Ceramic engines do not require a cooling system and hence allow a major weight reduction and therefore greater fuel efficiency. Fuel efficiency of the engine is also higher at high temperature, as shown by Carnot's theorem. In a conventional metallic engine, much of the energy released from the fuel must be dissipated as waste heat in order to prevent a meltdown of the metallic parts. Despite all of these desirable properties, such engines are not in production because the manufacturing of ceramic parts in the requisite precision and durability is difficult. Imperfection in the ceramic leads to cracks, which can lead to potentially dangerous equipment failure. Such engines are possible in laboratory settings, but mass-production is not feasible with current technology.
- Work is being done in developing ceramic parts for gas turbine engines. Currently, even blades made of advanced metal alloys used in the engines' hot section require cooling and careful limiting of operating temperatures. Turbine engines made with ceramics could operate more efficiently, giving aircraft greater range and payload for a set amount of fuel.

- Recently, there have been advances in ceramics which include bio-ceramics, such as dental implants and synthetic bones. Hydroxyapatite, the natural mineral component of bone, has been made synthetically from a number of biological and chemical sources and can be formed into ceramic materials. Orthopaedic implants made from these materials bond readily to bone and other tissues in the body without rejection or inflammatory reactions. Because of this, they are of great interest for gene delivery and tissue engineering scaffolds. Most hydroxyapatite ceramics are very porous and lack mechanical strength and are used to coat metal orthopaedic devices to aid in forming a bond to bone or as bone fillers. They are also used as fillers for orthopaedic plastic screws to aid in reducing the inflammation and increase absorption of these plastic materials. Work is being done to make strong, fully dense nano crystalline hydroxyapatite ceramic materials for orthopaedic weight bearing devices, replacing foreign metal and plastic orthopaedic materials with a synthetic, but naturally occurring, bone mineral. Ultimately these ceramic materials may be used as bone replacements or with the incorporation of protein collagens, synthetic bones.
- High-tech ceramic is used in watch-making for producing watch cases. The material is valued by watchmakers for its light weight, scratch-resistance, durability and smooth touch. IWC is one of the brands that initiated the use of ceramic in watch-making. The case of the IWC 2007 Top Gun edition of the Pilot's Watch Double chronograph is crafted in high-tech black ceramic.

Glass-Ceramics

Figure: *A high strength glass-ceramic cook-top with negligible thermal expansion.*

Glass-ceramic materials share many properties with both glasses and ceramics. Glass-ceramics have an amorphous phase and one or more crystalline phases and are produced by a so-called "controlled crystallization", which is typically avoided in glass manufacturing. Glass-ceramics often contain a crystalline phase which constitutes anywhere from 30% [m/m] to 90% [m/m] of its composition by volume, yielding an array of materials with interesting thermomechanical properties.

In the processing of glass-ceramics, molten glass is cooled down gradually before reheating and annealing. In this heat treatment the glass partly crystallizes. In many cases, so-called 'nucleation agents' are added in order to regulate and control the crystallization process. Because there is usually no pressing and sintering, glass-ceramics do not contain the volume fraction of porosity typically present in sintered ceramics.

The term mainly refers to a mix of lithium and aluminosilicates which yields an array of materials with interesting thermomechanical properties. The most commercially important of these have the distinction of being impervious to thermal shock. Thus, glass-ceramics have become extremely useful for countertop cooking. The negative thermal expansion coefficient (TEC) of the crystalline ceramic phase can be balanced with the positive TEC of the glassy phase. At a certain point (~70% crystalline) the glass-ceramic has a net TEC near zero. This type of glass-ceramic exhibits excellent mechanical properties and can sustain repeated and quick temperature changes up to 1000 °C.

Processing Steps

The traditional ceramic process generally follows this sequence: Milling → Batching → Mixing → Forming → Drying → Firing → Assembly.

- Milling is the process by which materials are reduced from a large size to a smaller size. Milling may involve breaking up cemented material (in which case individual particles retain their shape) or pulverization (which involves grinding the particles themselves to a smaller size). Milling is generally done by mechanical means, including *attrition* (which is particle-to-particle collision that results in agglomerate break up or particle shearing), *compression* (which applies a forces that results in fracturing), and *impact* (which employs a milling medium or the particles themselves to cause fracturing). Attrition milling equipment includes the wet scrubber (also called the planetary mill or wet attrition mill), which has paddles in water creating vortexes in which the material collides and break up. Compression mills include the jaw crusher, roller crusher and cone crusher. Impact mills include the ball mill, which has media that tumble and fracture the material. Shaft impactors cause particle-to particle attrition and compression.

- Batching is the process of weighing the oxides according to recipes, and preparing them for mixing and drying.
- Mixing occurs after batching and is performed with various machines, such as dry mixing ribbon mixers (a type of cement mixer), Mueller mixers, and pug mills. Wet mixing generally involves the same equipment.
- Forming is making the mixed material into shapes, ranging from toilet bowls to spark plug insulators. Forming can involve: (1) Extrusion, such as extruding "slugs" to make bricks, (2) Pressing to make shaped parts, (3) Slip casting, as in making toilet bowls, wash basins and ornamentals like ceramic statues. Forming produces a "green" part, ready for drying. Green parts are soft, pliable, and over time will lose shape. Handling the green product will change its shape. For example, a green brick can be "squeezed", and after squeezing it will stay that way.
- Drying is removing the water or binder from the formed material. Spray drying is widely used to prepare powder for pressing operations. Other dryers are tunnel dryers and periodic dryers. Controlled heat is applied in this two-stage process. First, heat removes water. This step needs careful control, as rapid heating causes cracks and surface defects. The dried part is smaller than the green part, and is brittle, necessitating careful handling, since a small impact will cause crumbling and breaking.
- Firing is where the dried parts pass through a controlled heating process, and the oxides are chemically changed to cause sintering and bonding. The fired part will be smaller than the dried part.

Forming Methods

Ceramic forming techniques include throwing, slipcasting, tape casting, injection moulding, dry pressing, isostatic pressing, hot isostatic pressing (HIP) and others. Methods for forming ceramic powders into complex shapes are desirable in many areas of technology. Such methods are required for producing advanced, high-temperature structural parts such as heat engine components and turbines.

Materials other than ceramics which are used in these processes may include: wood, metal, water, plaster and epoxy—most of which will be eliminated upon firing.

These forming techniques are well known for providing tools and other components with dimensional stability, surface quality, high (near theoretical) density and microstructural uniformity. The increasing use and diversity of speciality forms of ceramics adds to the diversity of process technologies to be used.

Thus, reinforcing fibres and filaments are mainly made by polymer, sol-gel, or CVD processes, but melt processing also has applicability. The most widely used speciality form is layered structures, with tape casting for electronic substrates and packages being pre-eminent. Photo-lithography is of increasing interest for precise patterning of conductors and other components for such packaging. Tape casting or forming processes are also of increasing interest for other applications, ranging from open structures such as fuel cells to ceramic composites.

The other major layer structure is coating, where melt spraying is very important, but chemical and physical vapour deposition and chemical (e.g., sol-gel and polymer pyrolysis) methods are all seeing increased use. Besides open structures from formed tape, extruded structures, such as honeycomb catalyst supports, and highly porous structures, including various foams, for example, reticulated foam, are of increasing use.

Densification of consolidated powder bodies continues to be achieved predominantly by (pressureless) sintering. However, the use of pressure sintering by hot pressing is increasing, especially for non-oxides and parts of simple shapes where higher quality (mainly microstructural homogeneity) is needed, and larger size or multiple parts per pressing can be an advantage.

The Sintering Process

The principles of sintering-based methods are simple ("sinter" has roots in the English "cinder"). The firing is done at a temperature below the melting point of the ceramic. Once a roughly-held-together object called a "green body" is made, it is baked in a kiln, where atomic and molecular diffusion processes give rise to significant changes in the primary microstructural features. This includes the gradual elimination of porosity, which is typically accompanied by a net shrinkage and overall densification of the component. Thus, the pores in the object may close up, resulting in a denser product of significantly greater strength and fracture toughness.

Another major change in the body during the firing or sintering process will be the establishment of the polycrystalline nature of the solid. This change will introduce some form of grain size distribution, which will have a significant impact on the ultimate physical properties of the material. The grain sizes will either be associated with the initial particle size, or possibly the sizes of aggregates or particle clusters which arise during the initial stages of processing.

The ultimate microstructure (and thus the physical properties) of the final product will be limited by and subject to the form of the structural template or precursor which is created in the initial stages of chemical synthesis and physical forming. Hence the importance of chemical powder and polymer processing as it pertains to the synthesis of industrial ceramics, glasses and glass-ceramics.

There are numerous possible refinements of the sintering process. Some of the most common involve pressing the green body to give the densification a head start and reduce the sintering time needed. Sometimes organic binders such as polyvinyl alcohol are added to hold the green body together; these burn out during the firing (at 200–350 °C). Sometimes organic lubricants are added during pressing to increase densification. It is common to combine these, and add binders and lubricants to a powder, then press. (The formulation of these organic chemical additives is an art in itself. This is particularly important in the manufacture of high performance ceramics such as those used by the billions for electronics, in capacitors, inductors, sensors, etc.)

A slurry can be used in place of a powder, and then cast into a desired shape, dried and then sintered. Indeed, traditional pottery is done with this type of method, using a plastic mixture worked with the hands. If a mixture of different materials is used together in a ceramic, the sintering temperature is sometimes above the melting point of one minor component – a *liquid phase* sintering. This results in shorter sintering times compared to solid state sintering.

Strength of Ceramics

A material's strength is dependent on its microstructure. The engineering processes to which a material is subjected can alter this microstructure. The variety of strengthening mechanisms that alter the strength of a material include the mechanism of grain boundary strengthening. Thus, although yield strength is maximized with decreasing grain size, ultimately, very small grain sizes make the material brittle. Considered in tandem with the fact that the yield strength is the parameter that predicts plastic deformation in the material, one can make informed decisions on how to increase the strength of a material depending on its microstructural properties and the desired end effect.

The relation between yield stress and grain size is described mathematically by the Hall-Petch equation which is

$$\sigma_y = \sigma_0 + \frac{k_y}{\sqrt{d}}$$

where k_y is the strengthening coefficient (a constant unique to each material), σ_o is a materials constant for the starting stress for dislocation movement (or the resistance of the lattice to dislocation motion), d is the grain diameter, and σ_y is the yield stress.

Theoretically, a material could be made infinitely strong if the grains are made infinitely small. This is, unfortunately, impossible because the lower limit of grain size is a single unit cell of the material. Even then, if the grains of a material are the size of a single unit cell, then the material is in fact amorphous, not crystalline, since there is no long range order, and dislocations can not be defined in an amorphous material. It has been observed experimentally that the microstructure with the highest yield strength is a grain size of about 10 nanometres, because grains smaller than this undergo another yielding mechanism, grain boundary sliding. Producing engineering materials with this ideal grain size is difficult because of the limitations of initial particle sizes inherent to nanomaterials and nanotechnology.

Theory of Chemical Processing

Microstructural Uniformity

In the processing of fine ceramics, the irregular particle sizes and shapes in a typical powder often lead to non-uniform packing morphologies that result in packing density variations in the powder compact. Uncontrolled agglomeration of powders due to attractive van der Waals forces can also give rise to in microstructural inhomogeneities.

Differential stresses that develop as a result of non-uniform drying shrinkage are directly related to the rate at which the solvent can be removed, and thus highly dependent upon the distribution of porosity. Such stresses have been associated with a plastic-to-brittle transition in consolidated bodies, and can yield to crack propagation in the unfired body if not relieved.

In addition, any fluctuations in packing density in the compact as it is prepared for the kiln are often amplified during the sintering process, yielding inhomogeneous densification. Some pores and other structural defects associated with density variations have been shown to play a detrimental role in the sintering process by growing and thus limiting end-point densities. Differential stresses arising from inhomogeneous densification have also been shown to result in the propagation of internal cracks, thus becoming the strength-controlling flaws. It would therefore appear desirable to process a material in such a way that it is physically uniform with regard to the distribution of components and porosity, rather than using particle size distributions which will maximize the green density. The containment of a uniformly dispersed assembly of strongly interacting particles in suspension requires total control over particle-particle interactions. Monodisperse colloids provide this potential.

Monodisperse powders of colloidal silica, for example, may therefore be stabilized sufficiently to ensure a high degree of order in the colloidal crystal or polycrystalline colloidal solid which results from aggregation. The degree of order appears to be limited by the time and space allowed for longer-range correlations to be established.

Such defective polycrystalline colloidal structures would appear to be the basic elements of submicrometre colloidal materials science, and, therefore, provide the first step in developing a more rigorous understanding of the mechanisms involved in microstructural evolution in inorganic systems such as polycrystalline ceramics.

Self-Assembly

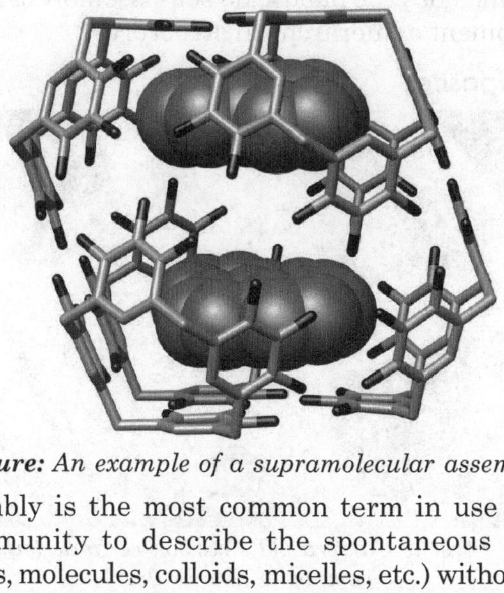

Figure: *An example of a supramolecular assembly.*

Self-assembly is the most common term in use in the modern scientific community to describe the spontaneous aggregation of particles (atoms, molecules, colloids, micelles, etc.) without the influence of any external forces. Large groups of such particles are known to assemble themselves into thermodynamically stable, structurally well-defined arrays, quite reminiscent of one of the 7 crystal systems found in metallurgy and mineralogy (e.g. face-centred cubic, body-centred cubic, etc.). The fundamental difference in equilibrium structure is in the spatial scale of the unit cell (or lattice parameter) in each particular case.

Thus, self-assembly is emerging as a new strategy in chemical synthesis and nanotechnology. Molecular self-assembly has been observed in various biological systems and underlies the formation of a wide variety of complex biological structures. Molecular crystals, liquid crystals, colloids, micelles, emulsions, phase-separated polymers, thin films and self-assembled monolayers all represent examples of the types of highly ordered structures which are obtained using these techniques. The distinguishing feature of these methods is self-organisation in the absence of any external forces.

In addition, the principal mechanical characteristics and structures of biological ceramics, polymer composites, elastomers, and cellular materials are being re-evaluated, with an emphasis on bioinspired materials and structures. Traditional approaches focus on design methods of biological materials using conventional synthetic materials. This includes an emerging class of mechanically superior biomaterials based on microstructural features and designs found in nature. The new horizons have been identified in the synthesis of bioinspired materials through processes that are characteristic of biological systems in nature. This includes the nanoscale self-assembly of the components and the development of hierarchical structures.

Ceramic Composites

Figure: *The Porsche Carrera GT's carbon-ceramic (silicon carbide) composite disc brake*

Substantial interest has arisen in recent years in fabricating ceramic composites. While there is considerable interest in composites with one or more non-ceramic constituents, the greatest attention is on composites in which all constituents are ceramic. These typically comprise two ceramic constituents: a continuous matrix, and a dispersed phase of ceramic particles, whiskers, or short (chopped) or continuous ceramic fibres. The challenge, as in wet chemical processing, is to obtain a uniform or homogeneous distribution of the dispersed particle or fibre phase.

Consider first the processing of particulate composites. The particulate phase of greatest interest is tetragonal zirconia because of the toughening that can be achieved from the phase transformation from the metastable tetragonal to the monoclinic crystalline phase, aka transformation toughening. There is also substantial interest in dispersion of hard, non-oxide phases such as SiC, TiB, TiC, boron, carbon and especially oxide matrices like alumina and mullite. There is also interest too incorporating other ceramic particulates, especially those of highly anisotropic thermal expansion. Examples include Al_2O_3, TiO_2, graphite, and boron nitride.

In processing particulate composites, the issue is not only homogeneity of the size and spatial distribution of the dispersed and matrix phases, but also control of the matrix grain size. However, there is some built-in self-control due to inhibition of matrix grain growth by the dispersed phase. Particulate composites, though generally offer increased resistance to damage, failure, or both, are still quite sensitive to inhomogeneities of composition as well as other processing defects such as pores. Thus they need good processing to be effective.

Particulate composites have been made on a commercial basis by simply mixing powders of the two constituents. Although this approach is inherently limited in the homogeneity that can be achieved, it is the most readily adaptable for existing ceramic production technology. However, other approaches are of interest.

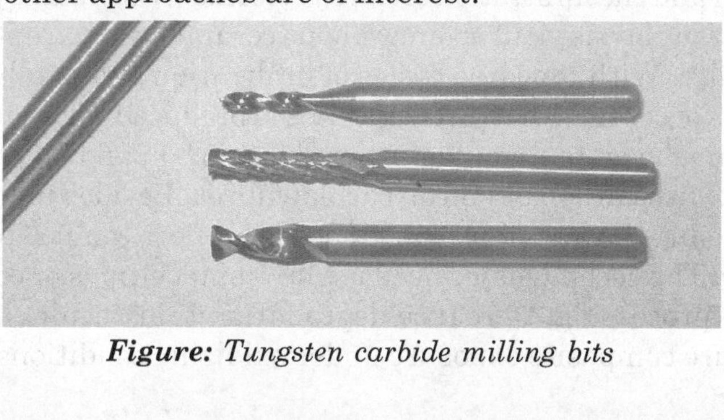

***Figure:** Tungsten carbide milling bits*

From the technological standpoint, a particularly desirable approach to fabricating particulate composites is to coat the matrix or its precursor onto fine particles of the dispersed phase with good control of the starting dispersed particle size and the resultant matrix coating thickness. One should in principle be able to achieve the ultimate in homogeneity of distribution and thereby optimize composite performance. This can also have other ramifications, such as allowing more useful composite performance to be achieved in a body having porosity, which might be desired for other factors, such as limiting thermal conductivity.

There are also some opportunities to utilise melt processing for fabrication of ceramic, particulate, whisker and short-fibre, and continuous-fibre composites. Clearly, both particulate and whisker composites are conceivable by solid-state precipitation after solidification of the melt. This can also be obtained in some cases by sintering, as for precipitation-toughened, partially stabilized zirconia. Similarly, it is known that one can directionally solidify ceramic eutectic mixtures and hence obtain uniaxially aligned fibre composites. Such composite processing has typically been limited to very simple shapes and thus suffers from serious economic problems due to high machining costs.

Clearly, there are possibilities of using melt casting for many of these approaches. Potentially even more desirable is using melt-derived particles. In this method, quenching is done in a solid solution or in a fine eutectic structure, in which the particles are then processed by more typical ceramic powder processing methods into a useful body. There have also been preliminary attempts to use melt spraying as a means of forming composites by introducing the dispersed particulate, whisker, or fibre phase in conjunction with the melt spraying process.

Other methods besides melt infiltration to manufacture ceramic composites with long fibre reinforcement are chemical vapour infiltration and the infiltration of fibre preforms with organic precursor, which after pyrolysis yield an amorphous ceramic matrix, initially with a low density. With repeated cycles of infiltration and pyrolysis one of those types of ceramic matrix composites is produced. Chemical vapour infiltration is used to manufacture carbon/carbon and silicon carbide reinforced with carbon or silicon carbide fibres. Besides many process improvements, the first of two major needs for fibre composites is lower fibre costs. The second major need is fibre compositions or coatings, or composite processing, to reduce degradation that results from high-temperature composite exposure under oxidizing conditions.

Applications

The products of technical ceramics include tiles used in the Space Shuttle program, gas burner nozzles, ballistic protection, nuclear fuel uranium oxide pellets, bio-medical implants, jet engine turbine blades, and missile nose cones. Its products are often made from materials other than clay, chosen for their particular physical properties. These may be classified as follows:

- Oxides: silica, alumina, zirconia
- Non-oxides: carbides, borides, nitrides, silicides
- Composites: particulate or whisker reinforced matrices, combinations of oxides and non-oxides (e.g. polymers).

Ceramics can be used in many technological industries. One application is the ceramic tiles on NASA's Space Shuttle, used to protect it and the future supersonic space planes from the searing heat of re-entry into the Earth's atmosphere. They are also used widely in electronics and optics. In addition to the applications listed here, ceramics are also used as a coating in various engineering cases. An example would be a ceramic bearing coating over a titanium frame used for an aircraft. Recently the field has come to include the studies of single crystals or glass fibres, in addition to traditional polycrystalline materials, and the applications of these have been overlapping and changing rapidly.

Aerospace

- Engines; Shielding a hot running aircraft engine from damaging other components.
- Airframes; Used as a high-stress, high-temp and lightweight bearing and structural component.
- Missile nose-cones; Shielding the missile internals from heat.
- Space Shuttle tiles
- Space-debris ballistic shields – ceramic fibre woven shields offer better protection to hypervelocity (~7 km/s) particles than aluminium shields of equal weight.
- Rocket nozzles, withstands and focuses the exhaust of the rocket booster.
- Unmanned Air Vehicles; Implications of ceramic engine utilisation in aeronautical applications (such as Unmanned Air Vehicles) may result in enhanced performance characteristics and less operational costs.

Biomedical

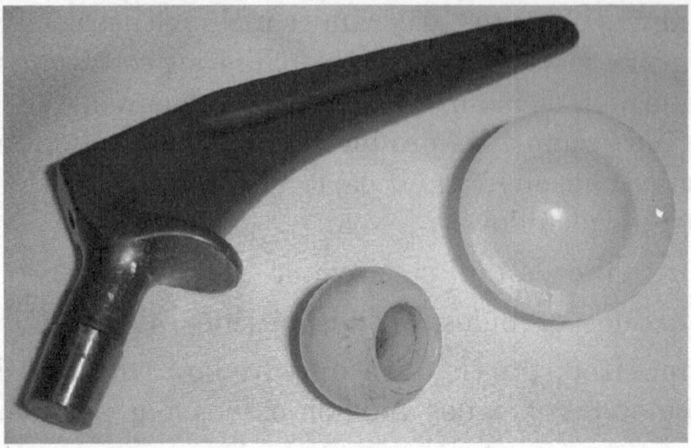

***Figure:** A titanium hip prosthesis, with a ceramic head and polyethylene acetabular cup.*

- Artificial bone; Dentistry applications, teeth.
- Biodegradable splints; Reinforcing bones recovering from osteoporosis
- Implant material

Electronics

- Capacitors
- Integrated circuit packages
- Transducers
- Insulators

Optical

Main article: Transparent ceramics

- Optical fibres, guided lightwave transmission
- Switches
- Laser amplifiers
- Lenses
- Infrared heat-seeking devices

Automotive

- Heat shield
- Exhaust heat management

Biomaterials

Silicification is quite common in the biological world and occurs in bacteria, single-celled organisms, plants, and animals (invertebrates and vertebrates). Crystalline minerals formed in such environment often show exceptional physical properties (e.g. strength, hardness, fracture toughness) and tend to form hierarchical structures that exhibit microstructural order over a range of length or spatial scales.

The minerals are crystallized from an environment that is undersaturated with respect to silicon, and under conditions of neutral pH and low temperature (0–40 °C). Formation of the mineral may occur either within or outside of the cell wall of an organism, and specific biochemical reactions for mineral deposition exist that include lipids, proteins and carbohydrates.

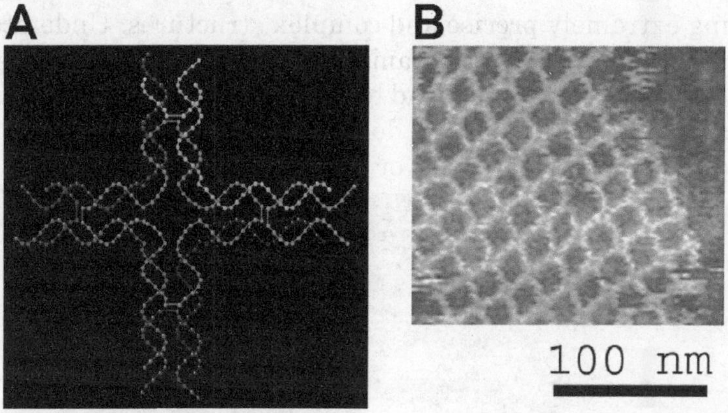

Figure: *The DNA structure at left (schematic shown) will self-assemble into the structure visualized by atomic force microscopy at right.*

Most natural (or biological) materials are complex composites whose mechanical properties are often outstanding, considering the weak constituents from which they are assembled. These complex structures, which have risen from hundreds of million years of evolution, are inspiring the design of novel materials with exceptional physical properties for high performance in adverse conditions. Their defining characteristics such as hierarchy, multifunctionality, and the capacity for self-healing, are currently being investigated.

The basic building blocks begin with the 20 amino acids and proceed to polypeptides, polysaccharides, and polypeptides–saccharides. These, in turn, compose the basic proteins, which are the primary constituents of the 'soft tissues' common to most biominerals. With well over 1000 proteins possible, current research emphasizes the use of collagen, chitin, keratin, and elastin. The 'hard' phases are often strengthened by crystalline minerals, which nucleate and grow in a biomediated environment that determines the size, shape and distribution of individual crystals. The most important mineral phases have been identified as hydroxyapatite, silica, and aragonite. Using the classification of Wegst and Ashby, the principal mechanical characteristics and structures of biological ceramics, polymer composites, elastomers, and cellular materials have been presented. Selected systems in each class are being investigated with emphasis on the relationship between their microstructure over a range of length scales and their mechanical response.

Thus, the crystallization of inorganic materials in nature generally occurs at ambient temperature and pressure. Yet the vital organisms through which these minerals form are capable of consistently producing extremely precise and complex structures. Understanding the processes in which living organisms control the growth of crystalline minerals such as silica could lead to significant advances in the field of materials science, and open the door to novel synthesis techniques for nanoscale composite materials, or nanocomposites.

Figure: *The iridescent nacre inside a Nautilus shell.*

High-resolution SEM observations were performed of the microstructure of the mother-of-pearl (or nacre) portion of the abalone shell. Those shells exhibit the highest mechanical strength and fracture toughness of any non-metallic substance known. The nacre from the shell of the abalone has become one of the more intensively studied biological structures in materials science. Clearly visible in these images are the neatly stacked (or ordered) mineral tiles separated by thin organic sheets along with a macrostructure of larger periodic growth bands which collectively form what scientists are currently referring to as a hierarchical composite structure. (The term hierarchy simply implies that there are a range of structural features which exist over a wide range of length scales).

Future developments reside in the synthesis of bio-inspired materials through processing methods and strategies that are characteristic of biological systems. These involve nanoscale self-assembly of the components and the development of hierarchical structures.

Chemical Process

In a scientific sense, a chemical process is a method or means of somehow changing one or more chemicals or chemical compounds. Such a chemical process can occur by itself or be caused by an outside force, and involves a chemical reaction of some sort. In an "engineering" sense, a chemical process is a method intended to be used in manufacturing or on an industrial scale to change the composition of chemical(s) or material(s), usually using technology similar or related to that used in chemical plants or the chemical industry.

Neither of these definitions is exact in the sense that one can always tell definitively what is a chemical process and what is not; they are practical definitions. There is also significant overlap in these two definition variations. Because of the inexactness of the definition, chemists and other scientists use the term "chemical process" only in a general sense or in the engineering sense. However, in the "process (engineering)" sense, the term "chemical process" is used extensively. The rest of the article will cover the engineering type of chemical process.

Although this type of chemical process may sometimes involve only one step, often multiple steps, referred to as unit operations, are involved. In a plant, each of the unit operations commonly occur in individual vessels or sections of the plant called units. Often, one or more chemical reactions are involved, but other ways of changing chemical (or material) composition may be used, such as mixing or separation processes.

The process steps may be sequential in time or sequential in space along a stream of flowing or moving material. For a given amount of a feed (input) material or product (output) material, an expected amount of material can be determined at key steps in the process from empirical data and material balance calculations. These amounts can be scaled up or down to suit the desired capacity or operation of a particular chemical plant built for such a process. More than one chemical plant may use the same chemical process, each plant perhaps at differently scaled capacities. Chemical Processes like Distillation and Crystallization, goes back to Alchemy in Alexandria.

Such chemical processes can be illustrated generally as block flow diagrams or in more detail as process flow diagrams. Block flow diagrams show the units as blocks and the streams flowing between them as connecting lines with arrowheads to show direction of flow.

In addition to chemical plants for producing chemicals, chemical processes with similar technology and equipment are also used in oil refining and other refineries, natural gas processing, polymer and pharmaceutical manufacturing, food processing, and water and wastewater treatment.

Unit Processing in Chemical Process

Unit processing is the basic processing in chemical engineering. Together with unit operations it forms the main principle of the varied chemical industries. Each genre of unit processing follows the same chemical law much as each genre of unit operations follows the same physical law.

Chemical engineering unit processing consists of the following important processes:

- Oxidation
- Reduction
- Hydrogenation
- Dehydrogenation
- Hydrolysis
- Hydration
- Dehydration
- Halogenation
- Nitrification
- Sulfonation
- Ammoniation
- Alkaline fusion
- Alkylation
- Dealkylation
- Esterification
- Polymerization
- Polycondensation
- Catalysis

Chemical Process Modelling

Chemical process modelling is a computer modelling technique used in chemical engineering process design. It typically involves using purpose-built software to define a system of interconnected components, which are then solved so that the steady-state or dynamic behaviour of the system can be predicted. The system components and connections are represented as a Process Flow diagram. Simulations can be as simple as the mixing of two substances in a tank, or as complex as an entire alumina refinery.

Chemical process modelling requires a knowledge of the properties of the chemicals involved in the simulation, as well as the physical properties and characteristics of the components of the system, such as tanks, pumps, pipes, pressure vessels, and so on.

Chemical Reactor

In chemical engineering, chemical reactors are vessels designed to contain chemical reactions. One example is a pressure reactor. The design of a chemical reactor deals with multiple aspects of chemical engineering. Chemical engineers design reactors to maximize net present value for the given reaction.

Designers ensure that the reaction proceeds with the highest efficiency towards the desired output product, producing the highest yield of product while requiring the least amount of money to purchase and operate. Normal operating expenses include energy input, energy removal, raw material costs, labour, etc. Energy changes can come in the form of heating or cooling, pumping to increase pressure, frictional pressure loss (such as pressure drop across a 90° elbow or an orifice plate) or agitation.

Chemical reaction engineering is the branch of chemical engineering which deals with chemical reactors and their design, especially by application of chemical kinetics to industrial systems.

Overview

Figure: *Cut-away view of a stirred-tank chemical reactor with a cooling jacket*

Figure: *Chemical reactor with half coils wrapped around it*

There are a couple of main basic vessel types:

- A tank
- A pipe or tubular reactor (laminar flow reactor(LFR))

Both types can be used as continuous reactors or batch reactors, and either may accommodate one or more solids (reagents, catalyst, or inert materials), but the reagents and products are typically fluids. Most commonly, reactors are run at steady-state, but can also be operated in a transient state. When a reactor is first brought into operation (after maintenance or inoperation) it would be considered to be in a transient state, where key process variables change with time.

There are three main basic models used to estimate the most important process variables of different chemical reactors:

- *batch reactor* model (batch),
- *continuous stirred-tank reactor* model (CSTR), and
- *plug flow reactor* model (PFR).

Furthermore, catalytic reactors require separate treatment, whether they are batch, CST, or PF reactors, as the many assumptions of the simpler models are not valid.

Key process variables include

- Residence time (τ, lower case Greek tau)
- Volume (V)
- Temperature (T)
- Pressure (P)
- Concentrations of chemical species (C_1, C_2, C_3, ... C_n)
- Heat transfer coefficients (h, U)

A chemical reactor, typically tubular reactor, could be a packed bed. The packing inside the bed may have catalyst to catalyze the chemical reaction. A chemical reactor may also be a fluidized bed. Chemical reactions occurring in a reactor may be exothermic, meaning giving off heat, or endothermic, meaning absorbing heat. A chemical reactor vessel may have a cooling or heating jacket or cooling or heating coils (tubes) wrapped around the outside of its vessel wall to cool down or heat up the contents.

Types

Figure: *Checking condition inside the case of a continuous stirred tank reactor (CSTR). Note the impeller (or agitator) blades on the shaft for mixing. Also note the baffle at the bottom of the image which also helps in mixing.*

CSTR (Continuous Stirred-Tank Reactor): In a CSTR, one or more fluid reagents are introduced into a tank reactor (typically) equipped with an impeller while the reactor effluent is removed. The impeller stirs the reagents to ensure proper mixing. Simply dividing the volume of the tank by the average volumetric flow rate through the tank gives the *residence time*, or the average amount of time a discrete quantity of reagent spends inside the tank. Using chemical kinetics, the reaction's expected percent completion can be calculated. Some important aspects of the CSTR:

- At steady-state, the mass flow rate in must equal the mass flow rate out, otherwise the tank will overflow or go empty (transient state). While the reactor is in a transient state the model equation must be derived from the differential mass and energy balances.
- The reaction proceeds at the reaction rate associated with the final (output) concentration, since the concentration is assumed to be homogenous throughout the reactor.

- Often, it is economically beneficial to operate several CSTRs in series. This allows, for example, the first CSTR to operate at a higher reagent concentration and therefore a higher reaction rate. In these cases, the sizes of the reactors may be varied in order to minimize the total capital investment required to implement the process.
- It can be demonstrated that an infinite number of infinitely small CSTRs operating in series would be equivalent to a PFR.

The behaviour of a CSTR is often approximated or modelled by that of a Continuous Ideally Stirred-Tank Reactor (CISTR). All calculations performed with CISTRs assume perfect mixing. If the residence time is 5-10 times the mixing time, this approximation is considered valid for engineering purposes. The CISTR model is often used to simplify engineering calculations and can be used to describe research reactors. In practice it can only be approached, particularly in industrial size reactors in which the mixing time may be very large.

PFR (Plug Flow Reactor)

In a PFR, one or more fluid reagents are pumped through a pipe or tube. The chemical reaction proceeds as the reagents travel through the PFR. In this type of reactor, the changing reaction rate creates a gradient with respect to distance traversed; at the inlet to the PFR the rate is very high, but as the concentrations of the reagents decrease and the concentration of the product(s) increases the reaction rate slows. Some important aspects of the PFR:

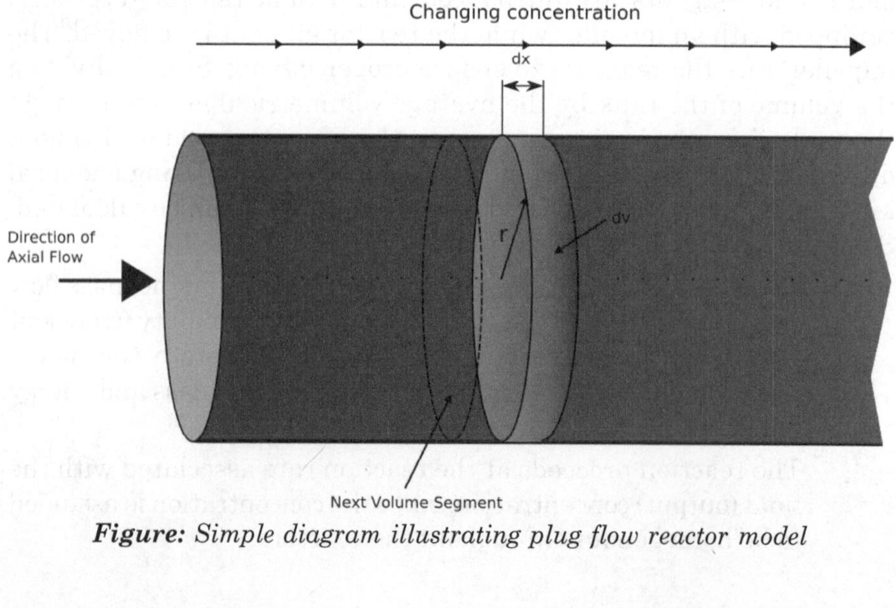

***Figure:** Simple diagram illustrating plug flow reactor model*

- All calculations performed with PFRs assume no upstream or downstream mixing, as implied by the term "plug flow".
- Reagents may be introduced into the PFR at locations in the reactor other than the inlet. In this way, a higher efficiency may be obtained, or the size and cost of the PFR may be reduced.
- A PFR typically has a higher efficiency than a CSTR of the same volume. That is, given the same space-time (or residence time), a reaction will typically proceed to a higher percentage completion in a PFR than in a CSTR. This is not always true for reversible reactions.

For most chemical reactions of industrial interest, it is impossible for the reaction to proceed to 100% completion. The rate of reaction decreases as the reactants are consumed until the point where the system reaches dynamic equilibrium (no net reaction, or change in chemical species occurs). The equilibrium point for most systems is less than 100% complete. For this reason a separation process, such as distillation, often follows a chemical reactor in order to separate any remaining reagents or byproducts from the desired product. These reagents may sometimes be reused at the beginning of the process, such as in the Haber process. In some cases, very large reactors would be necessary to approach equilibrium, and chemical engineers may choose to separate the partially reacted mixture and recycle the leftover reactants.

Continuous oscillatory baffled reactor (COBR) is a tubular plug flow reactor. The mixing in COBR is achieved by the combination of fluid oscillation and orifice baffles, allowing plug flow to be achieved under laminar flow conditions with the net flow Reynolds number just about 100.

Semi-Batch Reactor

A semi-batch reactor is operated with both continuous and batch inputs and outputs. A fermenter, for example, is loaded with a batch of medium and microbes which constantly produce carbon dioxide that must removed continuously. Analogously, driving a reaction of gas with a liquid is usually difficult, since the gas bubbles off. Therefore, a continuous feed of gas is injected into the batch of a liquid. One chemical reactant is charged to the vessel and a second chemical is added slowly (for instance, to prevent side reactions).

Catalytic Reactor

Although catalytic reactors are often implemented as plug flow reactors, their analysis requires more complicated treatment. The rate of a catalytic reaction is proportional to the amount of catalyst the reagents contact, as well as the concentration of the reactants. With a solid phase catalyst and fluid phase reagents, this is proportional to the exposed area, efficiency of diffusion of reagents in and products out, and efficacy of mixing. Perfect mixing usually cannot be assumed. Furthermore, a catalytic reaction pathway often occurs in multiple steps with intermediates that are chemically bound to the catalyst; and as the chemical binding to the catalyst is also a chemical reaction, it may affect the kinetics. Catalytic reactions often display the so-called *falsified kinetics*, i.e. the apparent kinetics differ from elementary chemical kinetics due to physical transport effects.

The behaviour of the catalyst is also a consideration. Particularly in high-temperature petrochemical processes, catalysts are deactivated by sintering, coking, and similar processes.

A common example of a catalytic reactor is the catalytic converter following an engine. However, most petrochemical reactors are catalytic, and are responsible for most of industrial chemical production in the world, with extremely high-volume examples such as sulfuric acid, ammonia, reformate/BTEX (benzene, toluene, ethylbenzene and xylene) and alkylate gasoline blending stock.

Bibliography

Anthony, H.: *The Gulf and the West: Strategic Relations and Military Realities,* Westview Press, Boulder, Colo., 1988.

Archibald, H. T.: *The First Kuwait Oil Concession Agreement: A Record of the Negotiations, 1911-1934,* Frank Cass, London, 1975.

Beizer, Boris: *Software Testing Techniques,* International Thomson Computer Press, Boston, MA., 2004.

Bernard S. Matisoff: *Handbook of Electronics Manufacturing Engineering,* Chapman & Hall, New York, 1997.

Chopey, Nicholas P.: *Handbook of Chemical Engineering Calculations,* McGraw-Hill, New York, 2004.

Conaway, Charles F.: *The Petroleum Industry: A Nontechnical Guide,* PennWell Pub. Co., Tulsa, OK, 1999.

Daniel Yergin, *The Prize: The Epic Quest for Oil, Money, and Power,* Simon and Schuster 1991.

Davis and Masten: *Principles of Environmental Engineering and Science,* McGraw-Hill Companies, Inc. New York, 2002.

Edward, B. Maxted: *Catalytic Hydrogenation and Reduction,* P. Blakiston's Son, Philadelphia , 1919.

Eisner, A.: *Hydrogenation and Petrography of Some Low-rank Coals from the Western United States,* Bureau of Mines, Washington, D.C., 1940.

Fenichell, Stephen: *Plastic: The Making of a Synthetic Century,* Harper Business, New York, 1996.

Frank, Alison Fleig: *Oil Empire: Visions of Prosperity in Austrian Galicia,* Harvard University Press, 2005.

George B. Hogaboom: *Principles of Electroplating and Electroforming (Electrotyping),* McGraw-Hill, New York, 1949.

Gomes, K.J., Braden, J.F.: *Mechanical Fastening of Plastics - An Engineering Handbook,* Marcel Dekker, New York, 1984.

Hala, E., I. Wichterle, J. Polak, and T. Boublik: *Vapor–Liquid Equilibrium at Normal Pressures,* Pergamon, Oxford, 1968.

Hill, J. W.; Petrucci, R. H.; McCreary, T. W.; Perry, S. S.: *General Chemistry*, Pearson Prentice Hall, Upper Saddle River, New Jersey, 2005.

Kenneth, E. Anderson: *Modern Petroleum: a Basic Primer of the Industry,* PennWell Books, Tulsa, OK, 1992.

Kondolf, G. Mathias; Hervé Piégay: *Tools in Fluvial Geomorphology*, Wiley, New York, 2003.

Laurence M. Harwood, Christopher J. Moody: *Experimental Organic Chemistry: Principles and Practice*, WileyBlackwell, Oxford, 1989.

Levinson, Paul: *Mechanism of the Synthesis of Hydrocarbons from Carbon Monoxide and Hydrogen,* Bureau of Mines, Washington, D.C., 1958.

Manthorp, P.A.: *Estuarine Circulation and Sediment Transport in the Saco River estuary*, Boston University, Boston, MA, 1995.

Matthew R. Simmons: *Twilight in the Desert The Coming Saudi Oil Shock and the World Economy*, John Wiley & Sons, NJ, 2005.

Schmidt, Lanny D.: *The Engineering of Chemical Reactions*, Oxford University Press, New York, 2005.

Selby, Michael John: *Earth's Changing Surface: an Introduction to Geomorphology*, Clarendon Press, Oxford, 1985.

Sheng, James: *Modern Chemical Enhanced Oil Recovery: Theory and Practice,* Gulf Professional Pub., Amsterdam, Boston, 2011.

Stewart Truswell: *Essentials of Human Nutrition*, Oxford University Press, Delhi, 2008.

Suarez-Ruiz, I., and J.C. Crelling: *Applied Coal Petrology: the Role of Petrology in Coal Utilization*, Academic Press, New York, 2008.

Thomas, L.: *Handbook of Practical Coal Geology*: John Wiley & Sons, Chichester, 1992.

Tissot, B., and D.H. Welte: *Petroleum Formation and Occurrence*, Springer-Verlag, New York, 1984.

Walter D.: *The Elements Beyond Uranium,* John Wiley & Sons, New York , 1990.

Wang, Y.: *Physical Chemistry and Photobiology of Nucleic Acids: Chemistry*, Athens: U of Georgia Press, 2005.

Waterman, H. I.: *Hydrogenation of Fatty Oils*, Elsevier, New York, 1951.

Wiley, J.L. and H.C. Anderson: *Bibliography of Pressure Hydrogenation,* J. L. Wiley, Washington, D.C., 1950.

William C. Lyons, and Ali Ghalambor: *Petroleum Production Engineering: A Computer-assisted Approach,* Gulf Professional Pub., Burlington, MA, 2007.

William, L. Leffler: *Petrochemicals in Nontechnical Language,* PennWell, Tulsa, Okla, 2001.

Index

❑❑❑